"How Japanese is cherry blossom? Jung Lee's vivid and subtle tale of interdependence and rivalry between botanists in Japan and Korea is an outstanding contribution both to the transnational history of modern science, and to our understanding of the dynamics of imperialism and nationalism."

Francesca Bray, *University of Edinburgh, UK*

"One of the most important books on modern East Asian science in recent years. *Renaming Plants and Nations in Japanese Colonial Korea* is an impressive examination of the entanglement of science (particularly botany), Japanese imperialism/colonialism, and Korean nationalism. This book also provides a fascinating case study in the global history of science. A must-read."

Fa-ti Fan, *Binghamton University, USA*

"This book makes an incisive analysis of how the mundane yet connected multi-local practices of classifying and naming plants, such as Japanese cherry and rose of Sharon, in Japanese colonial Korea (1910–1945) simultaneously shaped Korean and Japanese botany while also changing 'Western' principles and practices of botany."

Geun Bae Kim, *Jeonbuk National University, South Korea*

"This book provides a new view of the global history of botany from the local perspective of botanists in East Asia under Japanese imperialism. Lee sheds light on forgotten botanists, both colonizers and colonized, who tried their own practices under colonial culture and politics."

Akihisa Setoguchi, *Kyoto University, Japan*

Renaming Plants and Nations in Japanese Colonial Korea

This book studies a striking example of intensely negotiated colonial scientific practice: the case of botanical practice in Korea during the Japanese colonization from 1910 to 1945.

The shared aim of botanists who encountered one another in colonial Korea to practice "modern Western botany" is successfully revealed through analysis of their fieldwork and subsequent publications. By exploring the variations in what that term should mean and the politically charged nature of the interactions between both imperial and colonial players, it reveals how botanists of the region created a form of scientific practice that was neither clearly Western nor particularly modern. It shows how the botany that evolved in this context was a product of colonially resourced, globally connected practice, immersed in intertwined traditions, rather than simply a copy of "modern Western botany."

Utilizing extensive primary sources, this book will be of interest to students and scholars of the history of science, colonial Korean history, and environmental history.

Jung Lee is an Assistant Professor in the Institute for the Humanities at Ewha Womans University in Seoul, South Korea.

Needham Research Institute Series
Series Editor: Christopher Cullen

Joseph Needham's 'Science and Civilisation' series began publication in the 1950s. At first it was seen as a piece of brilliant but isolated pioneering. However, at the beginning of the twenty-first century, it became clear that Needham's work had succeeded in creating a vibrant new intellectual field in the West. The books in this series cover topics that broadly relate to the practice of science, technology and medicine in East Asia, including China, Japan, Korea and Vietnam. The emphasis is on traditional forms of knowledge and practice, but without excluding modern studies that connect their topics with their historical and cultural context.

Celestial Lancets
A history and rationale of acupuncture and moxa
Lu Gwei-Djen and Joseph Needham
With a new introduction by Vivienne Lo

A Chinese Physician
Wang Ji and the Stone Mountain medical case histories
Joanna Grant

Chinese Mathematical Astrology
Reaching out to the stars
Ho Peng Yoke

Medieval Chinese Medicine
The Dunhuang medical manuscripts
Edited by Vivienne Lo and Christopher Cullen

Chinese Medicine in Early Communist China, 1945–1963
Medicine of revolution
Kim Taylor

Explorations in Daoism
Medicine and alchemy in literature
Ho Peng Yoke

Tibetan Medicine in the Contemporary World
Global politics of medical knowledge and practice
Edited by Laurent Pordié

The Evolution of Chinese Medicine
Northern Song dynasty, 960–1127
Asaf Goldschmidt

Speaking of Epidemics in Chinese Medicine
Disease and the geographic imagination in Late Imperial China
Marta E. Hanson

Reviving Ancient Chinese Mathematics
Mathematics, history and politics in the work of Wu Wen-Tsun
Jiri Hudecek

Rice, Agriculture and the Food Supply in Premodern Japan
The Place of Rice
Charlotte von Verschuer
Translated and edited by Wendy Cobcroft

The Politics of Chinese Medicine under Mongol Rule
Reiko Shinno
Asian Medical Industries

Contemporary Perspectives on Traditional Pharmaceuticals
Edited by Stephan Kloos and Calum Blaikie

The Chinese Astronomical Bureau, 1620–1850
Lineages, Bureaucracy and Technical Expertise
Ping-Ying Chang

Bamboo in Vietnam
An Anthropological and Historical Approach
Đinh Trọng Hiếu and Emmanuel Poisson

Disability and Impairment in Early China
Other Bodies
Avital H. Rom

Renaming Plants and Nations in Japanese Colonial Korea
Jung Lee

https://www.routledge.com/Needham-Research-Institute-Series/book-series/SE0483

Renaming Plants and Nations in Japanese Colonial Korea

Jung Lee

LONDON AND NEW YORK

First published 2025
by Routledge
4 Park Square, Milton Park, Abingdon, Oxon OX14 4RN

and by Routledge
605 Third Avenue, New York, NY 10158

Routledge is an imprint of the Taylor & Francis Group, an informa business

British Library Cataloguing-in-Publication Data
A catalogue record for this book is available from the British Library

ISBN: 978-1-032-83625-6 (hbk)
ISBN: 978-1-032-84217-2 (pbk)
ISBN: 978-1-003-51175-5 (ebk)

DOI: 10.4324/9781003511755

Typeset in Times New Roman
by codeMantra

Contents

Figures

Notes on Transliteration and Third-Party Materials

1. All translations of Japanese and Korean texts are mine.
2. For the Romanization of Korean, I adopt the McCune-Reischauer system except when authors and institutes already have Romanized names. For the Romanization of Japanese, I adopt the Revised Hepburn system without macrons and apostrophes. For example, Tokyo instead of Tōkyō.
3. In the main text, Korean, Japanese, and Chinese names are in their original order, surnames followed by given names. For the works published in English by these authors that use the reverse order, I keep the order in the publications. When Korean given names are two syllables written separately, I have hyphenated them, like Paik Nam-June, instead of Paik Nam June. If they already have connected syllables, like Lim Yunchan, I do not add the hyphen.
4. To save space and ink, I use only English translations and the original language titles for the citation and bibliography. Specialists can easily locate the original language references with the original language titles and non-specialists can understand the meaning. For key terms, I provide transliterations, often both in Japanese and in Korea.
5. I mainly cite English works both for Japanese and Korean historiography, and my citations for Japanese and Korean language works are not complete because of their limited accessibility to the readers of this book in English.
6. I use parts of the following papers, though I have revised them for the construction of this book while correcting a few errors and reassessing some of the points. I thank the publishers for allowing me to reuse the content.

   Lee, J. "Provincializing Global Botany," in Helen Anne Curry, Nick Jardine, James Secord, and Emma C. Spary eds., *Worlds of Natural History*, Cambridge: Cambridge University Press, 2018: 433–446.

   Lee, J. "Mutual Transformation of Colonial and Imperial Botanizing: The Intimate and Remote Collaboration between Chung Tyaihyon and Ishidoya Tsutomu in Colonial Korea," *Science in Context* 29(2) (2016): 179–211.

   Lee, J. "Between Universalism and Regionalism: Universal Systematics from Imperial Japan," *The British Journal for the History of Science* 48(4) (2015): 661–684.

Acknowledgment

Looking back, I am truly grateful that I was given more than ten years for this book, thanks to quite a few publishers choosing to decline it. If I had not been forced to take the time to learn the practice of writing piece by piece, I would have published a manuscript with far more flaws than it has now. The fortunate postponement also allowed me to have the humbling experiences of learning from many generous and patient people. I am grateful to have been given this space to acknowledge them in some small way.

I have to thank all the institutes that supported my doctoral and postdoctoral studies and the great teachers and generous colleagues during my stays there. At Seoul National University, where I wrote the dissertation that became this book, Kim Yungsik, Kwon Tae-Eok, Kim Kiyoon, and Moon Manyong were excellent committee members, and the guidance of Hong Sungook and Chung Yong-Wook was also valuable. The Needham Research Institute in Cambridge, the Institute of History and Philology, Academia Sinica in Taipei, and the Max-Planck Institute for the History of Science in Berlin hosted me as a postdoctoral fellow, and Christopher Cullen, Mei Jianjun, John Moffet, Sue Bennet, Hasok Chang, James Secord, and Helen Curry during my stay in Cambridge; Chu Pingyi, Sean Hsiang-lin Lei, Szuwei Chai, Yun-ju Chen, Kuangchi Hung, Yawen Ku, Wen-Hua Kuo, Jender Lee, and Hsiu-yun Wang during my stay in Taipei; and Lorraine Daston and Dagmar Schafer during my stay in Berlin all greatly nurtured this project. Kim Dongwon at the D. Kim Foundation also gave generous support. My sabbatical stay at the Paris Institute for Advanced Study greatly enabled this publication, putting me under the excellent guidance of Christopher Cullen and Catherine Jami. Specifically, Hasok Chang generously read my dissertation and James Secord went through the ordeal of reading the first version of a not so readable full manuscript. Lorraine Daston and Dagmar Schafer read the proposal and introduction. They all provided very helpful comments. I cannot thank Francesca Bray and Sandy Robertson enough for their extremely helpful readings of the proposal and their kind support, and I also deeply appreciate Fa-ti Fan's generous reading of the entire manuscript and wise guidance, Georges Métailie's constructive critique on the proposal, and Karine Chemla's pertinent comments on the introduction. They taught me a lot about my hasty thinking and writing. Two anonymous reviewers' kind reading and very helpful comments are also deeply appreciated.

I thank Shin Chang-gon for the help and advice provided during my archival trip to Japan, and the Institute for Japanese studies at Seoul National University for funding the trip. I thank all the archivists, librarians, and staff at various institutes who helped me find and use those materials for this book. I thank all the colleagues at those institutes and the audiences at various talks. Though I may have forgotten some of them, the following people helped this project in many ways: Hakmun Byun, Hyundeuk Cheon, David Fedman, Hansun Hsiung, Tae-ho Kim, John Lee, Tae-Hee Lee, Victoria Lee, Elaine Leong, Joseph D. Martin, Takuya Miyagawa, Albert Park, Hyunhee Park, Victor Seow, Ann Sherif, Holly Stephens, Junghyun Won, Huiyi Wu, and Kiebok Yi. I thank my very reliable proofreader Jeremy McMillan. I cannot thank Christopher Cullen enough for his immense help and judicious advice throughout the publication process, and I am also grateful for the editorial guidance of Sue Bennet at the Needham and Andrew Leach and Stephanie Rogers at Routledge.

I thank all my colleagues at Ewha Institute for the Humanities, who provide a stimulating learning environment with their intellectual diversity, and my students at Ewha Post-Human Studies, whose ecological, social, and feminist concerns teach me a lot.

Several kind taxonomists in Japan, Taiwan, and Korea showed me their herbarium collections, shared materials, and answered my queries, which helped this project a lot. I do admire their love for plants and scholarship. While I met none of the botanists discussed in this book, I did meet some of their children and relatives. I thank them for kindly sharing their stories about them, and allowing me to use some of the materials. I endlessly thank my parents, whose love and guidance have sustained me, and my brother, whose caring support also enabled much of what I have done so far.

My greatest thanks for this book go to Lim Jongtae, who has read numerous versions of this manuscript throughout the years since he advised the dissertation submitted in 2013. Without his excellent guidance, exemplary scholarship, and great tolerance of my faults, this book, and myself as a researcher, could not exist. I dedicate this book to him, though he or anyone above mentioned should not be responsible for any flaws remaining in the work.

Introduction

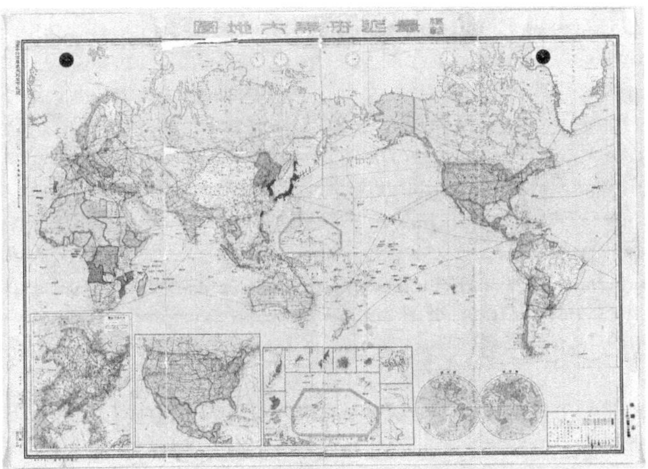

Figure 0.1 The Model Great Map of the Modern World (模範最近世界大地圖). This is a map authorized by Sato Tetsutaro (佐藤鉄太郎, 1866–1942) in 1935, retired Vice Admiral of the Imperial Japanese Navy and member of House of Peers. It marks Japan and its colonies of Korea and Taiwan in black (red), tiny but at the center, and Japan's other occupied territories, Manchukuo and Micronesia, respectively, in gray and in a heptagon (National Museum of Korea).

This book tells the story of those Japanese and Koreans at the turn of the twentieth century who joined the botanical practice of classifying and naming plants rooted in their regional soils in the name of science (科學, kagaku in Japanese, kwahak in Korean) (Figure 0.1). Its focus is on their practice in the Korean peninsula, where these Japanese and Koreans who identified themselves as botanists (植物學者, shokubutsu gakusha in J, shingmul hakcha in K) most extensively interacted with each other, starting with the Japanese colonization of Korea in 1910 to its end in 1945. While their new scientific names of plants were in Latin, the universal language of botany for the universal order of nature, these Japanese and Korean botanists contributed to not just the universal order of the plant world but also

DOI: 10.4324/9781003511755-1

regionally formulated meanings of science, botany, plants, and nations.[1] These all newly emerging terms dynamically shaped and were shaped by this very act of studying and naming plants in novel ways.

What is in the new names of plants and what is in the names "science," "botany," "botanist," "Japan," and "Korea"? Let me start with the names of two plants, rose of Sharon, *mugung hwa* (無窮花, 무궁화) in Korean, and Japanese cherry, *sakura* (櫻, さくら) in Japanese. They are beautiful garden plants cherished in many parts of the world outside of where they are originally from. Both have the names of places in their scientific binomials, too, *Hibiscus syriacus* and *Prunus yedoensis*, respectively.

Carl von Linné (1707–1788), the Swedish father of botany, saw a rose of Sharon in the garden of George Clifford III (1685–1760), one of directors of the Dutch East India Company. Clifford loved plants and imported many, including this one from Syria. He asked Linné to catalog them, which Linné did with his "natural system" of ordering plants from classes, orders, genera, and species. Linné claimed many new species from Clifford's garden in *Species Plantarum* (1753), including *Hibiscus syriacus*, noting its origin as Syria.[2] Matsumura Jinzo (松村任三, 1856–1928), a Professor at Tokyo Imperial University who studied in Germany, named the most beloved and widely planted cherry tree in Japan as *Prunus yedoensis* in an international botanical journal published in Tokyo in 1901. He believed it was native to Edo, the previous name of Tokyo.[3] To emerging Japanese botanists, this renaming of a familiar Japanese plant, which almost every Japanese person knew, in Latin, which only a few of them understood, symbolized Japan's successful modernization in studying plants. Japanese academia honored this naming with the Imperial Academy Award in 1904.

It resonates with the striking catchphrase of Meiji Japan (1868–1912), "Escaping Asia towards Europe (脫亞入歐)," Fukuzawa Yukichi's (福澤諭吉, 1835–1901) influential mantra for "all-out Europeanization (萬事歐化主義)." All things in Japan would become modern (近代, kindai in J, kŭndae in K) by replacing Japan's Asian past with "Western" (西歐, literally Western European, seio in J, sŏgu in K; 西洋, literally [the realm of the] Western oceans, seiyo in J, sŏyang in K) practice.[4] These Japanese fully embraced the imperial Western ideas of Western exceptionality in historical progress and their own status as benighted by denying their past, asserting a historicism that depicts history as a linear progression that culminates in European modernity.[5]

In terms of origins, Linné was wrong and Matsumura was about half right according to the Plants of the World Online at the Royal Botanic Gardens, Kew. The rose of Sharon is native to southern China, while the cherry was from a flower shop, albeit one in Tokyo, making its current binomial *Prunus x yedoensis*, with x noting its status as a hybrid cultivar.[6]

As making refutations on previous conjectures by new empirical findings is expected of science, and as science progresses, according to Karl Popper, by such refutations, it seems unobjectionable to provide somewhat mistaken names for a couple of plants with confused origins. Furthermore, origins, although often attributed in this flimsy way, had not been primary concerns in classification. Yet these

Japanese and Korean botanists responded to these mistaken names in shaping their respective botany. How they came to respond differently in their common pursuits of modern botanical practice sheds light on what it means to do botany and to name things in new ways in Japanese colonial Korea.

Matsumura's mistake received attention, possibly too soon. In 1912, less than a decade after the award of honor, a German botanist who had studied specimens of cherry sent by a French missionary based on the Korean island of Cheju claimed that *Prunus yedoensis* was native to that Korean island, which lies between Japan and the Korean peninsula.[7] In a sense, that Korean island was Japan. After an impressive victory over its giant neighbor Qing China in 1895 that secured Taiwan, Japan achieved an even more celebrated victory over White Russia in 1905, fought over Korea largely on Korean soil, designating Korea as its protectorate that year and ultimately colonizing it in 1910.[8] Yet the Japanese botanical establishment at the imperial university appeared crestfallen.

According to *Bushido: The Soul of Japan*, the bestselling introduction to Japanese culture in English published in 1899, the cherry symbolized the Japanese ethos. Its author Nitobe Inazo (新渡戸稲造, 1862–1933), an influential scholar and statesman, says the cherry "is ever ready to depart life at the call of nature," showing the samurai's "sense of honor which cannot bear being looked down upon." It was not like the roses loved by Westerners that rot on the stem, clinging to life.[9] Korea, its closest neighbor, however, represented the darkness of Asia's past in the heightened nationalistic building up of Japan. One Japanese congressman said in 1905: "Koreans are animals that walk upright rather than human beings."[10] Nitobe agreed in 1907: "Koreans are not the people of the twentieth nor of the tenth century. No, they are not even the people of the first century. They belong to the prehistoric era."[11] To have this Korea as the origin of the Japanese soul must have been inconceivable. Most Japanese botanists saw a single set of specimens from Cheju as insufficient to confirm the Korean origin of *Prunus yedoensis*[12] and kept quiet about it while Japanese governments transplanted this representative Japanese cherry, with seedlings from Tokyo, into Korea as into other colonies. It was "an effort to symbolically stamp occupied areas as spaces for Imperial Japan."[13]

Matsumura and his students were even less enthusiastic about the possibility of its creation in a Tokyo florist's. Somei Yoshino (染井 吉野), the Japanese name for *Prunus yedoensis*, had suggested it all along; Somei was an area packed with flower shops. E. H. Wilson (1876–1930), a British expert from the Arnold Arboretum of Harvard University, invited by Matsumura to find its habitats in 1916, made this possibility official by suggesting two species of native cherry that are likely to beget the beautiful cultivar.[14] With the x in the binomial, the naming's scientific status, as one that orders nature, diminishes, they thought.[15]

These Japanese botanists searched for an achievement more secure than the naming of *Prunus yedoensis*. Venturing out to Korea just across the strait, a "virgin land for botany" which had had no trade port before its opening to Japan in 1876, provided them with success. Holding onto sponsorship from the Government-General of Korea (朝鮮總督府, GGK hereafter) for 30 years, Matsumura's student Nakai Takenoshin (中井猛之進, 1882–1952), who later studied in the US and the

UK, produced 22 volumes of work on Korean flora from 1913 to 1942, providing numerous new scientific names. In 1926, he became the first non-Westerner to join the International Committee on Botanical Nomenclature for the International Congress of Plant Sciences.

In addition to this resounding success, Japan's imperial botanizing in colonial Korea produced contentious partners, colonized Koreans who dared to call themselves botanists, *shingmul hakcha*, refusing to remain mere informants and collectors for Japanese *shokubutsu gakusha*. Clearly, no Korean botanist had their names as authors in those 22 volumes of authoritative works on Korean flora by Nakai, written in a mixture of English, Latin, and Japanese. Yet some of them contributed to scholarly journals in Japan and in colonial Korea and produced several monographs about Korean flora on their own. Those were mostly in Japanese, the language of official publication in the colony, except for Latin names.

To share their new knowledge about plants, Korean botanists had their own venues, too, though not academic, the public media in Korean.[16] For example, Toh Bong-Syup (都逢涉, 1904–?), a pharmaceutics graduate from Tokyo Imperial University, while summarizing Nakai's findings on Korean flora for a Korean audience in 1936, wrote: "That the origin of cherry trees was not inner Japan but Cheju was declared four or five years ago."[17] While the origin was by no means a settled issue, continuing into the twenty-first century, he was uncharacteristically assertive in declaring Korea as the origin of the Japanese cherry tree.[18] It could restore the pride that Koreans had held, a transmitter of important things to Japan in their shared history.

In addition to saying the cherry tree was from Cheju, some Koreans also showed a special interest in roses of Sharon. It was not about their origin in Latin or English. It concerned the vernacular name *mugung hwa*, which literally means "endless flower." This name for a plant that blossoms, falls, yet endlessly blossoms again during the heat of the summer could be a symbol of Koreans' endless resistance to the Japanese colonizer. They excavated some old Korean texts written in classical Chinese, *Chibong's Classified Commentaries* (芝峯類說 Chibong yusŏl, 1614), one of the earliest encyclopedic works by a Korean polymath, and a medical work, *Treasured Mirror of Eastern Medicine* (東醫寶鑑 Tongŭi bogam, 1613). Chinese visitors in antiquity praised the abundance of the plant (木槿 mujin or 槿花 jinhua in Chinese) in the land, calling it the land of the rose of Sharon (槿花鄉), while *Treasured Mirror* confirmed *mujin*'s vernacular name, *mugung hwa*.[19]

Neither *mugung hwa* nor *sakura* had been a "national" flower before since nations themselves had existed differently in the mind of people. According to Kaibara Ekken's (貝原益軒, 1630–1714) *Japanese Materia Medica* (大和本草 Yamato honzo, 1709), both the Japanese and Koreans loved cherry trees.[20] As with many traditions like English afternoon tea, the seemingly eternal and distinctive traditions of celebrating falling cherry petals or endlessly blossoming roses of Sharon were of recent making in globally shaped interactions.[21] Historian Sheldon Garon notes that Nitobe's work became very popular, and was "widely reprinted in the United States and Britain, and soon translated into German, Swedish, Norwegian, and Polish," because it was "in dialogue with Western preoccupations about national reinvigoration." Nitobe simply presented virtues widely endorsed in these emerging nation-states, "loyalty to the nation, frugality, military valor, and an enduring tradition of 'chivalry,'" as Japanese[22] Nitobe's work was translated back into Japanese in 1908, to help secure the *sakura*'s national flower status while stimulating nation-less Koreans too to claim the *mugung hwa* as a national flower.[23]

Botanists in Japan and Korea, like their counterparts in other countries, played active roles not just in finding a universal order in nature but in this dynamic making of their respective nationhood. The various challenges that they encountered in ordering and naming their local plants shaped their respective botany as well as their knowledge practice's relations to their respective nations in the making. It may be ironic that these figures, who most openly embraced the modernity and universality of science in seeking a transnational education, also led these inventions of traditions and divisive meanings of plants. This book looks into how these botanists joined in colonial Korea studied and named their plants, while trying to appreciate the various and changing meanings that they attached to new terms like "science," "botany," and "nation."

This book follows more recognized figures like Nakai Takenoshin and Toh Bong-Syup, while examining other figures who worked with them, all obscure figures in the history of botany. In a sense, Korean and Japanese were not the languages of science, and their works in Korean or Japanese existed outside the proper world of botany, written mostly in English and Latin.[24] Yet this book contends that these obscure self-made botanists, who were sometimes modest schoolteachers in colonial schools and experts in colonial government institutes, were as important as internationally recognized figures like Nakai in globalizing botany. Above all, it was not particularly Linné's doing that they decided to celebrate *Prunus yedoensis* and *Hibiscus syriacus* as names indicating the value of their knowledge practices. In the words of British journalist Meredith Townsend in 1901:

> We are told every day how Europe has influenced Japan, and forget that the change in those islands was entirely self-generated, that Europeans did not teach Japan, but that Japan of herself chose to learn from Europe methods of organization, civil and military, which have so far proved successful. She imported European mechanical science as the Turks years before imported European artillery. That is not exactly "influence," unless, indeed, England is "influenced" by purchasing tea of China. Where is the European apostle or philosopher or statesman or agitator who has re-made Japan?[25]

These Japanese and Korean botanists indeed worked on their own from the ground up, actively joining this globalizing knowledge practice that simultaneously shaped modern botanies and nations across the globe. Like their Western companions, they actively relied on imperial or colonial interactions that shaped the modern world.[26] Yet thanks to the unique dynamics between the non-Western Empire of Japan and its adjacent colony, Korea, their encounters in colonial Korea simultaneously created two botanies, Japanese and Korean, between empire and colony, with distinctive botanical constructions, contents, and rules. In doing so, they portrayed themselves as practitioners of not simply Japanese and Korean botany but proper transmitters of "Western" botany too as they defined it.

By hearing these complex voices of Korean and Japanese botanists, we can reshape our understanding of the globalization of science, which has been heavily shaped by European or American experiences in metropoles and their historically and geographically remote colonies. Lorraine Daston has asserted that enlightenment was a process, to be revealed by asking "Not 'What is enlightenment?' but 'Who are the enlightened, and how did they get that way?'"[27] While many Westerners also struggled to define science as exceptionally "Western" in creating the new hierarchy of civilization or the "measure of men,"[28] there were many more non-Westerners who asserted themselves as practitioners of "Western" science, especially around this time. Newly emerging experts everywhere claimed to study and practice Western science for the enlightenment (啓蒙, kiemo in J, kyemong in K) of themselves and their nation.[29]

To understand how the globalized practice of science became Western and enlightened, it is crucial to follow these non-Western botanists to their non-Western outposts. This book thus makes concerted shifts, from European to non-European fields, from a few thinkers to many practitioners, and from outcomes to processes, as many in the field have suggested and carried out.[30] The often confused and muffled voices of Japanese and Korean botanists in their languages reveal much more about what it meant to practice a science like botany for these self-made botanists. How did it become scientific and enlightened to give Latin names for their local plants?

The dynamics between the non-Western Empire of Japan and its adjacent colony Korea renders this outpost most vibrant by making their botanical interactions especially massive and highly charged. While Japan did not become a settler colony like Australia or the United States, it did count almost a million Japanese settlers at the end of its rule, for about 20 million Koreans.[31] Japanese experts came in great numbers and stayed for long periods, making them colonially attached "go-betweens," at times against metropolitan authorities like Matsumura and Nakai.[32] Some Japanese settlers questioned what was lost in Japan's hasty Westernization, which seemed as vulnerable as *Prunus yedoensis*, while learning about the wealth of the shared tradition that was denied at home as being of a benighted Asian past. This tradition, renewed in colonial interactions, was alive materially too, providing financial support for Koreans' own botanizing. Toh came to have his own lab funded by a Korean pharmaceutics company that became prosperous in the traditional medicine trade. In these settings, it was not only the Japanese, whose model Westernization was widely recognized, but Koreans who also shaped their own botany, cataloging their plants themselves.

This book mainly covers these studies about ordering and naming plants, which was the botanical science carried out in colonial Korea most visibly, massively, and in close but complexly changing collaborations amongst Japanese and Korean botanists, while often significantly involving Western botanists like Wilson and the International Botanical Congress. Unlike laboratory botany like plant physiology and genetics, Japanese botanists in plant systematics, whether from Tokyo or settled in Korea, could not conduct their research without the help of colonized Koreans who knew the language and the land. As colonized Koreans mostly stopped at making a token success in other fields of science like physics and chemistry, this need for collaboration in botanical classification was crucial in simultaneously constructing distinctive botanies between empire and colony even within the unique colonial settings of colonial Korea.

Analytically, this book follows their interactively shaped research by paying attention to various vernacular works, including some personal reflections from these botanists, as well as internationally recognized academic works. Their entangled botanical practices culminated in the war-time collaboration for locally rooted (鄕土, literally local soil, kyodo in J, hyangt'o in K) botany involving many Japanese and Korean school teachers. Through their writings in Japanese and Korean, at times strategically chosen over English, Latin, and Japanese directed to outside audiences, these botanists revealed different thoughts about botany, nationhood, Western botanical rules and practices, and their uneasy collaborations.

How may the voices and actions of these obscure botanists change our understanding of terms like *Western science*, *modern systematics*, and *nation*, beyond simply the names of *Prunus yedoensis* and *Hibiscus syriacus*?[33] First, despite all their own initiatives to study abroad, translate Western books, and build their local institutes in connection with Western institutes as praised by Townsend, they themselves were the main force that sustained the much criticized "diffusion" narrative; they denied their own agency presenting themselves as mere transmitters. Many works in colonial, post-colonial, and global histories of sciences have effectively critiqued diffusion narratives. They have made many well-corroborated attempts to "provincialize" Europe and "relocate" modern sciences into colonial "contact zones" or "circulate" them in global networks.[34] Japan provides especially good examples to show that practitioners in every society were hardly ever "merely on the receiving end of innovations conceived elsewhere a century earlier." Japanese doctors who studied in Germany revealed that "'German' medicine was not an end product completed in Germany and then exported to Japan, but rather in a constant process of making and un making," and Japan, the earthquake nation, was the site that co-produced the "Anglo-Japanese science of seismology."[35] Yet the mistaken origins of incorrect names seem extremely resilient, not just in *Prunus yedoensis* and *Hibiscus syriacus*, but in the names of Western or modern science, because these practitioners of Western science everywhere kept evoking them with the hierarchical dichotomies between the West and the Rest, the center and the periphery, the imperial and the colonized, the modern and the traditional, and the scientific and the practical (or theory and practice). Their enlightened moves toward "Western" and "modern" science against their entrenched traditions, an ongoing process, seem to require and perpetuate these dichotomies embedded in the diffusion narrative.

Second, they nonetheless clearly display their own reasons and rules, often ignoring or reworking Western originals that they selectively brought in. Especially through their vernacular or non-scientific writings in Japanese and Korean, these pioneering modernizers in their societies expressed many dissatisfactions, doubts, and reformulations, addressing various imperfections and inadequacies of Western botany; they knew that theirs were "not local variants of or alternatives to a presumed original modernity" but their own original.[36] These vernacular discussions were in a sense the "hidden transcripts" of the global "rest," in the imperial era created by overwhelming European dominance.[37] On the one hand, these so far less-analyzed hidden thoughts reveal the "epistemic violence and the painful confinement" induced by Eurocentrism that these non-Western practitioners of modern science had internalized, which concealed their own agency and reason.[38] On the other hand, the voices even of colonized Koreans reveal many second thoughts about Western and Nakai's methods and classifications, showing the forceful reevaluations and reformulations of modernity that shaped all local ones.

The field of plant systematics, the focus of this book, is especially good in showing how there was no complete modernity to be diffused from Western centers. It delineates how even the foundational rules that had created mistaken names like *Hibiscus syriacus* were always in the process of being made by practitioners from scattered centers. As many have shown, and as will be discussed further in Chapter 2, Linné's—or European—rules had been in disarray, with many competing standards since the powerful critiques of Michel Adanson (1727–1836) and Antoine-Laurent de Jussieu (1748–1836), the emergence of the Kew botanists, and the challenges of evolutionary theory, which was only heightened by recent DNA analyses.[39] The local knowledge of plants rooted in regions, though dismissed in creating Linnaean centers of botany filled with dried specimens from around the world, always caused tensions. Forceful newcomers from regions detached from Europe, like America and Oceania, had all raised questions on rules shaped on dried specimens with their actual observations in local fields.[40] Botanists working at the impressive imperial center of botany at Tokyo Imperial University as well as those who joined in the practice from the colonial field by building up Japanese collections came up with their own rules to classify Japanese and Korean plants and raised fundamental questions about the insecurity of changing standards.

Third, partly thanks to their textual traditions, relatively easy to access for both imperial and colonial botanists, they show various crisscrossing of the traditional and the modern in new knowledge practices. Their specialized studies on plants were indeed unprecedented in the age-old and vital efforts of people studying these life forms. Many of their empirical methods and tools, like botanical explorations, collections, and microscopic and chemical analyses, as copied from Europe, seemed novel; both Japanese and Korean botanists emphasized this novelty of their practices, portraying theirs as overcoming an unscientific tradition.

However, tradition, entrenched in every way, intellectually, socially, and materially, did not just disappear. Tokugawa Japan (1603–1868) had an illustrious tradition in studying plants, highly shaped by transnational interactions. Several European scholars including Carl Linné's (1707–1778) pupil Carl Peter Thunberg (1743–1828) and Philipp Franz von Siebold (1796–1866) had entered Japan through the Dutch trading port in Nagasaki since the seventeenth century.[41] Siebold noted: "Outside of Europe, there is no place where botanical studies are more advanced than Japan and China."[42] Chosŏn Korea (1392–1910), in between Japan and China, also had its own knowledge practice about plants, both textual, as with *Treasured Mirror*, and material, as with herbal medicines, which sufficiently inspired Tokugawa Japan to carry out three botanical explorations in the peninsula between 1718 and 1722.[43] These were earlier than the first dispatches of Linné's disciples, showing that the novelty of the Linnaean approach was not one of kind but more in its systematic articulation and massive execution. This story from colonial Korea, where imperial and colonial practitioners were complexly intertwined with traditions, shows various forces of traditions that nurtured modernization.

Ian J. Miller says: "Until quite recently, historians of science have tended to treat the emergence of modern biology in Japan as a revolution, something akin to Thomas Kuhn's notion of a 'paradigm shift,' a transformation between two incommensurable systems."[44] Historians of Japanese botany indeed used to present Japan's modernization of botany as a break from the past, following the words rather than the actual practices of botanists. They claimed, endorsing their predecessors' words, that the Meiji generation broke new ground by adopting Western methods. It was like "grafting bamboo onto wood," one on top of the other but with a clear chasm between them.[45] Yet quite a few historians of science and technology noted early on the roles of tradition in the successful making of various Japanese sciences. [46] This book follows these works, joined by the growing efforts that variously appreciate and delineate these continuities to later periods.[47] For the history of Korean botany also, this book seeks to fill some gaps by showing the roles of tradition in the Korean transformation of knowledge practices, as already suggested with *Hibiscus syriacus* and Toh's pharmaceutical lab.

Fourth, through the uniquely parallel moves of Japanese and Korean botanists, this book seeks to reflect on the asymmetry in our understanding of imperial and colonial actors that presents colonial actors as mere hands supporting imperial theorists who did real science. In demonstrating how colonial hands were their own theorists in botanical science, inseparable from botanical practice in the field, I take insights from works that deepen our understanding of science through the tacit and embodied knowledge of practitioners.[48] In stories of global or colonial science where important European theorists had limited interactions with various assistants in remote lands, the asymmetry of power and knowledge appeared too great to make the assertion that both parties were equally significant knowledge producers. David Arnold once wrote, not with enthusiasm, I think,

> It would be a mistake, on the basis of the kind of evidence presented here, to presume that indigenous ideas and agency had an equal role or that some kind of open, mutually respectful, discourse existed between Indian science and its European counterpart. As far as botany was concerned, indigenous knowledge and agency was, even by the 1830s, assigned a secondary, subaltern status, and though both were undoubtedly taken up and utilized by European travellers and Indian-based naturalists, this was more often akin to exploitation than a recognition of epistemological equality.[49]

I think this sense of an unbridgeable gap between imperial and colonial actors shaped by Western cases created and justified a certain compartmentalization in the field. Important histories of colonial, global, and regional sciences tend to exist apart from those narratives shaped by scientific revolutions and other important theoretical breakthroughs. Or they matter only in terms of their contributions to this real science of Western centers.[50] As Francesca Bray suggests, and as some scholars have tried, we have to overcome this compartmentalization in the field that keeps these regional stories only as regional by making more symmetric connections.[51] This book tries to do so by following all regional botanies entangled in colonial Korea as equally in the making, and presenting the "multi-local" stories of many botanies, to borrow the term that Gregory Clancey suggested.[52]

Assuredly, colonial Korea was also a "contact zone," where "conditions of coercion, radical inequality, and intractable conflict" pervaded.[53] As shown in the words of the aforementioned Japanese congressman and Nitobe, Japanese ideologues tried to render Korea as remote as the faraway colonies of Western empires, reproducing Western orientalism. However, in this contact zone shaped by Japan's model Westernization, Japan was the West, imperial, center, metropole, and modern, but simultaneously a "self-colonized" Asian periphery anxious about the security of *Prunus yedoensis*.[54] Also, while Japan's military success over Qing China and Russia that startled the West surely impressed Koreans, making some of them eager for colonial tutelage, some still believed the strength of the sword was just the strength of the sword. They remembered all too well that they used to look down on the samurai of the archipelago within the Confucian hierarchy that placed the letter above the sword.[55] Korean botanists hardly questioned the authority of Japanese botanists in person, accepting discriminatory conditions during their collaboration. But they eventually built their own center of botany at Toh's lab for their own botanizing.

Finally, the parallel making of Japanese and Korean botanies focusing on their interactive practices shows how these Japanese and Korean botanists, like their Western counterparts, were not mere tools of imperial or colonial politics; they were no less creative than their global counterparts in their initiatives to connect their knowledge practice to their emerging nations, though it was indeed an "imagined community" for Korean botanists without a nation.[56] For Japanese botanists, too, there was no ready-made powerful empire that scientists could tap into. They made the state, overwhelmed with many all-important issues and lacking resources, pay attention to their botanies, through *Prunus yedoensis* and *mugung hwa*. Matsumura and Nakai reproduced the Western imperial posture, leading Japan's colonizing of Korea for their imperial center of botany. Some Koreans eagerly accepted limited roles as colonial informants advancing imperial projects, in the name of both nation and science, revealing the various limitations and tensions of a colonial nationalism that sought empowerment through imperial logic and guidance.[57] Nonetheless, since there was nothing natural about these power-laden couplings of science and nation, there was a wide and changing spectrum of these couplings that variegated the courses of their botanies.[58]

These relations between science and nation have been important in Japanese and Korean historiographies. In Japanese historiography, although works on Japanese botany have not yet significantly dealt with Japanese works on Korean or Taiwan flora, Japanese science's relation to imperial expansion has received increasing attention.[59] Still, they often treat empire as the static infrastructure that enabled the development of Japanese science. Or, like the works for Western imperial science, they discuss these imperial practices or field works in colonies mostly in terms of their contribution to Japanese science, although the roles of colonial sites, cultures, and assistants are much more appreciated than before.[60] These are multi-sited and multi-local stories, but hardly that of multi-sciences of the empire and the colony simultaneously forming in interaction.[61]

Korean historiography tends to take more interest in the imperial Japanese side even when it mainly discusses colonial practice. It agonizes over the coloniality of Korean practices that are always in comparison with Japanese and Western ones, though many historians have moved beyond such concerns, illuminating coloniality's encroachment into metropolitan modernity, the "true intimacy between coloniality and modernity" everywhere and showing strong politics from the ground.[62] Neither the imperial government in Japan nor the GGK, like all governments today, had the ability or the means to formulate policies that could foresee the next several decades, spanning several wars including the two World Wars. Whatever any government did was the result of various actors with conflicting ideas. The GGK especially suffered from perennial financial problems, making it more vulnerable to conflicting domestic, inter-colonial, and international politics within and beyond the empire.[63] Kwon Tae-Eok shows that the GGK came to promise the civilizing mission (文明化, bunmeika in J, munmyŏnghwa in K) and emphasize "scientific" colonialism, pressured by Koreans as well as Western competitors who doubted the legitimacy of its rule or civilized status.[64] David Fedman thoughtfully discusses how colonial Korean forestry, which at times travelled back to Japan to reshape Japanese forestry, was dynamically shaped by constant negotiations among these forces.[65] Albert Park shows how colonial Koreans, regardless of their disenfranchised status, eagerly "participated in determining how modernization would take shape and form in the peninsula."[66] They made all kinds of reforms and campaigns against, for, or with the GGK, showing more interest in certain areas and making its modernity different from that of Japan and its other colonies.

The case of botanical modernity in colonial Korea, which did not exist in Japan's other colonies like Taiwan or Micronesia, clearly demonstrates how all colonial modernity came from on the ground politics of the colonized.[67] Despite the promise of the civilizing mission, the GGK had a ready excuse not to carry it out: the primitive status of the Korean people. In practice, the GGK actively made various discriminatory measures against Korean education and careers to keep Koreans benighted. Although Koreans became quite unanimous about the need to accept Western science and technology for Korea's survival around this time, there were few practicing scientists in other fields as a result of these measures.[68] Botany became a field populated by Koreans because the colonized themselves could build up modest institutions for their education and research. The pharmaceutics industry based on traditional medicines, the need for colonial informants for the GGK's colonial development in forestry and agriculture, and the belief that classification based on the empirical observation of nature was a key to cultivate scientific nationhood all worked to help Korean botanists settle in colonial schools and modest research institutes.

In sum, using the specific dynamics of colonial Korea concerning plant systematics and hearing out diverse voices, articulating murmurs, and paying attention to changing interactions, this book delineates these concurrent, multi-local, and interactive makings of Japanese, Korean, and, in turn, "Western" botanies. These Japanese and Koreans celebrated their successful "diffusion" of Western botany as they defined it without detailing their own efforts and reformulations that shaped their distinctive botanies. Yet they were strong institution builders in their own society, as Linné was in his. Embroiled in multi-local interactions of the imperial era and nurtured by dynamic local politics, these Japanese and Koreans shaped their own botanies and our vulnerably connected world of nations.

Synopsis of Chapters

This book offers eight episodes that closely follow the interconnected intellectual trajectories of individuals and groups, displaying a spectrum of achievements and frustrations in the practicing of botany in colonial Korea. Chapter 1 starts in the eighteenth century, partly unraveling what was in the name of *Ginkgo biloba*, named by Linné. It examines the extensive earlier interactions between European, Japanese, and Korean knowledge practitioners concerning plants and their related traditions that inspired those interactions. Carl Peter Thunberg's and Philip Franz von Siebold's expeditions to Japan and Japanese herbalists' expeditions to Korea are among them. Chapter 2 follows the establishment of the imperial center of botany at Tokyo Imperial University. Korea, a "hermit country" unknown to Western botanists, became its strategic site for building the Japanese Kew with herbariums and gardens by launching collection expeditions almost a decade prior to Japan's actual colonization of the land in 1910, which produced the internationally recognized authority on Korean flora, Nakai Takenoshin. It reveals the divergent paths that Japanese imperial botanists explored, by comparing Nakai's work on Korean flora with his colleague Hayata Bunzo's (早田文蔵, 1874–1934) work on Taiwanese flora, and examining Nakai's suppression of Hayata's open critique of the classification rules and methods. Chapter 3 follows a Japanese settler forester-cum-systematist, Ishidoya Tsutomu (石戸谷勉, 1891–1958), who could not trust Nakai's classification of Korean plants and became a most vocal and influential critic of Japanese, and, in turn, Western botanical practices. Nakai's confused naming of a willow tree collected by him from *Salix rorida* Laksch to *Salix splendida* Nakai, and again to *Chosenia splendida* Nakai, and Western botanical rules that accepted all these changing names, disillusioned him. He left his 15-year career as a forester in 1925 to study traditional Korean medicines. Chapter 4 demonstrates how Koreans demanded the right to learn and practice Western botany as in Japan, notwithstanding such doubts raised from the colony concerning Japanese and Western botanical practices. They came to build their own centers of botany with encouragement for their scientific pursuit from the colonized collective, including serious sponsorship of the Korean pharmaceutical industry.

Chapter 5 examines the interactive making of nationhood and botany following two Korean botanists, Toh Bong-Syup and Chung Tyaihyon (鄭台鉉, 1883–1971), Nakai's interpreter and collector and the so-called "father of Korean botanical systematics." It examines how successful self-made botanists like them still suffered from colonial "double consciousness," at times trying to remove their Koreanness from their works and even from their appearance in order to overcome colonial discrimination against their bold wish to become botanists.[69] It was not just Koreans who were caught up in imperial politics. Japanese botanists also had to be or chose to be "Japanese" or imperial in their knowledge practice. Chapter 6 examines the making of a Japanese imperialist recovery of colonial traditions by following the tumultuous journey that Ishidoya, the disillusioned colonial expert, made after his transition to Korean traditional medicines. He made complex and changing negotiations with Korean and then Chinese traditions, eventually producing a German work on Chinese herbal medicines through and in the market. Chapter 7 looks into the organizing efforts of settled Japanese teachers who attempted a kind of professionalization by studying plants and animals in Korea. They, like Ishidoya, made their niche by emphasizing their expertise based on the colonial land, claiming locally rooted studies. Nonetheless, in their separate quarters, they only succeeded in Japanizing their studies against Nakai's work that they found "foreign oriented." Chapter 8 looks into what was in the vernacular names and knowledge that Korean botanists laboriously added to the Latin and English works of Nakai. It follows the collective botanizing of Koreans who gradually forged an independent intellectual agenda through their constrained identity as Koreans.

These concurrent and multi-local makings of modernities came not just with great and important achievements but also with many contradictions, scars, and lost opportunities. The current predicament with nature, with fragmented politics incapable of addressing it, is one of them. Mainly to help all moderns in the making see different perspectives than their own for more reflective interactions with their locally variegated but connected modernities, this book relates these stories of Japanese and Korean botanists who classified and named their plants in the names of science, botany, modernization as well as their interactively shaped nations. It makes a long reflective journey into many mistaken and divisive names, in order to move beyond them.

Notes

1 Benedict Anderson, *Imagined Communities: Reflections on the Origin and Spread of Nationalism*, revised ed. London: Verso, 1991; Carol E. Harrison and Ann Johnson, eds., *National Identity: The Role of Science and Technology*, Chicago, IL: University of Chicago Press, 2009.

2 Carl Linné, *Species Plantarum,* Holmiæ [Stockholm], Impensis Laurentii Salvii, 1753, vol. 2, 695; Gerard Thijsse, "A Contribution to the History of the Herbaria of George Clifford III (1685–1760)," *Archives of Natural History* 45(1) (2018): 134–48.

3 Matsumura Jinzo, "Cerasi Japonicæ Duæ Species Novæ," *The Botanical Magazine, Tokyo* (植物學雜誌, *BMT* hereafter) 15 (1901): 99–101.

4 From now on, "West" means this historical actors' term. Meiji Japan means Japan under the Meiji Emperor, who came to the power with the Meiji restoration in 1868. It is followed by the Taisho (大正, 1912–1926), and Showa (昭和, 1926–1989) periods. The

modern Japanese attitude was under this mantra of "Escaping Asia towards Europe," by which the Japanese tried "aggressively to render themselves similar to Europeans, not Asians," and even claimed the "Whiteness" of the Japanese race. David G. Wittner and Philip C. Brown, eds. *Science, Technology, and Medicine in the Modern Japanese Empire*, New York: Routledge, 2016, 12; Marius B. Jansen, *The Making of Modern Japan*, Cambridge, MA: Belknap Press of Harvard University Press, 2000.

5 For a penetrating analysis on the complexes in Japanese minds concerning their all-out Europeanization, see Komori Yoichi 고모리 요이치, *Postcolonial: Colonial Sub-Consciousness and Colonial Consciousness* (포스트콜로니얼: 식민지적 무의식과 식민주의적 의식), tr. Song Tae-uk, Seoul: Samin, 2002. For critiques of Euro-centric historicism and its relation to history writing, see Prasenjit Duara, *Rescuing History from the Nation: Questioning Narratives of Modern China*, Chicago, IL: University of Chicago Press, 1996; Dipesh Chakrabarty, *Provincializing Europe: Postcolonial Thought and Historical Difference*, Princeton, NJ: Princeton University Press, 2000.

6 Plants of the World Online (https://powo.science.kew.org/) accessed 2021.12.15.

7 Bernhard Adalbert Emil Koehne (1848–1918), an expert on the Rosaceae family, claimed it after examining the specimens sent in 1908 by a French missionary on Cheju Island, Emile Taquet (1873–1952), who collected plants to finance his missionary activities. It is suggested as a natural variety of *Prunus yedoensis*. Bernhard Koehne, "*Prunus* yedoënsis var. nudiflora nov. var. (Originaldiag.)," *Feddes Repertorium Novarum Specierum Regni Vegetabilis* 10 (1912): 507.

8 Japan colonized Karafuto (present day South Sakhalin) and the Kwantung Leased Territory in 1905, went on to occupy the equatorial Pacific Islands known as Nanyo (Micronesia) in 1914, and founded Manchukuo in 1931. Ramon H. Myers and Mark R. Peattie, eds., *The Japanese Colonial Empire, 1895–1945*, Princeton, NJ: Princeton University Press, 1984.

9 Nitobe Inazo, *Bushido: The Soul of Japan*, New York: G.P. Putnams Sons, 1905, 28, 33.

10 Requote from Chung Yountae 정연태, *Everyday Colonial Discrimination in Japanese Korea* (식민지 차별의 일상사), Seoul: Purŭnyŏksa, 2021, 237.

11 This was after his investigative trip to Korea in 1906. Nitobe Inazo, *Essays*, Tokyo: Teimi Shuppansha, 1907, 108.

12 Nakai Takenoshin, Matsumura's student, who came to study Korean flora, showed such a stance. Nakai Takenoshin 中井猛之進, *Reports on Plants of Cheju and Wan Islands* (済州島并莞島植物調査報告書), Seoul: The Government General of Korea, 1914, 53.

13 Emiko Ohnuki-Tierney, *Kamikaze, Cherry Blossoms, and Nationalisms: The Militarization of Aesthetics in Japanese History*, Chicago, IL: The University of Chicago Press, 2002, 122.

14 Ernest H. Wilson, *The Cherries of Japan*, Cambridge: The University Press, 1916.

15 This seems why some students of Matsumura tried to confirm Cheju as its natural habitat. Koidzumi Genichi 小泉源一, "The Habitat of the Somei Yoshino Cherry そめゐよしのざくらノ自生地," *BMT* 27(320) (1913): 395 Koidzumi Genichi , "Prunus Yedoensis Matsum Is a Native of Quelpaert! 染井吉野櫻の天生地 分明す," *Acta Phytotaxonomica et Geobotanica* (植物分類地理) 1(2) (1932): 177–79.

16 The Korean language media boomed in the 1920s as the GGK made conciliatory moves after the so-called March First movement. Koreans strongly revealed their dissatisfaction about Japanese rule with the March First movement, begun on March 1, 1919, in Seoul that reverberated throughout the peninsula that year despite the GGK's repression that killed thousands and arrested tens of thousands of peaceful marchers. For the complex makeup of Korean nationalism, see Michael Edson Robinson, *Cultural Nationalism in Colonial Korea, 1920–1925*, Seattle: University of Washington Press, 1988. Park Changsung 박찬승, *The Era of Nationalism* (민족주의의 시대), Seoul: Kyŏngin Ch'ulp'ansa, 2007.

17 Toh must be referring to Koidzumi's article in 1932 in note 12. Now some call *Prunus yedoensis* "Tokyo cherry" while giving another name, "King Cherry," to a natural hybrid in Cheju. Toh Bong-Syup, "Classification of Korean Plants: Five Genera and

about Five Hundred Species of Native Korean Plants (朝鮮産植物의 分類 (中) 朝鮮固有植物五屬五百餘種)," *Dong-a Daily* (동아일보), April 21, 1936.

18 Park Mankyu 박만규, "A Historological Survey on the *Prunus yedoensis* in Korea (韓國왕벚나무의 調査研究史)," *Journal of Botany* (식물학회지) 8(3) (1965): 12–15; Cho Ara, Seunghoon Baek, Goon-Bo Kim, Chang-Ho Shin, Chan-Soo Kim, Kyung Choi, Youngje Kang, Hee-Ju Yu, Joo-Hwan Kim and Jeong-Hwan Mun, "Genomic Clues to the Parental Origin of the Wild Flowering Cherry *Prunus yedoensis* var. *nudiflora* (Rosaceae)," *Plant Biotechnology Reports* 11 (2017): 449–59.

19 The Chinese texts *Shanhaijing* (山海經) and *Gujinzhu* (古今注), both attributed to the Jin period (265–420), were mentioned, among many others. U. Hoik 우호익, "About Hibiscus (무궁화고)," *Eastern Light* (동광) 13 (May 1927): 2–15, 12. Augustine Henry (1857–1930), British expert working at the time for the Kew Gardens, provided the current information that its origin was not Syria, but southern China. George Graves, "Hibiscus Syriacus," *Arnoldia* 1(8) (1941): 41–44, 41. Korean botanist did not or could not pay attention to the mistaken origin about their national flower, which was then discussed. Andrew C. Forbes, "Augustine Henry," *Empire Forestry Journal* 9(1) (1930): 129–31. Consensus that this was the rose of Sharon in the King James Bible (1611) came after much confusion in the twentieth century. Rose of Sharon was claimed to be a reed and an anemone, too. CWT, "What Was the Rose of Sharon?" *Science* 7(171) (1886): 439; Amos W. Butler, "What Was the Rose of Sharon?" *Science* 9(205) (1887): 13.

20 Requote from Federico Marcon, *The Knowledge of Nature and the Nature of Knowledge in Early Modern Japan*, Chicago, IL: University of Chicago Press, 2015, 95–96.

21 Eric Hobsbawm and Terence Ranger, eds., *The Invention of Tradition*, Cambridge: Cambridge University Press, 2012.

22 Sheldon Garon, "Transnational History and Japan's 'Comparative Advantage,'" *Journal of Japanese Studies* 43(1) (2017): 65–92, 77. For another case of this transnationally shaped tradition, see Victor Seow, "A Tradition of Invention: The Paradox of Glorifying Past Technological Breakthroughs," *East Asian Science, Technology and Society: An International Journal* 16(3) (2022): 349–66.

23 The cherry blossom is not the official national flower though used profusely in military uniforms and insignias to imbibe the way of samurai. Koreans made mugung hwa the national flower after independence. Ohnuki-Tierney, *Kamikaze, Cherry blossoms, and Nationalisms*.

24 Michael D. Gordin, *Scientific Babel: How Science Was Done Before and After Global English*, Chicago, IL: University of Chicago Press, 2015.

25 Meredith Townsend, *Asia and Europe*, Edinburgh: Archibald Constable & Co. Ltd., 1901.

26 For examples, Lucile H. Brockway, *Science and Colonial Expansion: The Role of the British Royal Botanic Gardens*, New York: Academic Press, 1979; Nicholas Jardine, James A. Secord and Emma C. Spary eds., *Cultures of Natural History*, Cambridge: Cambridge University Press, 1996; Lisbet Koerner, *Linnaeus: Nature and Nation*, Cambridge, MA: Harvard University Press, 1999; Emma C. Spary, *Utopia's Garden: French Natural History from Old Regime to Revolution*, Chicago, IL: University of Chicago Press, 2000; Richard H. Drayton, *Nature's Government*, New Haven, CT: Yale University Press, 2000; Londa L. Schiebinger, *Plants and Empire*, Cambridge, MA: Harvard University Press, 2004; Londa Schiebinger and Claudia Swan, eds., *Colonial Botany*, Philadelphia: University of Pennsylvania Press, 2005; Jim Endersby, *Imperial Nature: Joseph Hooker and the Practices of Victorian Science*, Chicago, IL: University of Chicago Press, 2008; Daniela Bleichmar, *Visible Empire: Botanical Expeditions and Visual Culture in the Hispanic Enlightenment*, Chicago, IL: University of Chicago Press, 2012.

27 Lorraine Daston, "Afterword: The Ethos of Enlightenment," in William Clark, Jan Golinski, and Simon Schaffer eds., *The Sciences in Enlightened Europe*, Chicago, IL: University of Chicago Press, 1999, 495–504, 495.

28 Michael Adas, *Machines as the Measure of Men: Science, Technology, and Ideologies of Western Dominance*, Ithaca, NY: Cornell University Press, 1990.

29 The early twentieth century witnessed the strong emergence of global practitioners aiming at "Western" science. As mentioned, this book uses "the West" as these actors' term. Fa-ti Fan, "Mr. Science," May Fourth, and the Global History of Science," *East Asian Science, Technology and Society: An International Journal* 16(3) (2022): 279–304; Marwa Elshakry, "When Science Became Western: Historiographical Reflections," *Isis* 101(1) (2010): 98–109; Benjamin A. Elman, *On Their Own Terms: Science in China, 1550–1900*, Cambridge, MA: Harvard University Press, 2005, xxi–xxxviii, 396–422; Benjamin A. Elman, "'Universal Science' Versus 'Chinese Science': The Changing Identity of Natural Studies in China, 1850–1930," *Historiography East and West* 1 (2003): 68–116.

30 Thomas Kuhn, *The Structure of Scientific Revolutions*, 2nd ed., Chicago, IL: University of Chicago Press, 1970; Jan Golinski, *Making Natural Knowledge: Constructivism and the History of Science*, Chicago, IL: University of Chicago Press, 1998; Bruno Latour, *Science in Action: How to Follow Scientists and Engineers through Society*, Cambridge, MA: Harvard University Press, 1987.

31 See Sidney Xu Lu, *The Making of Japanese Settler Colonialism: Malthusianism and Trans-Pacific Migration, 1868–1961*, Cambridge: Cambridge University Press, 2019, although Korea is not the focal point of the book. Among those one million Japanese in colonial Korea, 700,000 were civilians and the rest were military personnel. Excluding the military, it was also notable that almost one fourth of the civilians could be considered a part of the colonial bureaucracy, a stark contrast with the British colony of Kenya where only 80 English bureaucrats governed over 5 million Kenyans. Colonial Korea was remarkably well-staffed, even compared with French colonies. French Vietnam had 2,920 French administrative personnel for 17 million Vietnamese in 1937. Though the settlers had different opinions about colonial governance and development from the center, as those who greatly benefited from colonization, they mostly acted as "brokers" of the empire, as Uchida Jun's meticulous research on the Japanese settlers in Korea has shown. Jun Uchida, *Brokers of Empire: Japanese Settler Colonialism in Korea, 1876–1945*, Cambridge, MA: Harvard University Asia Center, 2011, 3, 23. For comparison with French Vietnam, Bruce Cumings, *The Origins of the Korean War*, vol. 1. Princeton, NJ: Princeton University Press, 1981, 11–12.

32 Simon Schaffer et al. eds., *The Brokered World: Go-Betweens and Global Intelligence, 1770–1820*, Sagamore Beach, MA: Science History Publications, 2009.

33 Mostly, I use 'systematics' to indicate the discipline, and 'classification' for the act.

34 Chakrabarty, *Provincializing Europe*; James A. Secord, "Knowledge in Transit," *Isis* 95(4) (2004): 654–72; Fa-ti Fan, *British Naturalists in Qing China*, Cambridge, MA: Harvard University Press, 2004; Harold Cook, *Matters of Exchange: Commerce, Medicine, and Science in the Dutch Golden Age*, New Haven, CT: Yale University Press, 2007; Kapil Raj, *Relocating Modern Science: Circulation and the Construction of Knowledge in South Asia and Europe, 1650–1900*, New York: Palgrave Macmillan, 2010; Bleichmar, *Visible Empire*; Fa-ti Fan, "Modernity, Region, and Technoscience: One Small Cheer for Asia as Method," *Cultural Sociology* 10(3) (2016): 352–68.

35 Sebastian Conrad, "Enlightenment in Global History: A Historiographical Critique," *The American Historical Review* 117(4) (2012): 999–1027, 1016; Sheldon Garon, "Transnational History and Japan's 'Comparative Advantage,'" *Journal of Japanese Studies* 43(1) (2017): 65–92, 72; Gregory K. Clancey, *Earthquake Nation: The Cultural Politics of Japanese Seismicity, 1868–1930*, Berkeley: University of California Press, 2006, 63; Kim Boumsoung 金凡性, *Meiji and Daisho Seismology: Beyond Local Science* (明治・大正の日本の地震学: ローカル・サイエンスを超えて), Tokyo: Tokyo Daigaku Shuppankai, 2007; Hoi-Eun Kim, *Doctors of Empire: Medical and Cultural Encounters*

between Imperial Germany and Meiji Japan, Toronto: University of Toronto Press, 2014, 10.

36 Fa-ti Fan, "Modernity, Region, and Technoscience: One Small Cheer for Asia as Method," *Cultural Sociology* 10(3) (2016): 352–68, 353.

37 James C. Scott, *Domination and the Arts of Resistance: Hidden Transcripts*, New Haven, CT: Yale University Press, 1990.

38 Sean Hsiang-lin Lei, *Neither Donkey nor Horse: Medicine in the Struggle over China's Modernity,* Chicago, IL: University of Chicago Press, 2014, 12. On the difficulty of finding another reason to challenge Western reason, see Gyan Prakash, *Another Reason: Science and the Imagination of Modern India*, Princeton, NJ: Princeton University Press, 1999.

39 Peter F. Stevens, *The Development of Biological Systematics: Antoine-Laurent de Jussieu, Nature, and the Natural System*, New York: Columbia University Press, 1994; David E. Allen, *The Naturalist in Britain*, Princeton, NJ: Princeton University Press, 1994; Christophe Bonneuil, "The Manufacture of Species: Kew Gardens, the Empire and the Standardisation of Taxonomic Practices in late 19th Century Botany," in M.-N. Bourguet, C. Licoppe, and O. Sibum eds., *Instruments, Travel and Science: Itineraries of Precision from the 17th to the 20th Century*, London: Routledge, 2002, 189–215. Carol Kaesuk Yoon, *Naming Nature: The Clash between Instinct and Science*, New York: W. W. Norton & Company, 2009. The success of Linné is attributed to Linné's dedication, its relative simplicity, and his globalizing strategy. Lisbet Koerner, *Linnaeus: Nature and Nation*, Cambridge, MA: Harvard University Press, 1999; Londa L. Schiebinger, *Plants and Empire*, Cambridge, MA: Harvard University Press, 2004.

40 Alix Cooper, *Inventing the Indigenous: Local Knowledge and Natural History in Early Modern Europe*, Cambridge: Cambridge University Press, 2007; Sharon E. Kingsland, *The Evolution of American Ecology, 1890–2000*, Baltimore, MD: Johns Hopkins University Press, 2005.

41 Federico Marcon, *The Knowledge of Nature and the Nature of Knowledge in Early Modern Japan*, Chicago, IL: University of Chicago Press, 2015; Maki Fukuoka, *The Premise of Fidelity Science, Visuality, and Representing the Real in Nineteenth-Century Japan,* Stanford, CA: Stanford University Press, 2012.

42 Philipp Franz Siebold, "On the Status of Japanese Botany (日本植物學ノ現狀ニツキテ)," in Philipp Franz Siebold ed., *Flora Japonica*, Tokyo: Shokubutsu Bunken Kankokai, 1937, 1–11.

43 Tashiro Kazui 田代和生, *An Investigation of Korean Medicines in the Edo Period* (江戸時代朝鮮藥材調査の研究), Tokyo: Keio Gijuku Daigaku Shuppankai, 1999.

44 Ian J. Miller, *The Nature of the Beasts: Empire and Exhibition at the Tokyo Imperial Zoo*, Berkeley: University of California Press, 2013, 50. The following would be an example. Suzuki Zenji 鈴木善次, *Biology-The Beginning* (バイオロジ-事始), Tokyo: Yoshikawa Kobunkan, 2005.

45 Oba Hideaki 大場秀章, a systematist retired from Tokyo University, and other systematist-cum-historians Kimura Yojiro 木村陽二郎 and Ueno Masuzo上野益三, who led the compilation of the biology volume of the compendium of the history of science in Japan in 1965, all expressed similar views. Meiji biology and agricultural science were disconnected from Tokugawa traditions. Oba Hideaki, *History of Botany/Cultural History of Botany* (植物学史・植物文化史), Tokyo: Yasakashobo, 2006, 139; Kimura Yojiro, *Naturalists of the Edo Era* (江戸期のナチュラリスト), Tokyo: Asahishimbunsha, 1988, 188. Ueno Masuzo, *A History of Natural History in Japan* (日本博物学史), Tokyo: Heibonsha, 1973, 137; The History of Science Society of Japan 日本科学史学会 ed., *A Compendium of History of Science in Japan* (日本科學技術史大系 , *Compendium* hereafter), vol. 15, Tokyo: Daiichi Hoki Shuppan, 1965, 15.

46 Hiroshige Tetsu 廣重徹, *Social History of Science* (科學の社會史), Tokyo: Iwanami Shoten, 1973/2002, 6–7. James R. Bartholomew, *The Formation of Science in Japan*,

New Haven, CT: Yale University Press, 1993; Tessa Morris-Suzuki, *The Technological Transformation of Japan: From the Seventeenth to the Twenty-First Century*, Cambridge: Cambridge University Press, 1994. For a thoughtful discussion of the Chinese making of industrial modernity with innovations also of traditions, see Eugenia Lean, *Vernacular Industrialism in China: Local Innovation and Translated Technologies in the Making of a Cosmetics Empire, 1900–1940,* New York: Columbia University Press, 2020.

47 Gregory K. Clancey, *Earthquake Nation: The Cultural Politics of Japanese Seismicity, 1868–1930*, Berkeley, CA: University of California Press, 2006; Boumsoung Kim 金凡性, *Meiji and Daisho Seismology: Beyond Local Science* (明治・大正の日本の地震学: ローカル・サイエンスを超えて), Tokyo: Tokyo Daigaku Shuppankai, 2007; Miyagawa Takuya, "The Meteorological Observation System and Colonial Meteorology in Early 20th-Century Korea," *Historia Scientiarum* 18 (2008): 140–50. Victoria Lee, *The Arts of the Microbial World: Fermentation Science in Twentieth-Century Japan*, Chicago, Il: University of Chicago, 2021. On the influence of the lingering "catch-up" model in Japanese historiography of science, see Kim Yung Sik, "Problem of Early Modern Japan in the History of Science in East Asia," *Historia Scientiarum* 18 (2008): 49–57.

48 These works also properly blur the line between science and technology. Harry Collins, *Changing Order: Replication and Induction in Scientific Practice*, Chicago, IL: University of Chicago Press, 1985; Pamela O. Long, *Artisan/Practitioners and the Rise of the New Sciences, 1400–1600,* Corvallis, OR: Oregon State University Press, 2011; Lissa Roberts, Simon Schaffer, and Peter Dear eds., *The Mindful Hand*, Amsterdam: Koninkliijke Nederlandse Akademie van Wetenschappen, 2007. Pamela Smith, *From Lived Experience to the Written Word: Reconstructing Practical Knowledge in the Early Modern World*, Chicago, IL: University of Chicago Press, 2022.

49 David Arnold, *The Tropics and the Traveling Gaze: India, Landscape, and Science, 1800–1856*, Seattle: University of Washington Press, 2006, 8.

50 Chakrabarty's attempt to provincialize Europe emphatically warns that "provincializing Europe cannot ever be a project of shunning European thought," and Arnold's thoughtful diffusion of diffusionism of European technology notes the undeniable significance of the great inventions like railroad and telegraph. Yet no matter how articulate and creative those ideas and inventions were, they existed thanks to local builders who had to put into them their equally ingenious and original thoughts in materializing those ideas in their different fields while necessarily reshaping the "originals." Dipesh Chakrabarty, *Provincializing Europe: Postcolonial Thought and Historical Difference*, Princeton, NJ: Princeton University Press, 2000, 255; David Arnold, "Europe, Technology, and Colonialism in the 20th Century," *History and Technology* 21(2005): 85–106, 91–93.

51 Francesca Bray, "Only Connect: Comparative, National, and Global History as Frameworks for the History of Science and Technology in Asia," *East Asian Science, Technology and Society* 6 (2012): 233–41. Notably, Mueggler treats his non-European actors as knowledge practitioners with independent agendas, though their concurrently emerged knowledge culture was shaped around their roles as plant collectors. Erik Mueggler, *The Paper Road: Archive and Experience in the Botanical Exploration of West China and Tibet*, Berkeley: University of California Press, 2011. See also, Gabriela Soto Laveaga, *Jungle Laboratories: Mexican Peasants, National Projects, and the Making of the Pill,* Durham, NC: Duke University Press, 2020.

52 Gregory Clancey, "Japanese Colonialism and its Sciences: A Commentary," *East Asian Science, Technology and Society* 1(2) (2007): 205–11.

53 Mary Louise Pratt, *Imperial Eyes: Travel Writing and Transculturation*, London: Routledge, 2008, 8.

54 All those were more or less how colonized Koreans or Japanese had felt about Japan, although terms like 'West' or 'metropolis' for Japan may look strange to non-specialists of

the region. Komori Yoichi 고모리 요이치, *Postcolonial: Colonial Sub-Consciousness and Colonial Consciousness* (포스트콜로니얼: 식민지적 무의식과 식민주의적 의식), tr. Song Tae-uk, Seoul: Samin, 2002.

55 For Korean transitions between the two powerful cultures of China and Japan through these negotiations, see Andre Schmid, *Korea between Empires, 1895–1919*, New York: Columbia University Press, 2002.

56 See note 1.

57 Takashi Fujitani, *Race for Empire: Koreans as Japanese and Japanese as Americans during World War II*, Berkeley: University of California Press, 2011; Frantz Fanon, *Towards the African Revolution: Political Essays*. Translated by Haakon Chevalier, New York: Grove Press, 1988, 31–44.

58 In Japan, too, though dwindling, there were those who said: "There is thus no need to endlessly repeat 'nation, nation' all the time," distancing themselves from the jingoistic appeal for a more and more powerful Japan. Tanaka Akira (田中彰), *The Idea of Small Japan: Re-Reading Japanese Modernity* (소일본주의 : 일본의 근대를 다시 읽는다) Kang Chin-a 강진아 tr., Seoul: Sohwa, 2002.

59 My engagement with recent works is admittedly limited. Research Group on Japanese Colonialism (日本植民地研究会) ed., *The Status and Task of Japanese Colonialism Studies* (日本植民地研究の現状と課題), Tokyo: Atenesha, 2008; Kawamura Yutaka (河村豊), et al., "The Status and Task for War-Time Science in Japan: The Report from the 2003 Annual Conference (日本戦時科学史の現状と課題(2003年度年会報告)," *Journal of History of Science, Japan. Series II* (科学史研究. 第II期) 43(229) (2004): 45–56; Togo Tsukahara, "Introduction to Feature Issue: Colonial Science in Former Japanese Imperial Universities," *East Asian Science, Technology and Society* 1(2) (2007): 147–52; Akihisa Setoguchi, "Control of Insect Vectors in the Japanese Empire: Transformation of the Colonial/Metropolitan Environment, 1920–1945," *East Asian Science, Technology and Society* 1(1) (2007): 167–81; Hiromi Mizuno, *Science for the Empire*, Stanford, CA: Stanford University Press, 2009; Daqing Yang, *Technology of Empire: Telecommunications and Japanese Expansion in Asia, 1883–1945*, Cambridge, MA: Harvard University Asia Center, 2010; Aaron S. Moore, *Constructing East Asia: Technology, Ideology, and Empire in Japan's Wartime Era, 1931–1945*, Stanford, CA: Stanford University Press, 2013; David G. Wittner and Philip C. Brown, eds. *Science, Technology, and Medicine in the Modern Japanese Empire*, New York: Routledge, 2016 Sakano Toru (坂野徹), *Investigating the Empire: A History of Science of Colonial Fieldwork* (帝国を調べる: 植民地フィールドワークの科学史), Tokyo: Keiso Shobo, 2016; Sakano Toru and Tsukahara Togo (塚原東吾), *Intellectual History of Science in Imperial Japan* (帝国日本の科学思想史), Tokyo: Keiso Shobo, 2018; Nakashima Koji (中島弘二) ed. *The Japanese Empire and Forests: Conservation and Development of Forest Resource in Modern East Asia* (帝国日本と森林——近代東アジアにおける環境保護と資源開発), Tokyo: Keiso Shobo, 2023. Aya Homei, *Science of Governing Japan's Population*. Cambridge: Cambridge University Press, 2023. There are many important works to be cited in chapters by Fujiwara Tatsushi (藤原辰史), Iijima Wataru (飯島渉), and Shin Chang-Geon (愼蒼健), etc.

60 The agency of colonial Taiwanese farmers can be seen greatly when it discusses mainly Taiwanese technology. Shuntaro Tsuru, "Irrigation Pumps in late Colonial Taiwan: Farmers' Utilization of Technology and the Transition to Rice Cultivation," *Modern Asian Studies* 57(6) (2023): 1–37.

61 These studies become multi-sciences more often when they deal with Japanese and Western sciences. Gregory K. Clancey, *Earthquake Nation: The Cultural Politics of Japanese Seismicity, 1868–1930*, Berkeley: University of California Press, 2006; Kim Hoi-Eun, *Doctors of Empire: Medical and Cultural Encounters between Imperial Germany and Meiji Japan*, Toronto: University of Toronto Press, 2014; Morris Low, *Science and the Building of a New Japan*, New York: Palgrave Macmillan, 2005; Lisa

Onaga, "Toyama Kametaro and Vernon Kellogg: Silkworm Inheritance Experiments in Japan, Siam, and the United States, 1900–1912," *Journal of the History of Biology* 43(2) (2010): 215–64.

62 In Korea also, the botanist-cum-historians produced earlier works on the making of Korean botany, which always discussed Japanese influence. Chŏng Yŏng-ho 정영호, *An Outline of History of Plant Classification in Korea* (한국식물분류학사개설), Seoul: Academy Books, 1986. *Compendium of History of Modern Culture in Korea* (한국현대문화사대계), vol.3, Seoul: Koryŏ Univ. Minjongmunhwayŏn'guwŏn, 1977; Lee Deok-Bong 이덕봉, "History of Recent Development of Korean Botany: The Institutes and Status of Botany in Colonial Korea (최근세조선 식물학연구사)," *The Journal of Asiatic Studies* (亞細亞硏究) 4(2) (1961): 101–49. These recent works that pay attention to the active development of natural history by Korean researchers during the colonial era importantly deal with interactions between Korean and Japanese scientists. Moon Manyong 문만용, "Butterfly-Taxonomy of 'the Korean Biologist,' Seok Joo myung ('조선적 생물학자' 석주명의 나비분류학)," *Journal of the Korean History of Science Society* (한국과학사학회지) 21 (1999): 157–93; Kim Sungwon 김성원, "The Context of a Korean Naturalist's Career-building in Colonial Korea: Cho Pok Sung as an Example of Colonial Entomologist (식민지시기 조선인 박물학자 성장의 맥락: 곤충학자 조복성의 사례)," *Journal of the Korean History of Science Society* 30 (2008): 353–82; Nayoung Aimee Kwon, *Intimate Empire: Collaboration and Colonial Modernity in Korea and Japan*, Durham, NC: Duke University Press, 2015, 9; Todd A. Henry, *Assimilating Seoul: Japanese Rule and the Politics of Public Space in Colonial Korea 1910–1945*, Berkeley: University of California Press, 2014; Lee Jeongseon 이정선, *Assimilation and Exclusion: Japanese Assimilation Policies and the Korean-Japanese Marriage* (동화와 배제: 일제의 동화정책과 내선결혼), Koyang: Yŏksabipyŏngsa, 2017; Yun Hae-Dong 윤해동, *Towards the East Asian History: Transnational History and Colonial Modernity* (동아시아사로 가는 길 : 트랜스내셔널 역사학과 식민지 근대), Seoul: Ch'aekwahamkke, 2018.

63 Ramon H. Myers and Mark R. Peattie eds., *The Japanese Colonial Empire, 1895–1945*, Princeton: Princeton University Press, 1984, 399–420; On the possible limitations that it caused to the government's ability to rule, particularly in comparison with the well-funded Taiwanese colonial government, see Moon Myung-ki 문명기, "A Comparative Study on the Annual Expenditures of Taiwan and Chosun Government General during the Japanese Colonial Period (일제하 대만 조선총독부의 세출구조 비교분석)," *Theses of Korean Studies* (한국학논총) 44 (2015): 409–54.

64 Kwon also says that "civilizing Korea" was self-civilizing for Japan, citing Sebastian Konrad. Kwon Tae-eok (권태억), *Japanese 'Civilizing mission' in Korea (1904–1919)* (일제의 한국 식민지화와 문명화), Seoul: Seoul Nat'l Univ. Ch'ulp'anmunhwawŏn, 2014, 53. For Japanese emphasis on the "scientific colonialism" shaped since its colonization of Taiwan, see Ming-cheng Miriam Lo, *Doctors within Borders: Profession, Ethnicity, and Modernity in Colonial Taiwan*, Berkeley: University of California Press, 2002.

65 David Fedman, *Seeds of Control: Japan's Empire of Forestry in Colonial Korea*, Seattle: University of Washington Press, 2020. In comparison, the following collection presents in the introduction a less dynamic perspective on Japanese imperial forestry, seeing it as the strategic response to secure colonies and mobilize their resources for the safety of the entire empire, guided by the sense of crisis that Japan felt as an island nation against Western imperial powers. While there certainly would have been such strategic minds asserting themselves more forcefully especially in the war-time regime after the Sino-Japanese War to the end of the World War II, tensions within domestic politics and competitions between colonies cannot be downplayed. Nakashima Koji (中島弘二) ed. *The Japanese Empire and Forests: Conservation and Development of Forest Resource in Modern East Asia* (帝国日本と森林——近代東アジアにおける環境保護と資源開発), Tokyo: Keiso Shobo, 2023, 4.

66 Albert L. Park, *Building a Heaven on Earth: Religion, Activism, and Protest in Japanese-Occupied Korea*, Honolulu: University of Hawai'i Press, 2015, 8.

67 As the following work suggests, the often mentioned differences between Japanese colonial experiences in Korea and Taiwan, as Korean resistance versus Taiwanese collaboration, should be more concretely tested. The famous Korean resistance explains so little about colonial Korean modernity with its botanists. Jina E. Kim, *Urban Modernities in Colonial Korea and Taiwan*, Leiden: Brill, 2019. For works delineating various Taiwanese resistance, see Miriam Lo, *Doctors within Borders*, Shuntaro Tsuru, "Irrigation Pumps in late Colonial Taiwan: Farmers' Utilization of Technology and the Transition to Rice Cultivation," *Modern Asian Studies* 57(6) (2023): 1–37

68 Kim Geun-Bae 김근배, *The Emergence of Modern Techno-Scientific Manpower in Korea* (한국 근대 과학기술 인력의 출현), Seoul: Munji, 2005; Noh Dae-hwan 노대환, *Studies on the Making of the Eastern Way and Western Means Idea* (동도서기론 형성 과정 연구), Seoul: Ilchisa, 2005.

69 The term coined by the American Du Bois also echoes Franz Fanon's analysis of the black soul in French Algeria, as it does for all those who are subjugated and discriminated against. William Edward Burghardt Du Bois, *The Souls of Black Folk*, New York: Penguin, 1903.

1 European Botany Made in Japan

The large green onion of Korea, an ordinary enough plant, is presented here in the elegant style of Chinese orchid painting, though superb in detail, as shown in the roots, and no less realistic (See Figure 1.1).[1] It was drawn by a Japanese painter, who was dispatched three times to Chosŏn Korea (1392–1910) between 1718 and 1722 as part of a serious mission that went on until 1751, based on the Japanese trading house in Pusan, a port city facing Japan. He was to draw things that could not be easily shipped back to Japan, while collectors in the team made dried and stuffed specimens of plants and animals or pots and cages of living ones. The enlightened shogun Tokugawa Yoshimune (徳川吉宗, 1684–1751) wanted to know more about the Korean plants and animals recorded in the 25-volume Korean medical text, *Treasured Mirror of Eastern Medicine* (東醫寶鑑 Tongŭibogam, 1613), then in wide circulation for its clinical efficacy in Japan and China. While he put forward an understanding of this Korean text as the primary purpose of the mission, the urgent task was to find a way to naturalize expensive Korean medicinal plants like ginseng, which, owing to its "strangely heated" popularity among

Figure 1.1 A depiction of the large green onion of Korea, in an elegant style of Chinese orchid painting. From "Animal and Plant Illustrations," *Rectification Records of Medical Material Survey, Documents in Tsushima Souke Bunko*, vol. 5. (National Institute of Korean History, http://db.history.go.kr/id/ts_005_$1ill, accessed 2024.01.07).

DOI: 10.4324/9781003511755-2

Japanese including the shogun himself as a cure-all energizer, had been draining the Tokugawa government's silver holdings. Niwa Seihaku (丹羽正伯, 1691–1756), a *honzo* (本草 materia medica, bencao in Chinese, bonch'o in Korean) scholar in charge of this mission, effectively answered his sponsor's needs. The domestic cultivation of ginseng that the Tokugawa government had tried since 1637 finally succeeded around 1738. This Japanese ginseng came to meet the huge demand, though people immediately noticed its low efficacy.[2]

Niwa also added Japanese names to 1,250 medicines in the *Treasured Mirror*, missing just 132 out of 1,392 of the entire list. He could do this because of similar country-wide investigations of Japanese plants and animals that he had orchestrated by sending directions and surveys to all prefectural governments, heightening the empirical turn in Japanese *honzo* studies modeled after his 30 year investigation in Korea.[3] Although it is not a well-known episode in the Japanese study of plants, it is not an exception. In 1784, several Japanese visited a state medicine shop in Beijing. This time they carried illustrations and specimens, seeds, and roots of those illustrated plants that they had collected in the Ryukyu Kingdom (1429–1879, present day Okinawa). They wanted to learn about the identities of those plants from Chinese medicinal dealers.[4]

Around the same time, Japan also received several European visitors, who wanted to learn more about Japanese plants through the Dutch trading house in Dejima, Nagasaki, which was about 35 times smaller than the Japanese one in Pusan. The last one of them, Philipp Franz von Siebold (1796–1866), who stayed longest in Japan from 1823 to 1829 before its opening, enthusiastically reported to Europe in 1829:

> Except Europe, there is no place where studies on plants are more advanced than Japan and China. Also, both Japan and China have used immense varieties of botanical riches collected from mountains and fields for food, clothes, houses, and amusement for more than a thousand years. Both Japan and China have imported the most valuable plants that each other had, and collected useful plants from hinterlands, neighbors, and archipelagos. They carefully acclimatized these plants in cultivating numerous exquisite and perfect varieties.[5]

In the early twentieth century, Japanese botanists made these kinds of transnational studies of plants an all-out passion. Japanese botanists and collectors showed themselves in all of its newly integrated territories of Okinawa and Hokkaido, parts of China, and newly colonized territories like Taiwan, Korea, and Karafuto (present day Sakhalin). They actively communicated with European botanists, sending and requesting specimens and publishing in Latin and English. It looks as if Japan effectively used its adventurous tradition and vibrant exchanges with other cultures in plant studies to develop a modern approach.

However, no Japanese botanist at the time appears to have thought so. In fact, until recently hardly any historical work even mentions the early expeditions of Tokugawa scholars to their Asian neighbors, or Meiji botany's substantial connections to Tokugawa tradition. The emphasis was on past failures to profit from earlier European contacts. What was also said and believed about the Japanese tradition of studying plants was that this diversely open knowledge tradition was insular, stuck in the detrimental tradition of China. Japan was said to have failed to develop full-blown natural history, and to have remained at the stage of *honzo* studies, merely investigated medicinal plants bound by Chinese *bencao* tradition.[6]

Like every story, this story of Japanese traditions with selected, excluded, and modified memories did not emerge out of nowhere; it had its persuasive storytellers. This chapter introduces the revolutionaries who created this story, to see how they led Japan's remarkable transformation about studying plants upon it. These conscious pioneers of Japanese botany eagerly accepted the European claim that there was one modern science, singularly European yet universal, which was a necessary acquisition for every nation wishing to be considered enlightened and modern.[7] By preemptively accepting this European claim, they dismissed the worth of Japanese traditions, calling for drastic changes toward the "European" path. They especially made transnational studies of plants as a characteristically European approach that the new nation had to support if it was to be serious about its goal.

Unmodern Interactions and European Botany Made in Japan

The Meiji revolutionaries who established a new regime upon the turmoil surrounding Japan's opening to the West, promised to renew Japan, all in European fashion. The striking catch-phrase "Escaping Asia towards Europe (脱亞入歐)" seized the imagination of these revolutionaries, denying their past in favor of thorough-going Europeanization, ranging from food and clothes to all kinds of institutions such as military, education, tax, legal, and administrative systems. However, Japanese botanists leading this transformation found a certain past tradition that they could not ignore. For it was properly European as they saw it, yet appeared to hamper their effort to Europeanize Japanese botany.

This tradition consisted of the fruitful explorations of Japanese flora made by several Europeans based in the Dutch trading post in Nagasaki. For example, Engelbert Kaempfer (1651–1716) had visited between 1690 and 1692; Carl Linné's (1707–1778) pupil Carl Peter Thunberg (1743–1828) had done so from 1775 to 1776; and Philipp Franz von Siebold (1796–1866) from 1823 to 1829. All three produced works on Japanese flora, not least the 30 volumes of *Flora Japonica* (1835–1870) that resulted from Siebold's final visit. Japanese scholars like Matsumura Jinzo who began their investigations of Japanese flora at the end of the nineteenth century shared the feeling that foreigners "had greedily collected Japanese plants since the eighteenth century" and "descriptions for most Japanese plants were [already] finished."[8] Their own investigations appeared pointless. These Japanese botanists presented themselves as passive victims of the European botanical explorations, forgetting their own ambitious explorations in Asia and the vibrant plant studies traditions that informed those explorations as well as European ones. The vast number of classical Chinese and Japanese works were not their concerns; instead, a few impressive Western-language works on Japanese plants became all their significant past.

Their memory of those European expeditions in Japan left out much detail about the actual interactions. As recent studies have demonstrated, the so-called "Closed Country (鎖國 sakoku)" system of Tokugawa Japan was not a system that closed off Japan from the outside world. It rather allowed Japan to set its own rules for international relations.[9] For example, with the exception of Siebold, European scholars, allowed into Nagasaki as doctors for Dutch traders, did not have much freedom to go outside the small island of Dejima, an island artificially built exactly for the purpose of controlling their free access to the mainland. Siebold's outings were also mostly confined to the vicinity of Nagasaki, except one trip to Edo, during which he met the shogun and some important naturalists. Given the limitations of their movements, what could be seen during those European expeditions depended largely on what their Japanese counterparts would share with them.

Their Japanese counterparts were generally generous but more so when they had a greater return in mind. Kaempfer, who stayed in Japan for about two years and made those ritualistic trips to Edo twice, had achieved less than Thunberg, who stayed about a year and made the trip once, although Kaempfer's fortune and legacy were significant too. When Kaempfer had arrived in 1690, his Japanese counterparts had little interest in what Kaempfer might be able to offer to them. They perceived "the Dutch" mostly as red-haired barbarians, and "the Dutch Learning (蘭学 rangaku)" group had yet emerged. The famous translation of the *New Text on Anatomy* (解体新書 Kaitai shinsho) did not appear until 1774.[10] Still, Kaempfer had an excellent interpreter, a learned young man who obtained all the books that he wanted to see, and explained and translated the passages indicated, as Kaempfer noted in his *Amoenitatum Exoticarum* (delightful exotics, 1712). Nakamura Tekisai's (中村惕斎, 1629–1703) *Illustrated Encyclopaedia for Everyone* (訓蒙図彙 Kinmozui), heavily used in Kaempfer's various works, seemed a key to their fruitful communication, as he could point out the illustrations and ask for explanations. Thanks to these exchanges, Kaempfer described about 500 Japanese plants with about 200 illustrations in *Amoenitatum Exoticarum.*[11] Thunberg and Linné all utilized Kaempfer's introduction to Japanese flora and added much more. Thunberg's monograph on Japanese flora, *Flora Japonica* (1784) described about 1,000 Japanese plants.[12]

Thunberg, who came in 1775 to help Linné's grand scheme to build a universal system of natural classification, was fortunate enough to meet many Japanese scholars and doctors, including two capable Dutch Learning scholars, Nakagawa Junan (中川淳庵, 1739–1786) and Katsuragawa Hoshu (桂川甫周, 1754–1809), who just finished the translation of the *New Text on Anatomy* and searched for more Dutch knowledge to learn. They offered specimens and illustrated Japanese works such as *Collection of Flowers* (花彙 Kai, 1765) by the so-called Japanese Linné Ono Ranzan (小野蘭山, 1729–1810), in exchange for European information and books. Linné, who named hundreds of Japanese plants in *Species Plantarum* (1753–), added some more in *Mantissa Plantarum Altera* (1771), including the famous *Ginkgo biloba* L., confirming Kaempfer's illustration and description (See Figure 1.2).[13]

Figure 1.2 A botanical illustration depicting a branch of a ginkgo tree, the shape of leaves, divided at the center, reflected in Linné's species epithet biloba, to be discussed further in the conclusion. Engelbert Kaempfer, *Amoenitatum Exoticarum*, Lemgoviae: Typis & impensis Henrici Wilhelmi Meyeri, aulae Lippiacae typographi, 1712, 813 (The Biodiversity Heritage Library).

Three other disciples of Linné dispatched earlier to China, Christoffer Tärnström (1711–1746), Olof Torén (1718–1753), and Pehr Osbeck (1723–1805), failed to secure any specimen or information about this tree widely cultivated in China. Thus, its first western name was *ginkgo*, a misspelled form of the Japanese name *ginkyo* (銀杏), not *yinxing*, the Chinese pronunciation of the same Chinese characters.[14] Ginkgo, a name globalized for this fossil tree in and outside the scientific communities, shows the imprint of strong Japanese naturalist tradition, which, with many other such traditions, had shaped the European understanding of the plant world. Although Linné himself did not note on this contribution, like for most other helps from other cultures that shaped his "natural system," Thunberg's monograph clearly noted the knowledgeable help of Nakagawa and Katsuragawa.[15]

Siebold, by some accounts the most well-liked foreigner in Japan, benefited the most from the strong initiative of the growing Japanese Dutch Learning communities. For both Siebold and his Japanese counterparts, the changing international milieu of the early nineteenth century greatly helped. Korea and China were losing importance to Japan, in part because of Japan's production of ginseng and silk, while the frequent appearances of Western ships near its shores increased the Tokugawa government's need to know about this new force.[16] When the Dutch learning scholars attempted an unprecedented interaction through the Dutch trading port, the Tokugawa government readily supported them.[17]

Changing dynamics within Europe pushed Japan's European partner, too. The Netherlands, cornered by heightened imperial competition, tried to strengthen its special relationship with Japan. It wanted to fully open up Japan for more lucrative trade that would restore its glory.[18] Siebold, an enthusiastic youth having just graduated from medical school, seized upon this mutually growing interest in 1823. He would collect information about Japanese resources and goods for the Dutch East India Company while answering the Dutch Learning scholars' wish to have a physician well versed in natural history in the trading port.[19] Siebold could regularly visit Nagasaki due to Japanese scholars' efforts to increase their contact with him. These scholars established a small school called Narutaki-jyuku in a Nagasaki suburb, where the inexperienced Siebold "taught," and dozens of mature Japanese scholars from all over Japan "learned." The Company gave substantial support to maintain this promising relationship, by providing scientific devices, instruments, and books that Siebold requested for his students.[20]

As in the previous interaction with Thunberg, it was a mutually beneficial learning experience. Siebold collected large quantities of material, especially by mobilizing his Japanese students; even after donating and selling many items to various European countries, he was still able to build a museum and publish three huge tomes about Japan: *Nippon* (1832), *Fauna Japonica* (1833–1850), and *Flora Japonica* (1835–1875). Except for *Nippon*, both were coauthored to compensate for his limited skills as a naturalist. Japanese naturalists also improved their knowledge of European natural history a great deal through this contact. An immediate result was the *Commentary on the Honzo Naming of the Great West* (泰西本草名疏 Taisei honzo meiso, 1829), the first full application of the Linnaean system to Japanese plants by a Japanese, Ito Keisuke (伊藤圭介, 1803–1901), a student at Narutaki-jyuku.[21]

Siebold was enthusiastic about the Japanese botanical tradition, which thoroughly enriched his *Flora Japonica*. Adding to the above high praise, he claimed that the "Japanese had investigated Japanese floral zones more fully than Europeans had done for theirs" by the time of Thunberg's arrival, and Europeans were then only just starting to investigate their local floras, of which only some general garden plants and herbs had been studied. He also thought that the Japanese plant names in Chinese characters, which showed the relationships between plants like Linnaean binomials, were sufficiently standardized throughout Japan through its book culture to accelerate the development of botanical knowledge.[22] Siebold took some pride in the Linnaean system and the wonderful hothouses filled with tropical plants in Europe, but on practical aspects, he believed Japan had developed much advanced knowledge about plants.[23] His remarks indicate who it was who completed the study of Japanese plants by the nineteenth century, which was neither himself nor Thunberg nor any other Europeans, but Japanese scholars working within their own transnationally shaped traditions.

Siebold knew that European encounters with Japan and Japanese encounters with Europe were not one party's superior and entirely different knowledge system flowing into the other culture and transforming it. Just as the arrival of the name *Ginkgo biloba* did not indicate the inflow of Japanese knowledge into Europe, driving Europe's modernization, neither were these encounters to be seen as the fortunate early influx of Western knowledge that suddenly shocked and enlightened a Japan stuck in its Asian traditions. The transformation of knowledge systems and practices in any society with its own intellectual and social inertia could not be that simple. One clear thing to note here is that Japanese scholars and their strong knowledge tradition had much more agency in these exchanges than has so far been presumed.

Naturally, no Japanese scholars at the time saw themselves as passive victims of European exploitation. They just used the interactions for their own knowledge practice and social recognition, which at least provided them with pupils and employment, as their exchanges with Chosŏn Korea and Ryukyu also did. The European expeditions to Japan sponsored by the Dutch East India Company were driven by the inchoate nationalism of the Netherlands, just as Japanese expeditions were driven by their own inchoate nationalism and economic interest. However, these Japan-Europe interactions display characteristics that differ from Japan's immediate relationship with its neighbors or from European imperial powers' relationship with their remote colonies: three characteristics to mull over. First, no parties, even Linné, seems to have conceived the plants that they explored as being a natural bounty left untouched by primitive people lacking culture. They explored these lands for more fully cultured natural riches to study and cultivate. Second, these interactions that began with curiosity about an unknown culture led to increased mutual respect for some as learning deepened. Third, based on this apparent mutual respect, neither party sought to impose their own ways of learning on the other. That is, there was no 'civilizing mission,' and its absence allowed cross-fertilization. While no one would describe these interactions as traditional, these were different from the 'modern' interactions that would be soon shaped around colonial Korea.

Old Ghosts in the Enlightened Nation

Tsurumi Shunsuke, an influential intellectual historian of pre-war Japan, claimed that the Japanese felt themselves isolated in their archipelago, "far away from the more advanced and more universal culture," which generated a "sense of cultural inferiority, curiosity about other cultures, and the desire to learn novel things."[24] Tokugawa Japan also had their own biases about other cultures in developing this sense of isolation. They felt isolated mostly because Middle Kingdom China, the universal and advanced culture that imposed a civilizational order that rendered everyone else barbarian, appeared out of reach. Ming and Qing China did not even bother to receive Japanese envoys, only allowing commercial exchanges. Possibly, the interactions with those 'minor' cultures like the Dutch, Korea, and Ryukyu could be more open because they were not China. What is less acknowledged is that those diverse interactions did not reflect Japan's reliance on other cultures but its strong knowledge market, in the making by interactively emerging material and print culture. Much more than passing exchanges, these social and intellectual institutions built by curious knowledge practitioners in Japan undergirded Japan's 'Europeanization.'

The Tokugawa government was the first Japanese government that seriously employed Confucian scholars like Hayashi Razan (林羅山, 1583–1657), who served the first four shoguns. Replacing Buddhist monks as the most important knowledge practitioners, these Confucian advisers created private and publicly funded schools, cultivating the next generations of government advisers while Confucianizing more and more samurai who did not have many battles left to fight. Yet Japanese Confucianization had unique features which distinguished it from the situation in China or Korea, where Confucian bureaucrats qualified to enter the ruling class through success in civil service examinations. The knowledge practitioners serving the samurai ruling class in Japan had no exams to take, hence no canons to memorize. Instead, it was better for them to distinguish themselves with special expertise.[25] Hayashi Razan's example and sanction of *honzo* studies made the investigation of natural things a widely shared specialty among Confucian scholars, whose livelihood often involved medicine. Hayashi famously offered the Chinese work, Li Shizhen's (李時珍, 1518–1593) *Systematic Materia Medica* (本草綱目 Bencaogangmu, 1596), in 1607 to the first shogun, and annotated it.[26] Intellectually inclined samurai and merchants also joined this trend. Florists and horticulturalists, thriving in the urbanizing culture of Japan, participated through writing and practice, supporting the famous Japanese love for gardening and bonsai. They published their own treatises helping their customers in the proper cultivation and appreciation of virtuous 'Confucian' plants like bamboos and chrysanthemum, jumping into the rising print market.

In this vibrantly evolving tradition, Li Shizhen's vast work, empirically rich and infused with his unique Confucian cosmology, was the authority.[27] The Japanese Linné Ono Ranzan, whose work had much enriched that of Linné himself, completed its Japanese localization by his extensive collection trips and philological investigations. His most influential work was his 48 volumes of *Enlightenment of Systematic Materia Medica* (本草綱目啓蒙 Honzo komoku keimo, 1803–1806), the compilation of his lectures on Li Shizhen's work. Yet, Ono had no reason to limit himself to Li's work in expanding his knowledge, freely citing all available works including a considerable number of Korean texts, including the *Treasured Mirror*. Although his work was not flawless,[28] his expansive work on the *honzo* tradition was what garnered the most admiration from Japanese naturalists in that period, including those Dutch Learning scholars who interacted with Siebold.[29]

Lots of these studies relied on the vibrant knowledge market that allowed Ono Ranzan to concentrate on his studies while just lecturing at his private school without even practicing medicine. Yet, these scholars did not mind the calls from the bakufu or local governments, early on establishing the relationship between knowledge practice on nature and governments.[30] Those scholars more receptive to the bakufu's or local governments' practical concerns fashioned their work as 'Studies on Things and Products' (物産學 butsusan gaku). Though similar to the 'Studies on Names and Things' (名物學 meibutsu gaku), or the 'Studies on Broad Things' (博物學 hakubutsu gaku) that were mainstays of the Confucian tradition, *butsusan* studies more clearly defined a practical and empirical orientation, seen as an investigation of natural resources and products to help strategize the economic pursuits of a region or the central government. Niwa Seihaku's expeditions to Chosŏn Korea and country-wide surveys reflected and strengthened such a practical and empirical turn. His detailed questionnaires addressed to each local government helped further spread this interest in nature studies among local samurai and merchants. They often formed groups for studying local riches, and exhibited them through specimens and paintings, etc., utilizing visual techniques like microscopes, photographs and *ukiyo-e*.[31]

As Siebold enthusiastically noted, all this constituted an impressive and well-entrenched tradition of plant studies. These scholars with various expertise and established working relations with local and central governments naturally had much to contribute to the modernization of Japanese botany, and they gave it generously. The Bureau of Natural History (博物局 Hakubutsu kyoku), established in 1871 under the Ministry of Education and staffed by naturalists of solid Tokugawa lineage, symbolizes the importance of the Tokugawa tradition for Japanese modernization of science and technology in general. Tanaka Yoshio (田中芳男, 1838–1916), a central figure in the bureau, was a naturalist who learned *honzo* and medicine from Ito Keisuke. Ono Motoyoshi (小野職愨, 1838–1890), a great-great-grandson of Ono Ranzan also joined the bureau.[32]

These people well immersed in various Tokugawa traditions left a lasting mark on Japanese modern botany, through the institutional base that they had built up and the ideological foundation on which it was constructed. The bureau's role in setting up science education is one example. Established a year before the drafting of the "Education Decree (學制, 1872)," the bureau promoted natural history as the central subject for modern science education, in its traditional term of *hakubutsu*, not in the later modernized term *shizenshi* 自然史, natural history.[33] At first, natural history failed to receive the highest attention from Meiji reformers. The primary choices shown in the Decree were physics, chemistry, and mathematics. They believed that these hard sciences represented the "mechanical civilization of the West" and were thus fit to "propagate the modern scientific view of nature and the empirical attitude." Physics, chemistry, and mathematics, not natural history, would be the core subjects in building a strong and prosperous Japan. However, this "extremely ambitious plan that arranged every curriculum for science" failed to take effect, due in part to the lack of trained teachers in those 'foreign' fields.[34]

Tanaka's Bureau led the redesigning of the science curriculum for modern Japan, with natural history at its core.[35] In addition to the wealth of trained personnel, there was another motive for giving it a central place: to correct a certain dangerous element in the scientific enlightenment of the Japanese people. For some in the Meiji government encountering the Freedom and People's Rights Movement, which objected to its tyranny and "all-out-Europeanization," it was science education that was responsible for fostering the emergence of such a critical and free spirit amongst the Japanese people. The reform answered such worries by reducing the number of science classes and strengthening the ethics curricula nurturing Confucian virtues (修身 *shushin*, cultivating oneself), as set out in the first amendment of the education law in 1880. Yet, science education could not be neglected if Japanese were to compete with Westerners. Placing natural history at the center of the reshuffled science curricula (理科 *rika*, literally disciplines about principles) could contain the risk by displaying the order and harmony of nature. It was a science to produce conforming citizens.[36]

Not all welcomed this change. There had been constant efforts to redirect this natural history turn in science education in pre-war Japan, including three quite organized attempts.[37] Nevertheless, natural history maintained its central place in primary and secondary science education until the end of the Second World War. The consensus among those who study the history of science education in Japan seems to be that the ideology behind science education in prewar Japan never moved beyond what was stipulated in the Imperial Edict on Education of 1890, the production of attentive and order-observant citizens.[38]

Tanaka's Bureau of Natural History emphasized the practical value of natural history as well, broadening its institutional base. In inheriting the seasoned practice of Tokugawa scholars, it promised natural history's service in promoting industry and trade, primary concerns for the Meiji nation. It was not empty talk, since the agriculture-related sector to which natural history promised and carried out its service never lost its importance in industrializing Japan. Meiji Japan was an agricultural society, with 80 percent of its population involved in agriculture. Notwithstanding urbanization and industrialization, agriculture maintained its economic primacy until the early twentieth century, and reforesting the quickly receding forests and securing more precious raw materials all involved natural history.[39] To this end, the Bureau organized the Product Exposition in 1872 and the first Domestic Exposition for the Promotion of Industries in 1877. These expos, inheriting the Tokugawa exhibitions on things and products, became a model for many similar industrial expos throughout Japan.[40]

Upon this ideological and practical role offered to the government, the Bureau built up related educational and research facilities. It established the Komaba Agricultural School (駒場農學校) in 1877, combining the Agricultural Experiment Station founded in 1872 with the subsidiary agricultural school in 1874. The Komaba School became the Tokyo Agricultural and Forestry School (東京農林學校) in 1886 by merging with the Forestry Experiment Station (樹木試驗場), which was established in 1882. And in 1890, in spite of determined resistance from the Tokyo Imperial University, which argued that no prestigious university in the world had an agricultural faculty, it became a part of the Tokyo Imperial University. The Hokkaido development project and subsequent colonial expansion provided another impetus. More and more government research institutes in "applied botany," as Tanaka put it, came into being as well.

Symbolic of this successful institution building in botany was its inclusion in the science faculty of Tokyo University at its foundation in 1877. Initially, the Meiji government, which in 1871 had displayed its intention to sever itself from the past by abolishing the state school that had taught Confucian classics, was not interested in having a faculty of botany in Japan's first university. Biological sciences did not appear modern enough. However, it finally accepted the rationale of the assiduous and quiet institution-builders in the Bureau; a field that had had such an important role in developing Tokugawa Japan should also empower the science faculty of its first national university.[41] From the primary to the highest level, the influence of the old-school naturalists was clear. The modernization of Japanese botany was in tried and trusted hands.

The Imperial Spirit and the Rise of the New

The foundational work of the old hands did not satisfy all the Japanese botanists emerging in new Japan. One of its key beneficiaries, Yatabe Ryokichi (矢田部良吉, 1851–1899), the first professor of botany at Tokyo University, was the most active in framing his own revolution. Yatabe, who graduated from Cornell University in the US, was the sole Japanese professor amongst an otherwise foreign scientific professoriate at the university's foundation owing to his fully Western education. He was a strong supporter of "All out Europeanization," joining the 'Roman Society' that proposed the Romanization of the Japanese writing system.[42] In that spirit of "Escaping Asia towards Europe," Japanese botany had to shift from its previous track, which smelled too much of the Tokugawa spirit. He had a new story to write about that illustrious past to start the new generation's institution building.

Japanese biologists, not entirely unlike their Western counterparts, used to deplore that biology, considered to be the least 'Western' science, occupied the lowest level of the intellectual hierarchy, in comparison with more properly 'Western' and cutting edge disciplines like physics and chemistry.[43] The growth of botany in higher education was indeed slow. There was no addition of a botanical department in the newly established imperial universities in Kyoto in 1897 and in Tohoku in 1907; the botanical department at Tokyo University was the only one in the country until the Kyoto Imperial University established one in 1918.[44]

Yatabe's botanical section, which would soon become a department, moved to change the low prestige of botany. The first move was to sever its ties to the Tokugawa tradition. It had a new partner in mind, the powerful and impressive Western botanical establishment. Practicing Western botany without any adulteration from traditions and joining the international botanical community solely consisting of Western botanists would show botany to be a proper Western science. Yatabe taught "modern botany" in English, as he had learned it in the US. He made Ito Keisuke's professorship at the university botanical gardens in Koishikawa (小石川植物園) into a mere sinecure for a respected elder, who deserved credit only for his introduction of the Linnaean system; Ito was not allowed to teach.[45] Yatabe also pushed Makino Tomitaro (牧野富太郎, 1862–1957), the beloved father of Japanese botany in the Tokugawa naturalist tradition, who had made botanizing popular among Japanese, out of the university.[46] Neither the old spell nor the popularity associated with tradition could heighten the prestige of botany as he envisioned.

In putting Japanese botany back on the right path of progress initiated by the West, Yatabe drew another powerful concept, that of the emerging nation.[47] If such a relation with the governing body was a mainstay from the Tokugawa period, his approach was somewhat different. He did not promise to help govern unruly subjects nor to aid industrialization with practical knowledge. He aimed at a more prominent role, by suggesting that botany had the capacity to answer Meiji revolutionaries' shared need to recovering national pride. Internationally displaying Japan's ability to do Western science like them would prove Japan's equality with Western nations, and help Japan redress the humiliating and unequal treaties that treated Japan as if it were just another Asian nation. As the home of the nation's only college-level department in the field, his department could fulfill this ideological and "foreign-oriented" task of proving Japan's equality to the West if the necessary support was given.[48] The Tokugawa government's herbal gardens in Koishikawa, where Tokugawa naturalists had experimented with domestication of ginseng, needed renovation and expansion to become a Japanese Kew with herbariums, laboratories, and greenhouses. He launched a serious collection project throughout Japan and initiated specimen exchanges with Western botanical centers. His Tokyo Botanical Society (1882) began publishing its journals and publications largely in Western languages, while equipping its library with Western books on the plant world. He created an international presence for his department within the international botanical network that had so far consisted only of nations of the West.

Yet, something in the position of his Japanese center in the network was wrong. It was not just that the "European" past in Japanese botany had exhausted his task of naming, classifying, and cataloging Japanese plants. All the type specimens of Japanese plants, against which any new specimens had to be checked, were found not in Japan, but in the European centers of botany. The standard literature on Japanese flora, published in European languages, was also more to be found in Europe. Japan still seemed an isolated periphery in the international network of botany, even for the study of Japanese plants. Moreover, Western botanical centers appeared sabotaging his attempt to study Japanese plants from his center. Specimen verification through European centers was especially frustrating. He commented:

> [From] some of [those to] whom I sent many valuable specimens, I have been so unfortunate as to have received no answer whatever even after the lapse of several years. Nothing, it will be admitted, is more trying and disappointing than this to an earnest worker.[49]

Yatabe's way of overcoming this disappointing situation was to strengthen his Japanese center with enough duplicate specimens of Japanese plants and references. He finished his end of the job to make his center the center for studying Japanese plants and decided to let the Western counterparts know it. In 1890, he announced that upon those new foundations, he "decided to begin to give new names to those plants which [he] consider[ed] as new, without attempting in many cases to consult with European specialists."[50] It was his declaration of independence from them.

Yatabe, resigned from his post that year without having been able to realize his independent Europeanization of Japanese botany.[51] Matsumura Jinzo, who had been collecting Japanese plants with Yatabe and duly finished his study abroad in Germany, succeeded him in the chair. As discussed in the introduction, Matsumura, by naming *Prunus yedoensis* in 1901, showed his proper inheritance of his predecessor's task, Europeanization of Japanese botany. Yet Matsumura placed a more ambitious strategy at the core of his task. The Japanese center would not just duplicate what European centers owned but re-enact their ways of securing those things: expeditions to "peripheries." He embarked on serious botanical explorations to Japanese peripheries that still lacked their own botanical institutions. Matsumura himself explored, or sent his associates and students to explore all of the newly acquired territories of Japan. These included Taiwan (1896), Okinawa (1897), Hokkaido (1899), Korea (1900, 1902), and Karafuto and the Kwantung Leased Territory (1906).[52]

At first sight, Matsumura's path appears to simply follow the successful imperial march of his nation, which colonized Taiwan in 1895, Karafuto in 1905, the Kwantung Leased Territory in 1905, and Korea in 1910.[53] However, Matsumura's most prescient move was not a mere response to Japan's military and political advances. His exploration of Korea, the "virgin land of botany" in 1900 and 1902 predated Japan's colonization of that country by almost a decade. As was the case for Europe, Japanese imperial expansion was not just a maneuver by politicians and the military, but a collective social movement involving the participation of entrepreneurs, intellectuals and professionals.[54] The new kind of botanists at the Imperial University, with all kinds of different experts navigating the peninsula since the late nineteenth century, became eager participants in the making of the Japanese empire, determined to complete their Europeanization through that imperial path.

Matsumura published the fruits of these expeditions in his series on "East Asian Plants" in the *Botanical Magazine, Tokyo* from 1901 and compiled the *Index Plantarum Japonicarum* (1904–1912) and *Icones Plantarum Koisikavenses* (1911–1921). As with his naming of *Prunus yedoensis*, these works were widely celebrated. According to one Japanese account, these works based on Japanese-collected specimens in the Koishikawa Botanical Gardens displayed "the name of Koishikawa to the world as the Japanese center of botany."[55] Displaying this Japanese center with its peripheries to the world, the modernization of Japanese botany appeared to have come to fruition.

But Matsumura himself was not yet so sure. He appears to have searched for a more distinctive imperial mark on the Japanese Kew, and possibly a less ambiguous success than *Prunus yedoensis*. He gave the accumulated specimens from Taiwan and Korea to two of his promising students, Hayata Bunzo (1874–1934) and Nakai Takenoshin (1882–1952), in 1903 and 1906, respectively, as thesis materials.[56] Matsumura saw that more solid success was to be secured by a more systematic engagement with colonial flora by Japanese botanists. Hayata and Nakai, who later inherited his professorship, successfully met his expectations by becoming international authorities on Taiwanese and Korean flora, respectively.

Conclusion: Changing Stories for the Japanese Imperial Center

Japanese botany, at the turn of the twentieth century, was very strong in that it had already collected, identified, described, and classified thousands of Japanese plants, secured most of their standard specimens in Japan, and had a variety of natural and cultivated plants that the Japanese people could eat, make cloth from, build houses with, enjoy the beauty of, and satisfy other emotional and industrial needs. They had a broad institutional foundation, with natural history education starting at the elementary school level, and a strong talent base for both intellectual and practical pursuits, much indebted to the long traditions on which it was built.

Yet, there emerged a new kind of botanist, with degrees from Western universities. They found little strength in these seasoned achievements and foundations, while imagining virgin land for Japanese botany in Korea and beyond. They presented the shift to imperial botanizing in Japanese peripheries as the only way to properly Europeanize Japanese botany, by forgetting Japan's illustrious past in plant studies. They barely remembered Japan's earlier encounters with other cultures, from which they were distracted by their current frustrations and ambitions. By means of a new story that denied and rather victimized Japan's illustrious past, they created a powerful political role for their knowledge practice, proving Japan's equality to Western imperial powers. They showed their capacity to do so, producing a new center that indeed resembled Western centers, from its physical facilities to its imperial strategy, and even to the language that it used for its newly produced knowledge. This proved to be a strategy that smoothly secured Japanese botanists' place in the international botanical community, which had hitherto consisted solely of botanists from Western imperial powers. Japanese botany overcame its alleged Asianness, simultaneously relying on and pushing away its Asian pasts and its Asian peripheries. It was a European botany made in Japan by new Japanese botanists.

Notes

1 "Animal and Plant Illustrations (동식물 도록)," *Rectification Records of Medical Material Survey* (藥材質正紀事), 對馬島宗家文書資料集 (Documents in Tsushima Souke Bunko), vol. 5. National Institute of Korean History, Korean History Database (http://db.history.go.kr/id/ts_005_$1ill) accessed 2024.01.07.
2 Tashiro Kazui 田代和生, *An Investigation of Korean Medicines in the Edo Period* (江戸時代朝鮮藥材調査の研究), Tokyo: Keio Gijuku Daigaku Shuppankai, 1999.
3 Tashiro Kazui田代和生, *An Investigation of Korean Medicines in the Edo Period* (江戸時代朝鮮藥材調査の研究), Tokyo: Keio Gijuku Daigaku Shuppankai, 1999.
4 Takatsu Takashi 高津孝, *East Asia of Natural History and Texts: Maritime Exchange between Satsuma and Ryukyu* (博物学と書物の東アジア ： 薩摩・琉球と海域交流), Okinawa: Yoju Shorin, 2010. Both of these episodes told by Japanese economic historians, despite their importance, do not receive attention in Japanese and recent English works on plant studies of Tokugawa Japan. Federico Marcon, *The Knowledge of Nature and the Nature of Knowledge in Early Modern Japan*, Chicago, IL: University of Chicago Press, 2015; Maki Fukuoka, *The Premise of Fidelity Science, Visuality, and Representing the Real in Nineteenth-Century Japan,* Stanford, CA: Stanford University Press, 2012.
5 Philipp Franz von Siebold, "On the Status of Japanese Botany (日本植物學ノ現狀ニツキテ)," in Philipp Franz von Siebold ed., *Flora Japonica*, Tokyo: Shokubutsu Bunken Kankokai, 1937, 1–11, 3.
6 For this historiography emphasizing the rupture between the tradition and the modern, see notes 43 and 44 in Introduction.
7 As insightfully discussed by Duara, having a national history fitting Western historical tropes and standards was an important part of non-Western modernization, where history was something that follows a constantly progressing linear path, by an overarching agent of the nation. In some sense, in order to become a nation and a true member of the modern world, a nation had to have such a progressive history, which culminated in the known destination of that progressive journey: becoming modern, i.e., Western. Prasenjit Duara, *Rescuing History from the Nation: Questioning Narratives of Modern China*, Chicago, IL: University of Chicago Press, 1996.

8 As retold by one of Matsumura's disciples in the popular biography of him, revealing how common and long-lasting this sense of loss was. Nagakubo Hen'un 長久保片雲, *The Life of the World-Class Botanist Matsumura Jinzo* (世界的植物学者松村任三の生涯), Tokyo: Akatsuki Inshokan, 1997, 3.

9 Japan allowed and tried various cultural exchanges on their own terms under this system that designated China and Korea as nations to correspond with and the Dutch and Ryukyu as nations to trade with. The characterization of this complex yet open system simply as "closed" emerged only in the nineteenth century. Mitani Hiroshi discusses how the visit of Commodore Perry was not the signaling of Japan's opening to the West but part of a longer development of Japan's engagement with foreign cultures. Mitani Hiroshi 三谷博, *The Visit of Commodore Perry* (ペリー来航), Tokyo: Yoshikawa Kobunkan, 2003.

10 The beautifully illustrated books on European flora and fauna, Rembert Dodoens's (1517–1585) work, given as a gift to the shogun in the mid-century were translated almost a century later in 1750. It was commissioned by Tokugawa Yoshimune, who sponsored the Chosŏn expeditions. Kimura Yojiro, *Naturalists of the Edo Era* (江戸期のナチュラリスト), Tokyo: Asahishimbunsha, 1988, 164–70, 188–94. Serious Dutch learning was embarked on only in the late eighteenth century except for some individual cases like that of Arai Hakuseki (1657–1725). Grant K. Goodman, *Japan and the Dutch, 1600–1853*, Richmond: Curzon, 2000.

11 Toshiyuki Nagata et al., "Engelbert Kaempfer, Genemon Imamura and the Origin of the Name Ginkgo," *Taxon* 64 (2015): 131–36, 133; Petra-Andrea Hinz, "The Japanese Plant Collection of Engelbert Kaempfer (1651–1716) in the Sir Hans Sloane Herbarium at the Natural History Museum, London," *Bulletin of the Natural History Museum. Botany Series* 31(1) (2001): 27–34; Engelbert Kaempfer, *Amoenitatum Exoticarum*, Lemgoviae: Typis & impensis Henrici Wilhelmi Meyeri, aulae Lippiacae typographi, 1712.

12 Thunberg lost his living plant collections just before arriving at the Dutch port, while Kaempfer managed to bring back hundreds of specimens. Much more careful studies of these interactions would be illuminating. Ueno Masuzo, *A History of Natural History in Japan* (日本博物学史), Tokyo: Heibonsha, 1973, 5, 92; Petra-Andrea Hinz, "The Japanese Plant Collection of Engelbert Kaempfer (1651–1716) in the Sir Hans Sloane Herbarium at the Natural History Museum, London," *Bulletin of the Natural History Museum. Botany Series* 31(1) (2001): 27–34.

13 Linné's species epithet, biloba, indicates the fan shaped leaf divided into two, notched at the center, as in Kaempfer's illustration.

14 Toshiyuki Nagata et al., "Engelbert Kaempfer, Genemon Imamura and the Origin of the Name Ginkgo," *Taxon* 64 (2015): 131–36, 133. Ginkgo has different names or nicknames in various languages, such as Icho in Japanese, Gongsunshu in Chinese, Maidenhair tree in English, etc., but the main name in any language is very stable and widely known, seemingly thanks to its very recognizable fan-shaped leaves and yellow fall foliage.

15 Lisbet Koerner, *Linnaeus: Nature and Nation*, Cambridge, MA: Harvard University Press, 1999, 113–63; Carl Peter Thunberg, *Flora japonica*, Lipsiae: Bibliopolio I.G. Mülleriano, 1784, xv–xviii. By the early nineteenth century, much translation of European botany was carried out. Udagawa Yoan (1798–1846), more famous for his translation of *Introduction to Chemistry* (1808), translated several books of botany, *Rules of Botany* (1822), *Introduction to Natural Science: Guide to Botany* (1835). Linnaeus's *Systema Natura* was translated by Nomura Ryuei (野村立栄, 1751–1828). Kimura Yojiro, *Naturalists of the Edo Era* (江戸期のナチュラリスト), Tokyo: Asahishimbunsha, 1988, 188–94.

16 Japan stopped the expensive visit of the Korean Delegation to Japan in 1811, losing an important conduit of inter-Asian cultural exchanges. As no such diplomatic delegation existed between China and Japan, Japanese scholars used to flock to meet these Korean delegates in their attempt to learn about Confucian culture; many Japanese literati would collect poems by Korean delegates and many Japanese medical practitioners published

their notes from these encounters. Hur Kyung Jin 허경진, "Travels of Chosŏn Medicine to Japan: On Publications of Written Conversations with Korean Doctors (조선 의원의 일본 사행과 의학필담집의 출판 양상)," *Korean Journal of Medical History* (의사학) 19 (2010): 137–56.

17 The Tokugawa government was at first so alarmed by Dutch learning that it instituted a ban on "barbarian learning" in 1790. Yet, with this changing situation it established the translation bureau in 1811 for the systematic translation of Dutch books. As Japan came to know more about the economic and social institutions, geography, and imperial competitions of Western countries, the realization was deepened that Japan could not pursue its selectively isolated existence. Mitani Hiroshi says that nineteenth century Japan was in contradiction by strengthening the 'closed country' system while actively pursuing Western knowledge. Mitani Hiroshi, *Meiji Restoration and Nationalism* (明治維新とナショナリズム: 幕末の外交と政治変動), Tokyo: Yamakawa Shuppansha, 1997; and *The Visit of Commodore Perry* (ペリー来航), Tokyo: Yoshikawa Kobunkan, 2003.

18 The Netherlands lost its colonies in the turmoil of the Napoleonic Wars (1803–1815), during which its Japanese trade was paused. It lost all its African colonies and was losing influence over its South Asian colonies. As an attempt to restore its power, it launched strategic investigations to South Asia and Japan. Andreas Weber, "Encountering the Netherlands Indies: Caspar G. C. Reinwardt's Field Trip to the East (1816–1822)," *Itinerario* 33 (2009): 45–60.

19 Siebold, who wanted to become a Humboldt for Japan by introducing this "terra incognita" to a Western audience, secured this job through a family connection and was introduced as a physician from a famous medical family to the Japanese. He made a second visit between 1859 and 1861 for his self-appointed diplomatic mission for Japan. Arlette Kouwenhoven and Matthi Forrer, *Siebold and Japan. His Life and Work*, Leiden: KIT Publishers, 2000. Studies of the relationship between Siebold and Japan have been continuous. For example, Kure Shuzo 呉秀三, *Dr. Siebold: His Life and Achievement* (シーボルト先生　その生涯及び功業), Tokyo: Tohodo Shoten, 1896/1926; Oba Hideaki 大場秀章, *The Man of Flower, Siebold* (花の男シーボルト), Tokyo: Bungeishunju, 2001; A. Thiede, Y. Hiki and G. Keil, eds., *Philipp Franz Von Siebold and His Era: Prerequisites, Developments, Consequences and Perspectives*, Berlin: Springer, 1999; Herbert E. Plutschow, *Philipp Franz von Siebold and the Opening of Japan: A Re-Evaluation*, Folkestone: Global Oriental, 2007.

20 The Dutch also sent pharmacists and illustrators to help his research. Siebold, a German, almost failed to enter Japan due to his inadequate Dutch. However, these Japanese scholars decided to learn from this youth in his poor Dutch, showing their determination. Kure Shuzo呉秀三, *Dr. Siebold: His Life and Achievement* (シーボルト先生　その生涯及び功業), Tokyo: Tohodo Shoten, 1896/1926, 90; Arlette Kouwenhoven and Matthi Forrer, *Siebold and Japan. His Life and Work*, Leiden: KIT Publishers, 2000, 9.

21 Despite what is often assumed, it is not accurate to say that Siebold taught the Linnaean system to Japanese scholars. As Siebold recorded after his meeting with Ito's teacher Mizutani Hobun (水谷豊文, 1779–1833), who sent his disciple Ito to Siebold, some Japanese scholars' understanding of the Linnaean system was quite mature already. In fact, better botanists than Siebold, they greatly contributed to Siebold's classification of Japanese plants. Mizutani showed Siebold his note that he tried the Linnaean classification, in which Siebold found few mistakes. "There were many not reported by Kaempfer or Thunberg and two or three of them were new to me," Siebold said. Siebold reported all species in Mizutani's *Names of Things* (物品識名, 1809) and *Names of Things Supplementary* (物品識名拾遺, 1825) that Mizutani failed to find Chinese names as new Japanese species in his *Flora Japonica*. Those were Mizutani's discoveries in Siebold's name. Kimura is the only Japanese historian who points out the mutuality of this interaction or Japan's contribution to European botany by detailing these encounters, see Kimura Yojiro, "Siebold in Japanese Botany (日本植物學におけるシーボルトと)," in Omori Minoru 大森實 ed., *Ph. Fr. Von. Siebold and the Modernization of Japan*

(PH.FR.VONシーボルトと日本の近代化), Tokyo: Hosei Daigaku, 1992, 275–86, 281–83; Ueno Masuzo, *A History of Natural History in Japan* (日本博物学史), Tokyo: Heibonsha, 1973, 81–103.

22 Needham also notes the convenience of the ideographic Chinese characters in showing the relationship between plants. Joseph Needham, *Science and Civilisation in China*, vol. 6, Biology and Biological Technology, Botany, Cambridge: Cambridge University Press, 1986, 178. Philipp Franz von Siebold, "On the Status of Japanese Botany (日本植物學ノ現狀ニツキテ)," in Philipp Franz von Siebold ed., *Flora Japonica*, Tokyo: Shokubutsu Bunken Kankokai, 1937

23 Philipp Franz von Siebold, "On the Status of Japanese Botany (日本植物學ノ現狀ニツキテ)," in Philipp Franz von Siebold ed., *Flora Japonica*, Tokyo: Shokubutsu Bunken Kankokai, 1937, 3–5.

24 Tsurumi Shunsuke 鶴見俊輔, *Conversion: Lectures on Japanese Intellectual History during the War Periods* (전향 : 쓰루미 슌스케의 전시기 일본정신사 강의 1931–1945), Ch'oe Yŏngho tr., Seoul: Nonhyong, 2005, 38.

25 Japanese intellectuals were not the ruling class of the samurai-ruled society. They had to compete to secure sponsorship for their studies. Ian J. Miller, *The Nature of the Beasts: Empire and Exhibition at the Tokyo Imperial Zoo*, Berkeley: University of California Press, 2013, 18.

26 Most works on Japanese naturalist tradition still begin with this episode. Federico Marcon, *The Knowledge of Nature and the Nature of Knowledge in Early Modern Japan*, Chicago, IL: University of Chicago Press, 2015.

27 On Li Shizhen's work, see Georges Métailie, "The Bencao gangmu of Li Shizhen: An Innovation in Natural History?" in Elisabeth Hsu ed., *Innovation in Chinese Medicine*, Cambridge: Cambridge University Press, 2007, 221–61; Carla Nappi, *The Monkey and the Inkpot*, Cambridge, MA: Harvard University Press, 2009.

28 Some modern day scholars noted many mistaken identifications of plants in Ono's works citing Korean and Chinese texts. Hong Munhwa 홍문화, "The Influence of Korean Transliterated Names of Local Medicines on Japanese Bencao Studies (우리의 이두향약명이 일본의 본초학에 미친 영향)," *Korean Journal of Pharmacognosy* (생약학회지) 3 (1972): 1–10. For a Japanese critique of similar confusion about Chinese names, Ueno Masuzo, *A History of Natural History in Japan* (日本博物学史), Tokyo: Heibonsha, 1973, 152.

29 Mizutani, who had mastered the Linnaen system by himself and impressed Siebold, also lectured on Ono's *Systematic Materia Medica* to his students. Kimura Yojiro, *Naturalists of the Edo Era* (江戸期のナチュラリスト), Tokyo: Asahishimbunsha, 1988, 194–200.

30 Ono Ranzan also received the bakufu's sponsorship at the age of 71, mostly as an acknowledgment of his achievement.

31 Kimura Yojiro, *Naturalists of the Edo Era* (江戸期のナチュラリスト), Tokyo: Asahishimbunsha, 1988, 55–58. Maki Fukuoka, *The Premise of Fidelity Science, Visuality, and Representing the Real in Nineteenth-Century Japan,* Stanford, CA: Stanford University Press, 2012.

32 Tanaka was a bakufu delegate to the International Exposition in Paris in 1867, a year before the Meiji Restoration. He presented his exquisite insect specimens there. Ueno Masuzo, *A History of Natural History in Japan* (日本博物学史), Tokyo: Heibonsha, 1973, 146.

33 The term *shizenshi* came to have greater usage only around the 1920s. The History of Science Society of Japan 日本科学史学会 ed., *A Compendium of History of Science in Japan* (日本科學技術史大系, *Compendium* hereafter), vol. 8, Tokyo: Daiichi Hoki Shuppan, 1965, 109–255.

34 Physics, chemistry, and mathematics all had Tokugawa lineages but those fields were hardly as populated as natural history, a field which was filled with practicing doctors and cultured hobbyists. *Compendium*, vol. 8, 199.

35 This political project with new conceptualization of nature is well analyzed. Julia Adeney Thomas, *Reconfiguring Modernity: Concepts of Nature in Japanese Political Ideology*, Berkeley: University of California Press, 2002.

36 The bureau had been promoting this moral value of natural history through its publications, such as leaflets and a popular science magazine for children, *Natural History Magazine* (博物雑誌). *Compendium*, vol. 8, 401–430, 438, 109–255.

37 One historian marks this change as a "regress" in Japanese science education. The first attempt was to have more physics and chemistry by the Association for Science Education Research (理科教育研究會), an organization of science educators founded during the science boom after the First World War. The second was part of the "freedom education (自由教育)" movement in the 1920s to reform teaching methods from the 'totalitarian' presentation, explanation, and summary to that enabling students' discovery through observation and experimentation. The third, the locally rooted (郷土) education movement of the 1930s, criticized science education as meaningless memorization even though they valued natural history as part of knowing local society. *Compendium*, vol. 9, 377–430; Kano Masanao 鹿野正直, "Cultural Movements under the 'Crossroads of Reform' (改造の十字路'下の文化運動)," *The Popular Movement and its Ideas in Modern Japan* (近代日本の民衆運動と思想), Tokyo: Yuhikaku, 1977, 139–72.

38 The edict's "Purposes of Science Education" says, "Science education enables the precise observation of nature and phenomena, and an understanding of nature's general inter-relationships and her connection to human lives, in addition to cultivating the love of nature." *Compendium*, vol. 9, 22.

39 Agriculture accounted for 45.1 percent of GNP in 1914. The integration of the Ministry of Industry, established in 1870, into the Ministry of Agriculture and Commerce in 1881 may symbolize the continued importance of agriculture. The Natural History Bureau itself moved to the Ministry of Agriculture and Commerce (農商務省) in the 1880s. Chalmers Johnson, *MITI and the Japanese Miracle: The Growth of Industrial Policy, 1925–1975*. Stanford, CA: Stanford University Press, 1982, 86–90. The study of insects orchestrated by the bureau was crucial in increasing agricultural produce. See Setoguchi Akihisa 瀬戸口明久, *The Birth of Insect Pests: Japanese History through Insects* (害虫の誕生: 虫からみた日本史), Tokyo: Chikuma Shobo, 2009, 7–95.

40 *Compendium*, vol. 15, 17–54; Ueno Masuzo, *A History of Natural History in Japan* (日本博物学史), Tokyo: Heibonsha, 1973, 145–48; Yoshida Mitsukuni 吉田光邦, *History of Science in Japan* (日本科學史), Tokyo: Kodansha, 1987, 152–57; Federico Marcon, *The Knowledge of Nature and the Nature of Knowledge in Early Modern Japan*, Chicago, IL: University of Chicago Press, 2015, 161–227, etc.

41 The influence of well-connected traditional Dutch Learning scholars like Ito Keisuke must have been significant here. Tokyo University became Imperial University in 1886 and Tokyo Imperial University in 1897. *Compendium*, vol. 1, 14.

42 That Yatabe was educated in the US and could teach in English like foreign professors such as Edward S. Morse (1838–1925) was a main reason for his employment. Oba Hideaki 大場秀章, *A History of Botanical Research in Japan: 300 Years of Koishikawa Botanical Gardens* (日本植物研究の歴史: 小石川植物園三〇〇年の歩み), Tokyo: Tokyo Daigaku Sogo Kenkyu Hakubutsukan, 1996, 84–89; *Compendium*, vol. 15, 15.

43 Biologists argued that there had been a disciplinary hierarchy of mathematics, physics, chemistry, biology, and geology since the Meiji Era. *Compendium*, vol. 15, 14.

44 Tohoku Imperial University opened its botanical department in 1921. Soon others followed. Kyushu Imperial University in 1923, Taihoku Imperial University in 1928, Tokyo and Hiroshima University in 1929, Hokkaido Imperial University in 1930. *Compendium*, vol. 15, 265–89, 290.

45 Yatabe also hindered the career of Ito Tokutaro (伊藤篤太郎, 1865–1941), Ito Keisuke's grandson who studied at Cambridge University, because he took away from him the honor of being "the first Japanese who gave a scientific name of a plant" from him. *Compendium*, vol. 15, 98.

46 Makino worked for the botanical department as a lecturer at first but Yatabe pushed him out. Makino, who became a naturalist after reading Ono Ranzan's work, greatly contributed to Japanese modern systematics while leading the publication of special journals and guiding the Japanese public to natural history. His illustrated flora of Japanese plants is still in print and the popularity of natural history in modern Japan owed much to him. Kamimura Minoru 上村登, *Loving Flowers: Biography of Makino Tomitaro* (花と恋して: 牧野富太郎伝), Kochi: Kochi Shinbunsha, 1999.

47 It was certainly one way to answer the founding creed of the university, providing "knowledge and technology that answers the demand of the nation." *Compendium*, vol. 15, 50.

48 Some later historians criticized that Japanese botany answered neither its own scholarly questions nor the practical domestic needs of the country, seeking recognition only from Europeans. *Compendium*, vol. 15, 16.

49 Yatabe Ryokichi, "A Few Words of Explanation to European Botanists," *The Botanical Magazine, Tokyo* (植物學雜誌, *BMT* hereafter) (1890): 355–56, 355. The *BMT* is the journal of the Tokyo Botanical Society.

50 Yatabe Ryokichi, "A Few Words of Explanation to European Botanists," *The Botanical Magazine, Tokyo* (植物學雜誌, *BMT* hereafter) 4(44) (1890): 355–56, 355. The *BMT* is the journal of the Tokyo Botanical Society.

51 The reason for Yatabe's early retirement was not fully known. "Conduct unbecoming of a university professor" was mentioned. Oba Hideaki 大場秀章, *A History of Botanical Research in Japan: 300 Years of Koishikawa Botanical Gardens* (日本植物研究の歴史: 小石川植物園三〇〇年の歩み), Tokyo: Tokyo Daigaku Sogo Kenkyu Hakubutsukan, 1996, 89.

52 Nakai Takenoshin, "An Outline of Dr. Matsumura Jinzo's Achievement (理學博士松村任三氏植物學上ノ事績ノ概略)," *BMT* 29(346) (1915): 342–48.

53 Ramon H. Myers and Mark R. Peattie eds., *The Japanese Colonial Empire, 1895–1945*, Princeton, NJ: Princeton University Press, 1984.

54 Bernard Semmel's classic discusses the "social" nature of imperialism in Britain. *Imperialism and Social Reform: English Social-imperial Thought, 1895–1914*, London: G. Allen & Unwin, 1960. For an excellent account of the Japanese build-up of "social" and "total" imperialism, though focused on the years after 1930, see Louise Young, *Japan's Total Empire: Manchuria and the Culture of Wartime Imperialism*, Berkeley: University of California Press, 1999. On the roles of small entrepreneurs, see Peter Duus, *The Abacus and the Sword: The Japanese Penetration of Korea, 1895–1910*, Berkeley: University of California Press, 1998; Jun Uchida, *Brokers of Empire: Japanese Settler Colonialism in Korea, 1876–1945*, Cambridge, MA: Harvard University Asia Center, 2011. See also David Fedman, *Seeds of Control: Japan's Empire of Forestry in Colonial Korea*, Seattle: University of Washington Press, 2020, 23–46.

55 Oba Hideaki 大場秀章, *A History of Botanical Research in Japan: 300 Years of Koishikawa Botanical Gardens* (日本植物研究の歴史: 小石川植物園三〇〇年の歩み), Tokyo: Tokyo Daigaku Sogo Kenkyu Hakubutsukan, 1996, 89–97, 93.

56 Nakai Takenoshin, *A Synoptical Sketch of Korean Flora*, Tokyo: The National Science Museum, 1952, 1.

2 Japanese Botany Made in Korea

Japanese botany, fully within its mature and diverse knowledge traditions and plant loving and exploring culture, rather systematically launched botanical interactions with other cultures to modernize Japanese botany. It was not surprising that Japanese botany made quick successes in its modern pursuits. The Japanese botanical establishment at the imperial center chose to make one of its successes, the success made in Japan's Asian colonies, more significant. As the mantra "Escaping Asia towards Europe" was not forced upon Japan, this imperial path was a choice consciously made by its founders Yatabe Ryokichi (1851–1899) and Matsumura Jinzo (1856–1928), as discussed in Chapter 1. It was to prominently position the Japanese botanical center in the increasingly globalizing world of botany, which first became more than European through these active participations not only from Japan but also from the US, Oceania, and beyond.

This Japanese center in the early twentieth century seemed to be in complete agreement about pursuing this path prescribed by its founders, and this chapter illuminates on those "imperial" characteristics of its botanical interactions with Asian neighbors through the work of Nakai Takenoshin (1882–1952), the investigator of Korean flora and representative of Japanese systematics to the international botanical society until his death in 1952. Yet, by looking into Nakai's interactions with Western botanical authorities and by comparing his practices with those of Hayata Bunzo's (1874–1934), the explorer of Taiwanese flora, it also sheds light on doubts and frustrations raised about strategies of the Japanese imperial center as well as the botanical rules of Europe that justified its choice.

Hayata, who had assumed both the professorship in plant taxonomy and the directorship at the Koishikawa Botanical Gardens before Nakai, rather came to be prevented from coming to the forefront of the Japanese botanical center since his illness in 1922. His questions about certain Japanese colonial research seemed not unrelated to his obscurity. Nonetheless, his research activities with global implications continued; at least, two of his works were read at the French Academy in 1933. He above all made significant challenges to the rules of the Western botanical establishment, which had been struggling with the "Linnaean" system that existed only in the name. By following the intertwined paths and struggles of Nakai and Hayata, this chapter reveals the complex negotiations of botanists at the Japanese imperial center about these imperfect botanical rules in the making.

DOI: 10.4324/9781003511755-3

Japanese Botany Made in Taiwan

Showing the wisdom of Matsumura's strategy, Hayata, a mature student who entered the university at the age of 26 (he had lacked financial means to study due to his parents' early death) after about ten years of self-education in botany, left a Japanese botanist's mark on Taiwanese flora quite quickly, by reporting a new genus of conifer that he named *Taiwania* through the prestigious journal of the Linnean Society in 1906.[1] Then, under the aegis of the colonial government in Taiwan, he produced a ten-volume monograph on Taiwanese flora, *Icones of the Plants of Formosa* (1911–1920), entirely in English and Latin. It was an indelible achievement, to which Matsumura had carefully guided him.

In this usual form of summarization, his was an achievement that looks almost exactly like that of Nakai right on the imperial and successful path that Matsumura had set for them. However, it was just the surface of Hayata's achievement. His deep concerns about what it meant to study nature and how to do it had different orientations, reflecting his religious inclination nurtured in his early loss of parents and Buddhist upbringing. He began his study of plants to understand life and its vicissitude. "That death was not annihilation but change" in relation of matters and that life continues through those matters in another life was his "life's understanding of the question," he said in 1933, a year before his death.[2] He examined his scientific practice in line with his pious search for the truth of life's vicissitude, resulting in his fundamental critique on the standard phylogenetic systematics of Heinrich Gustav Adolf Engler (1844–1930). To that end, he made globally engaged botanical practice immersed in tropical, sub-tropical, and temperate fields. Yet, his concerns and questions did not have much hearing in Japanese historiography. His colleague Nakai's intervention seems significant, who said:

> The dynamic theory of Prof. Hayata is a theory of excellent designation, but is an elaborate modern explanation of old incredible theory of spontaneous generation. I am one of the Japanese naturalists who keenly regret for having had the publication of such theory.[3]

It was a kind of official statement of the then head of the botanical department, published simultaneously in English and Japanese in 1930. It implied that Hayata's theory should not have been published in the first place. He regretted it as one of "Japanese naturalists."

Hayata indeed showed discomfort about core tools and approaches that the Japanese center had laboriously worked for. How Japanese center interacted with the object of his study, plants, bothered him foremost. Hayata was a field naturalist who had already twice been to Taiwan when he received his thesis material from Matsumura. Years of experience in the field watching big and small vegetative beings like conifers, moss, and ferns made him mindful about the partiality of information that dried specimens contained.[4] He did not share his predecessors' strong attachments to dried plant specimens, who believed the plants in the field without specimens could not make Japan the center for the study even of Japanese plants. They thus shipped those dried specimens from Japanese peripheries for the Japanese center's imperial transformation. Yet, Hayata seemed hardly comfortable about that strategy that allowed him to study Taiwanese plants at the comfortable Tokyo herbarium, so far away from plants in the field. Throughout his botanical career, he tried many ways of compensating for the partiality of information on dried specimens, and the standard systematics quite fixed upon them.

Hayata began his quest by working with those in the field, the collectors in Taiwan. His identification of his new conifer *Taiwania* details his cooperation process. At first, what he received from his collector was "but a few dry branches with cones." For a big tree like a conifer, the information that the dried specimens fixed on them was quite limiting. He wanted to know in what kind of habitat, and from which part of the tree these branches were taken. He desired to secure "some alcoholic materials for its histological study, and, if possible, to get a photograph of the whole plant." When he secured all these materials from his collectors, Hayata finally felt confident about the tree's singularity.[5]

Hayata's awareness of the limitation of specimens was important in his collaboration with European centers, too. While Yatabe, the first professor of botany in the university, declared that he would stop sending his specimens for European specialists' confirmation, such confirmation with standard and related specimens was necessary in identifying plants, whether it was more studied Japanese plants or less studied Taiwanese plants. Hayata sought direct confirmation for his first important finding from the leading expert on conifers, Maxwell T. Masters (1833–1907) of the Linnean Society. He did not simply send the specimens but all the subsidiary materials such as the photograph that showed its peculiar branching, trunk color, and its habitat. Hayata did not have the frustrating experience that Yatabe had described in 1890. Fully informed, Masters immediately acknowledged its newness and helped Hayata introduce the new plant through the journal of the Linnean Society.[6]

Such direct confirmation was, however, still a time-consuming process that Hayata could not go through every time. He referred to books, now readily available in Tokyo most of the time, to secure a more reliable information. Only he soon learned that working with European botanical references was also quite challenging. Some Western botanists accused Hayata of claiming too many new species without fully studying already reported plants.[7] He denied the accusation. Whenever he claimed a new taxon, he made a careful investigation of related existing species. He explained that some descriptions in the standard references from European centers were simply too inadequate, sometimes "less than three lines" long, to enable him to relate his new plant to an existing one.[8] European botanists appeared to think that others "may go to see the types themselves" if they found descriptions insufficient, which was almost impossible for those who were not in Europe.[9] The standard specimens undergirding the standard classification were too concentrated in Europe, with very limited circulation to outside Europe, creating a disparity in the practice and relations of European and non-European botanists.

Hayata was not the only one who acutely felt about this limited and geographically skewed availability of standard dried specimens. Other non-Europeans in the expanding botanical communities such as American and Oceanian botanists also shared this concern. They were making strong demands that European centers should make readier distributions of type specimens possible by producing more duplicate specimens. Hayata's approach was different. He did not think such measures that would necessitate more collections in peripheries and more travels of specimens to European centers were necessary. He wanted to make this growing concern about specimens a chance to reduce the reliance on them in botanical practice. He proposed to make "exhaustive descriptions of plants" compulsory so that all botanists could obtain clear ideas about that plant "without taking the trouble to look at types." While it was not uncommon to make botanical illustrations with specimens, he proposed sketches and photos taken "from the field," to accompany those exhaustive descriptions.[10] Given that most botanists merely focused on increasing the accessibility of dried specimens, the reliable "immutable mobiles," his was a unique approach reflecting his unusual qualm about dried specimens.[11]

Hayata's serious doubt and lack of enthusiasm about the plant specimens, the basic tool of modern classification, reflected his realization that it did not in fact have such power to bring consensus about classification. Joining in the globalization of European botany in their lands, Japanese botanists, like others in Americas and Oceania, all learned that there was no such thing as the "natural system," the universal systematics that they could just apply to their plants. The alleged universal systematics had always been on dispute with different criteria and rules contending with each other on every level. He noted that while the systems of Linné, Antoine de Jussieu, De Candolle, Bentham-Hooker, and now Engler had obtained many followers and functioned as standards in turn, no one system had ever succeeded in showing why its criteria were indeed natural and should become the ultimate standard. Hayata argued that Engler's system, "at present regarded as the most natural," revealed just how botanists had inadequate means to demonstrate relationships between plants, instead of capturing the "natural affinity" of plants as Engler claimed. What Engler showed in claiming natural affinity was just "constitutional resemblance instead of blood-relationship," which was much less obvious to demonstrate in plants than in animals. For example, although Hayata agreed that it was neat to divide the dicotyledons into subclasses of *Archichlamydeae*, plants with flowers without petals or separate petals, and *Metachlamydeae*, plants with flowers with fused petals, as Engler did, he could not see the natural ground for that division.[12]

In 1917, Hayata decided to look to nature to resolve this fundamental issue. Adding to his experience in temperate and sub-tropical fields in Japan and Taiwan, he chose to immerse himself in a tropical field this time. He would "start for Indo-China where [he] will stay for one year to accomplish [his] study in the midst of the tropical forests, observing every changing aspect of the flora through all season." In 1921, Hayata made a report on his observations, after spending three years thinking them through, in the last volume of his work on Taiwanese flora. "Much against [his] will," his conclusion was revolutionary, in that he denied any fixed order to plant relationships, and refuted not just Engler's system but more generally the possibility of finding phylogenetic relations among plants—the holy grail of plant taxonomy since the publication of Darwin's work.[13]

The dynamic and creative transformations of plants that he experienced day and night in the field did not justify Engler's attempt to establish one phylogenetic tree of life, especially the hierarchical one with "serial progressions" that Engler was drawing. Without enough paleontological or embryologic evidence to suggest plants' order of appearance on earth or ancestral relationships, Engler's arrangement of species and genera "according to their degree of advancement, or according to their simplicity or complexity," etc., only revealed a certain idea of evolution that Hayata had serious doubts about. Furthermore, if one agreed that one characteristic might be more primitive than another, there was also a doubt about whether such a thing meant an evolutionary divergence. That is, if bisexual flowers were more primitive than unisexual flowers, Hayata wondered whether that meant that no plants with bisexual flowers had affinity with those with unisexual flowers. Finding a "blood-relationship" between plants was an extremely complicated affair, he pointed out, given plants' ability to cross-pollinate, at times even beyond their species boundaries. He also believed that mechanisms of crossing, mutation, and adaptation, the mechanisms claimed for species formation, would be "by no means independent, but closely inter-related," increasing complexity in plant phylogeny.[14]

In understanding the dynamic transformations of plants that he observed in sub-tropical and tropical fields, where plants lightly crossed all sorts of boundaries in appearance, such as the one between trees and grasses, his major reference was Johann Wolfgang von Goethe's (1749–1832) "Metamorphose der Pflanzen," a text that "seeded a revolution in thought that would transform biological science during the nineteenth century," according to some account.[15] It was ironic that Hayata used this text to deny a certain revolution allegedly begotten by it.[16]

Hayata read Goethe and his followers very carefully and created a new way to interpret Goethe's ideas about plant transformation. For Hayata, the contemporary interpretations appeared to misrepresent Goethe, in whom he found his "own idea." What Goethe did was planting seeds of many annual plants and observe their growth by tracing their constant transformations, from their first cotyledons to fruits. Goethe discussed the transformative power of these first leaves, which he called "Blatt," and their unity and similarity with other organs like petals, pistils, and stamens of the plants. While this transformative power of "Blatt," literally "leaf," was interpreted as something enabling speciation by evolutionists, in Goethe's words, the metamorphosis of plants was more of the "process by which one and the same organ appears in a variety of forms."[17] Hayata agreed that Goethe noted various transformations or "divergence of characters" that one plant could display, and suggested "some idea of the theory of descent" by "Urpflanze (literally, primordial plants)." However, if one German biologist presented "Urpflanze" as a smoking gun that Goethe antedated Darwin by 70 years, Hayata thought it was a model plant with a perfect form that Goethe had in mind, rather than some ancestral plant.[18]

Hayata gave a new term to Goethe's "Blatt" and a new theory that he believed to be able to capture Goethe's ideas more accurately than the theory of evolution. The crucial idea of Goethe where he found "singular agreement" with his own idea was:

the constant variability of things and the "unity of everything" that nevertheless holds up the natural world.

> Nature creates ever new forms; what is there has never been, what is now never comes back; everything is new yet always old. It is an eternal life, becoming and moving within it. It evolves forever, without a moment of resting… In each she appears in a separate form. She hides in a thousand names and terms, and is always the same.[19]

While Goethe called this one and the same organ that transforms into everything "Blatt," he had also noted the need for "a general term to describe this organ that metamorphosed into such a variety of forms."[20] Hayata decided to give this very versatile organ with generative power, common in all things, a newly circulating term "gene." While it was an ambiguous term without much materiality in genetics as of yet, he gave it full materiality too. His gene would have "generating qualities," by working in various aggregates. It was governed by "the law of substance, i.e. the conservation of energy and the indestructibility of matter." These gene compounds created different yet similar things in different stages of growth within and across different organisms through "mutual participation" and the "mutual sharing" of genes by some forces like "chemical affinities or physical gravity." Although untested and abstract, this idea would come to find much resonance in the age of the double helix and DNA sequencing.[21]

This new way of seeing the variability of plants leads to a new natural way to classify them, Hayata's "dynamic system." Given the dynamic and various associations that plants have with other plants, any "static system" assuming the fixity of natural relationships cannot be natural, Hayata asserted. It was not that Linné's or Engler's classification criteria had no natural ground, but rather that their attempts to claim only one natural relationship was unnatural. For example, many botanists noted that plants in *Ranunculaceae*, a genus in dicotyledon, are very similar to plants in *Alismataceae*, a genus in monocotyledon. If one dynamically adopted different criteria for comparison than the one used in dividing dicotyledon and monocotyledon for such similar plants, their natural affinities could be expressed. Since such crisscrossing relations abound in the plant world, a system that fails to show those multiple lineages cannot be natural.[22] Hayata used Engler's system as a baseline in building his dynamic system; replacing the current standard with his own was not at all his aim. He wanted all contending systems to open up for other reasonable perspectives for a flexible and plural system serving different purposes and ever-changing nature.

Hayata had another concern in devising his revolutionary system, which he made public at the first international scientific congress held in Japan, in 1926. While this third Pan-Pacific Science Congress was a showcase to display the maturity of Japanese science for most of his colleagues, Hayata used it to share his concern about Japanese society, and a certain science undergirding it. He specifically deplored that an idea of "natural selection," flattering those who considered themselves selected, was taught in every Japanese school as "a veritable truth," promoting "a ruthless struggle for existence."[23] Such an idea of natural selection seemed to create a false idea of racial hierarchy, as if subjugation of other races was the proof of one's racial superiority.[24] He found an example that being subjugated did not mean inferiority, from the Japanese-colonized atoll off Taiwan, the Pescadores. Although they indeed lived "very poor in every respect" in the land that was too barren to produce enough food for its inhabitants, it was "as populated as any one of the most thickly peopled provinces in the Japanese Empire." They were thriving by living in "the spirit of mutual friendship" in which even the poor were "willing to share their food with others." They were also "quite strong, much stronger than the ordinary Chinese," capably carrying out their tough lives.[25]

Although Hayata did not say so explicitly, his insinuation was that such people were subjugated by the Japanese, who encouraged pride in such an act of subjugation while pretending "to join in the work of promoting peace and brotherhood" among their neighbors. It was a concern that he refrained from writing in Japan for a long time, he said, for fear of ridicule and suppression. He thought this international scientific congress that aimed at scientific cooperation as well as "the bonds of peace among Pacific peoples" would allow him, a devout Buddhist, to voice such concerns and to devote himself "to the cause of peace and truth."[26]

Hayata was certainly an uncommon figure with his idealism and originality. However, he was not alone in raising such fundamental doubts about phylogenetic systematics and a certain idea of evolution. He in fact utilized them in his works. In fact, criticism of phylogenetic systematics became so great that Engler finally admitted in 1926 that his system was just "constituted so as rather to show morphological steps." Phylogeny was yet an ideal rather than a reality, he added. In making such a concession, Engler must have noted Hayata's objection, too, as his dynamic system, originally in English, was introduced even in German, as a summary translation for *Die Naturwissenschaften* in 1923.[27] However, this concession did not stop the rule of Engler's system or the search for one phylogenetic tree.

Nonetheless, those who shared Hayata's concern could find inspiration from Hayata's work to unsettle the status quo. For example, a Swedish botanist Gustaf Einer Du Rietz (1895–1967) predicted in 1930 that Hayata's participation theory and the dynamic system, "a most remarkable revolt against the traditional phylogenetic method of taxonomy [...] will prove a most excellent working base for the great reform of higher botanical taxonomy that is so greatly needed." Du Rietz argued that Hayata's "imposing taxonomical knowledge and wide field-experience of tropical, subtropical, and temperate vegetation certainly ought to make European botanists read his contributions with greater attention than appears to have been the case until now."[28]

Matsumura seemed to have believed that the imperial path was the best way to modernize Japanese botany and Hayata followed that guidance in leaving his mark on the study of Taiwanese flora. His study of Taiwanese flora, however, prematurely ended in 1920 as the colonial government stopped funding his work. Although Hayata's last volume on Taiwanese flora contained his complaints about the situation, it was also a chance for him to expand his study bound by Japan's imperially shaped relations with Taiwanese flora.

Of course, including his exploration in south Asia funded by Japan's expansionist interest in the region, his intellectual expansion was also inconceivable without those imperial forces that shaped all ambitious botanical explorations. Yet, further aided by his isolation after his illness in 1922 created by rather generous and hasty exemptions from teaching and departmental duties till his premature death in 1934, his later intellectual journeys since his appearances at the Pan-Pacific Science Congress became more and more contemplative, searching for botanical practices not bound by imperial and asymmetric relations. Hayata, in close communication with other critical minds beyond Japan, and by open and deep explorations of scientific and religious ideas, thus made an internationally recognized revolt on the botanical standards, shaped and sustained by many imperial practitioners who had been too successful in those rules and relations to question them.

Japanese Botany Made in Korea

It is surprising that Hayata's contemporaries, who were so eager for Japanese botanists' international achievement, did not particularly appreciate his contribution to Japanese and global botany. Such seems the effect of the dismissal of Nakai, mentioned above. This silence about Hayata's contribution shows that Nakai's statement, given as the representative of Japanese botanists, indeed came to represent their opinions. Nakai further related his denunciation of Hayata to a non-academic issue: "a scholar should be able to communicate with the world."[29] What Nakai meant by these intriguing words would show in Nakai's communication with the world.

Nakai was a typical "armchair botanist," who went on field trips from time to time, but more as a duty or an excursion than as a way to study plants, it seems.[30] The immaculate white suit he wore on 1 of his 18 field trips in Korea, almost all much less than a month long and taken during his summer vacation from the university, shows how he perceived this duty (See Figure 2.1). When Nakai received an immense number of dried specimens of Korean plants in 1906, collected by Matsumura's assistant in 1900 and 1902, he showed no worries about carrying out his task of studying Korean plants in his Tokyo lab. He just showed a solemn acknowledgment of his teacher's wish. "It must have been [Matsumura's] long desire," Nakai reflected, "that the Korean plants should be studied by Japanese botanists."[31] Nakai lost no time in fulfilling his task. Finishing his thesis on Korean plants in three years, i.e., about a thousand days, without ever setting foot in Korea, he was remarkably productive, reporting 1,970 species, and 183 varieties of Korean plants, including about 200 new species and varieties in 149 families and 661 genera. Nakai's *Flora Koreana* (1909) far surpassed the most comprehensive work then extant on Korean flora, written by the Russian botanist Ivan Vladimirovich Palibin (1872–1949).[32]

Unlike Hayata, who had constantly asked questions about how to properly practice botany, Nakai hardly asked questions while en route to his quick success. He often looks like the white man depicted in George Orwell's perceptive and

Figure 2.1 A photo featuring Nakai Takenoshin at the center in a white suit. *Report on the Vegetation of the Island Ooryongto or Dagelet Island, Corea*, Seoul: GGK, 1919, 10 (The Biodiversity Heritage Library).

unsparing self-reflection of his experience as a police officer in imperial Burma, "Shooting an Elephant" (1936).

> Here was I, the white man with his gun, standing in front of the unarmed native crowd–seemingly the leading actor of the piece; but in reality I was only an absurd puppet pushed to and fro by the will of those yellow faces behind. I perceived in this moment that when the white man turns tyrant it is his own freedom that he destroys. He becomes a sort of hollow, posing dummy, the conventionalized figure of a sahib. For it is the condition of his rule that he shall spend his life in trying to impress the "natives," and so in every crisis he has got to do what the "natives" expect of him. He wears a mask, and his face grows to fit it. I had got to shoot the elephant. I had committed myself to doing it when I sent for the rifle. A sahib has got to act like a sahib; he has got to appear resolute, to know his own mind and do definite things. To come all that way, rifle in hand, with two thousand people marching at my heels, and then to trail feebly away, having done nothing–no, that was impossible. The crowd would laugh at me. And my whole life, every white man's life in the East, was one long struggle not to be laughed at.[33]

Like Orwell's hero, Nakai appears resolute. During the three years he was learning the trade by carrying out his first major task as a botanist, he already had an answer about the proper practice of plant classification that he decided to share with other botanists, especially "the venerable J. D. Hooker," a former director at the Kew Gardens. It was Nakai's response to what Joseph D. Hooker (1817–1911) called "a task awaiting the labour" of "a very judicious botanist."

In the two-part article about *Aconitum* in 1908, written in English and addressed to Hooker, Nakai made his approach and ambition very clear. He meant this article to be a clear guideline for those who were "at a loss how to classify a specimen," declaring classifying specimens, not real plants, as the task of a botanist. Providing a solution to Hooker's task was its main purpose. Nonetheless, while Hooker's task was "the reduction of the species and varieties of *Aconitum*," which Hooker believed had been recklessly increased up to 300 where 30 would suffice, Nakai did the opposite by adding about 30 new taxa of Japanese *Aconitum*.[34]

Their difference involved an old debate in botanical classification between "lumpers," who tend to combine similar plants into one taxa, and "splitters," who split genera and species by small differences. While Hooker was a well-known lumper, Nakai already established himself as a firm splitter.[35] In proposing his solution, Nakai implied that lumpers like Hooker had mistaken worries about classification criteria. As shown in Hooker's dictum that "species vary in a state of nature more than is usually supposed," lumpers pointed out the endless creativity of nature that rendered every individual plant distinct as a reason to ignore small variations. Nakai thought such worries about nature were unnecessary, if botanists could do the proper job. He said, "[t]he fault is perhaps due to the indefinite descriptions of it by different authors," which were too "comprehensible [sic] and varied," ending up giving rise to worries about diffused and overlapping characteristics of plants. If botanists could simply pay more attention to their *specimens* in accordance with his guidelines, no such confusion would occur. He prepared a series of dichotomous keys based on numerous distinctive single characteristics such as hood shape, flower color, stem shape, indentation of leaves, shape of leaves, and carpel numbers, etc. and claimed 30 more taxa in *Aconitum*.[36]

This answer shows how Nakai resolved the difficulties and confusions of classification that all botanists including Hayata had encountered in their practice. Dried specimens, where Hayata's questions began, became an answer for Nakai. Although Hayata pointed out that dried specimens hardly resolved issues about classification standards, Nakai found legitimacy of dried plant specimens in European centers' reliance on them. The venerable Hooker or Linné or Engler made authoritative final calls for controversial issues, relying on the systematically accumulated specimens at their centers. He, with the incomparable amount of dried plant specimens on Korean plants at his center, could make the final calls on local plants based on what he saw in them.[37]

Nakai's strong and firm predilection to split based on small morphological detail was a key to his productivity, seemingly reflecting his determination to use specimens from Korea as a foundation to advance Japanese botany.[38] He must have felt discouraged at the beginning. Korean plants were too similar to those of neighboring territories, Japan and China, the more studied and better-known territories in European botany. It was not surprising that European works on world plants did not recognize Korea as a distinctive botanical region, as shown in Siebold's comments on Japanese botany that only mentioned China and Japan.[39] Nakai had to admit that about 2,000 out of the 3,600 species of Korean plants that he recorded by 1926 were identical to Japanese ones. Another 1,100 species were identical to Chinese ones.[40] Nakai overcame this situation by paying attention to small details, which produced substantial new indigenous taxa from Korea. To have Korea acknowledged as a distinctive floral region, he needed to split.

Nakai's success with Korean flora, however, could not have been maintained if the virtual monopoly had not been secured, both by insignificance of Korea and through his active guarding. Korea, which had made little direct contact with the West before its opening, still did not attract many Western botanists, although it invited more animal collectors.[41] The world-leading authority Engler's remark reflects a sentiment among Western botanists. After his train trip passing through the peninsula in 1913, Engler told Matsumura that "there would be nothing for botanical research in Korea as it seemed just a barren land with a few willow trees."[42] Matsumura mused over this remark with Nakai, who had just visited Korea to secure the support of the GGK for his study, missing his chance to meet Engler. They were not discouraged. They seemed to know that Korea's low profile would maintain Japan's monopoly over the region, thereby providing a boon for Japanese botany. Nakai secured funding from the GGK, by appealing to that idea of "understudied yet academically important" Korean plants.[43] In this contested colony, the colonial regime was also responsive to such an ideological role of science, asking Nakai to "demonstrate that Japan makes advances in the scholarly fields as well as in the colonial governance" in Korea.[44] By presenting 22 volumes of *Flora Sylvatica Koreana*, written in a mixture of Latin, English, and Japanese, with numerous illustrations, between 1915 and 1940, Nakai maintained this sponsorship for 30 years until his retirement in 1942.[45]

Nakai's monopoly on Korean flora staved off a lot of dispute, but there was still some. His attention to detail was quite extreme to go unnoticed by his colleagues and competitors.[46] Nakai used many tactics to defend his own classification. The status of many native Korean plants and in turn his authority as a botanist were at stake. One constant ally was the dried specimens of Korean plants that he almost monopolized since he made sure to have ample specimens as the objective evidence for his classification.[47] Although minute, all the detailed characteristics that he used in classification, such as the shapes of veins and edges of leaves as well as the shapes of hairs on the backs of the leaves, were identifiable on his specimens.

In addition, he used two other tactics to defend his method. One was summoning the authority of European botany by claiming his method's identity with it. The other was to claim the specific regionality of his method, admitting his method's difference from the average European way. While this flexibility of Nakai may look curious, if not self-contradictory, he certainly appeared to know how to communicate with the world, effectively alternating two contradictory positions according to the situation.

In his early days, Nakai displayed a rather deep confidence in his method's identity with the best European one. In the preface to his first *Flora Sylvatica Koreana*, Nakai asserted:

> In such places like Germany, England, Russia, Austria, France and Sweden where the botany was most advanced, they now weightily consider small

> details like vein shapes and edges of leaves, which systematists once thought to do away with.[48]

After thus aligning his extreme splitter method with that of the leading botanists in Europe, he further explained that this renewed concern with small details was due to the new genetic understanding. The new genetics implied that any morphological characteristics, being "complex mixtures of genetic materials within cells," did not vary by chance and only within a very limited range.[49] Written only in Japanese, this authoritative-sounding preface targeted a domestic audience.

It is not known whether his domestic audience was convinced by this defense. On the other hand, it soon became quite obvious that his European counterparts did not think that his method was identical with the most advanced among theirs. European botanists dismissed Nakai's belabored new classification of *Aconitum* by choosing another proposal by one of their own, which completed the anticipated lumping. In 1917, in defending his classifications at least for Japanese *Aconitum*, Nakai said that his method was something for "Asiatic species." Characteristics that European botanists thought "too trivial for the European species" made for a "good and precise [distinction] for the Asiatic species," he said.[50]

While Nakai used relatively clear geographical term "Asia" in this regional defense targeting a Western audience, his regional categories were often more ambiguous. In 1914, showing his earlier awareness of his method's discrepancy with the usual European method, he evoked the "Orient."

> It was the age-old custom to classify Oriental (東洋) plants in the Great Western (泰西) classification method, but as plants living in the East and the West of the great ocean largely differ in their species, the classification for Oriental Plants must be the one that was proper to them. That is why I dare to use my original method.[51]

Although it is hard to locate the Orient and the Great West divided by the unnamed great ocean, it was obvious that he desired some region defined specifically by its separation from the "Great West." The next attempt was "East Asia (東亞 to-a)." In 1927, he contended, in front of a Japanese audience honoring his research on Korean flora with the prestigious Imperial Academy Award, that he had been using "East Asian Plant Systematics" because he needed a way to express the differences between East Asian and Euro-American (歐美 obei) plants. That the differences between East Asian and Euro-American plants were "so subtle as to be often overlooked" was his reason for a separate method.[52]

What was being defended seems remarkably consistent: his splitting of taxa based on small differences between plants. He did not entirely lack the consistency in his claims for various regions either. These all made good sense as oppositional concepts set against "Great Western" or "European" or "Euro-American" centers. The political nature of these oppositions became most clear in his last and most lasting attempt to associate his method with a specific region, "East Asia." For one thing, while its oppositional "Euro-America" could be quite definite as a geographical region, as a botanical region it was unusual to combine two large and separate regions. Euro-America made sense as a botanical region only if one considered what were, for Nakai, the most familiar forms of plants: the dried specimens. Euro-America was the region where numerous dried specimens were stored at its authoritative centers.[53]

Similarly, his choice of "East Asia," *To-a* seems to have been political, although the political connotation of "East Asia" has mostly been forgotten with its wide circulation. As insightfully analyzed by Koyasu Nobukuni, the term was originally coined in the fields of history and arts to replace the then usual "Chinese" epithet for describing "Japanese" things. The new term answered the needs of the expansionists to claim the primacy of Japan, not China, within the eastern part of Asia. It became more common in the 1930s, in response to Japan's more direct competition with the West. It helped transform Japanese imperial ideology, which had previously rested mostly on Japan's earlier acquisition of Western modernity, by stressing Japan's commonality with its *To-a* brothers. The claim that Japan had the unique ability to combine *To-a* traditions with good things from the West was the key.[54] Nakai astutely adopted this newly circulating term *To-a* to explain the value of his research on Korean flora. He explained that his imperial research on Korean plants allowed him to learn the "East Asian" characteristics of Korean and Japanese plants and thereby to establish his distinctive systematics.

With growing Japanese influence on the region, "East Asia" was also a plausible choice as a territory to assert the authority of his "original method." Nakai made his most serious attempt to define East Asia as a viable botanical region, testifying to his belief in its potential. His most remarkable attempt was his attempt to connect Japan to the rest of the Asian continent. It was his effort to overcome the problem of incorporating Japan's island flora into an otherwise continental "East Asia." A geological argument about the ancient continent between Korea and Japan was the inspiration.[55] He argued that the flora on the volcanic island of Ullŭng—which lay between the two lands—made it a place to "imagine the forest in the lost continent." The evidence for this was that it contained many native Japanese plants such as hemlock spruce, Japanese white pine, and beech, whose seeds were too heavy to be transplanted without a land connection. Moreover, the beech was a fossil tree, showing the antiquity of the island.[56] Nakai's attempt to connect Japan to the rest of the continent received political but not scientific support. The most convincing counter-evidence was suggested by Japanese scientists in Korea, including Nakai's closest associate for his colonial botanizing Ishidoya Tsutomu. By pointing out the island's distinctive fauna—the lack of land animals, insects, and freshwater fishes—they were able to strongly suggest its isolated emergence.[57]

Nakai did not give up on demarcating East Asia so easily. When he made his second attempt to define East Asia in 1935, he chose a completely different tactic. If his previous attempt had relied on nature, his new definition was critical of naturalistic approaches, now identified as characteristic of Western attempts at plant geography.[58] He singled out for criticism the approach of the then authority on plant geography, Engler. Though this direct charge against a Western authority was intended only for a Japanese audience (it was written in Japanese), Nakai's criticism targeted the fundamental assumption underlying Engler's approach:

> Although this division looks sophisticated at a glance, the method of division and union has many seriously unnatural elements. It is absolutely impossible to consent with for a scholar who is actually in contact with East Asian flora all the time.[59]

"Unnatural" might mean something more like "strange" than "not natural" here. Engler's was, in fact, a seriously naturalistic approach. It was the result of an exclusive utilization of elements of nature, such as climate and geography, based on numbers obtained through well-standardized instruments such as the thermometer, barometer, altimeter, and so on. As discussed in historiographies of European science, these "natural" elements adopted by European scientists became tools to demarcate the natural world as separate from the social one, allowing them to establish their autonomous authority for all things in "nature."[60]

Nakai found this exclusively naturalistic demarcation unnatural, although he did not specify why he thought so. Instead he just provided his alternative 'natural' division. His East Asia was demarcated simply by the governmental borders of China, Korea, Taiwan, and Japan, each of which, in turn, was composed of separate botanical regions. For example, the botanical region of Japan was further divided into several regions in chronological order of each entity's inclusion into Japanese territory. The botanical region of Korea was likewise divided into five regions—the Northern, Middle, and Southern regions within the peninsula, plus the Ullŭng and Cheju islands.[61] One might say that Nakai considered 'natural' categories like latitude in deciding Northern and Southern regions and used geographical entities like an island as a category of division. However, the guiding factor for Nakai's natural division seems to be his consideration of political dominance. If Engler's linkage of the Russian Maritime Territory to the adjacent Chinese Heilongjiang Province was unnatural to Nakai, what made it so seems to be that the domain was divided by different political regimes. Hence the Ullŭng and Cheju islands, under the rule of the colonial Korean government, could not be combined with any part of Japan in spite of Nakai's own claim of their botanical affinities with the Japanese Archipelago.

Nakai's labeling of European scientists' careful exclusion of cultural categories as 'unnatural' is significant. It suggests that the European attempt to impose a separation between nature and culture, and its concomitant claims of universality, may not have been always clear to non-European botanists. Nakai was not impressed; neither the global interconnectedness of nature nor these scientists' concern to define nature solely based on 'nature' left an impression on him. He could not see that European scientists had the naturally endowed right to apply their science everywhere, by virtue of having successfully "identified" this one grand natural world first. He saw science as inseparable from politics: the intellectual authority of Western botany came from Western imperial power that had conquered many territories in the non-West.

Nakai apparently perceived that Western power was nonetheless insurmountable. He scarcely revealed his criticisms to Western audiences, notwithstanding his inability to follow European classificatory norms and his defiant regionalist stance shown to Japanese audiences. Rather, in 1926, when he had the chance to display his stance about European rules to the international botanical community, he acted as a most adamant guardian of the Euro-centric status quo. The occasion was his proud appointment to the International Committee on Botanical Nomenclature. He was the sole non-Western member appointed to this important committee.[62] It was not a specific recognition for Nakai's taxonomic talent but seems to be a recognition for Japan's increasingly impressive imperial botanical center, of which Nakai had been in charge since Hayata's illness in 1922.[63] It was a momentous event for Japanese botany, achieving what modern Japanese science and society had long pursued: becoming a part of Western authority and thus being able to set the universal rules as a powerful insider.[64]

The international botanical community decided that it needed a special committee to meet at its next international congress in 1930 to address an issue that had been a thorn in their side for a long time: the instability of botanical nomenclature.[65] As displayed in Hayata's critique, once one moved beyond the superficial consensus about the Linnaean binomial system, the creation of a universal standard had been an elusive goal for modern systematics. Since Linné, there had been many systematists who criticized the artificial nature of Linnaean classification by claiming a real, "natural" system. Thus, even within Europe, actual practice varied with allegiances to different natural systems.[66] Increasing non-European participation in this practice, such as that of Americans and Oceanians, beginning around the late nineteenth and early twentieth centuries, was further diversifying opinions regarding established genera and species, creating an abundance of new names for similar plants.[67]

Newcomers did not merely heighten the nomenclatural instability that had been on systematists' agendas since the mid-nineteenth century. They had different solutions to the problem. They knew that the Priority Rule, proposed by Alphonse Pyramus de Candolle (1806–1893) and adopted at the first, all-European International Botanical Congress in 1867, was in shambles. European botanists had realized that its simple solution (keeping only first-given scientific names) caused too many familiar names to disappear. They had already unraveled it with various subsidiary measures like the Berlin and Kew codes, which had provisions for such names to be preserved. In 1907, some American botanists decided to make their voices heard as well, proposing an "American code." This code suggested having a representative type or types within each group whose name would be retained; other names in the group would be modified according to this. Moreover, its demand to scrap Latin as the stipulated language for botanical description was a shot at European dominance in systematics.[68]

Neither the integrity of the allegedly universal standard of botany nor the authority of European centers was safe. What finally produced the much awaited "international" consensus over this division was the 1930 congress, to which Nakai was invited. The consensus was, in principle, to accept the American code: revolution indeed.[69] However, Nakai's report sent to the committee in 1926 demonstrates that he neither expected nor wanted European centers to make such compromises. He was ready to fight for the authority and integrity of the Euro-centered standard against this American challenge.

Proclaiming the universality of the European standard was Nakai's first shot:

> The rules now supposed to be in force, seem too general and too susceptible of individual interpretation. To remedy this, the Rules of Nomenclature should be defined more accurately yet practicably in clear scientific terms, so that there should be the least possibility for human arbitrariness or traditional custom to ignore the real meaning of the rules and oblige them to observe them strictly.[70]

Nakai apparently found it problematic for the profession that botanical rules were in essence just voluntary guidelines.[71] For the unity of the profession, any unauthorized interpretation of the rules should not be allowed and the rules should be enforced universally. He argued that the much-maligned Priority Rule would allow for unambiguous application, so long as no exceptions were condoned. He added that "[t]he family name when not written in Latin should not be accepted under any circumstances, even if its description is given accurately."[72] His attempt to distance himself from the troublemaking newcomers was clear.

Nakai's proposition about color terminology further displays his strategy to perfect the rules to render them more enforceable. It was remarkably in line with the efforts being made in other soft sciences at the time in their pursuit of exactness and standardization.[73] He proposed, "[t]he terminology of shades should be standardized and uniformed, otherwise we may never be able to know the exactness of the natural colors described or defined by our colleagues." One of his references for a uniform color terminology thus offered names for distinguishing 1,115 color shades. And, he added, when these suggestions were not heeded and descriptions were still ambiguous and confusing, it should be clear that the standard specimens were to have the last word.[74] He presented himself, and by extension Japan, as anything but a trouble-making newcomer. He was one of the standard-bearers, who worked to maintain and strengthen the standard, as if to make the European standard truly universal.

Although he acted almost like a European botanist, his final act, where his mask would become most truly European, was yet to come. It also showed that making the European standard truly universal was not his concern. He wanted to make *his* troubled method the universal standard and he saw the chance, naturally, in Japan's further imperial expansion. Just ahead of Japan's deeper invasion of China in 1937, Nakai made his first attempt at applying his classification to the plants of the entire world. In his Japanese preface for volume 21 of *Flora Sylvatica Koreana*, published in 1936, he said that he decided to "try not just the *Flora Sylvatica* of Korea but comprehensive monographs on the seven families of plants." He was, he wrote, "inspired by extraordinary praise from botanists, dendrologists, foresters from all over the world" for his previous volumes.[75] It was ambitious of him to attempt to fit monographs on seven families with hundreds of genera and thousands of species into a single volume.

Nakai soon found a more plausible strategy. He came to focus on families and genera known to be particularly thriving in "East Asia," such as tea trees and honeysuckle families. This strategy was timely because he could then obtain many of specimens from mainland China, a fact about which he did not hide his satisfaction. "The natural world of China was open only to Euro-American scholars for a long time, leading Japanese, even scholars, to be in the dark about its secret." Owing to the war, "this artificial and unnatural situation disappeared." He promised to work day and night to "enlighten our knowledge on the Chinese natural world" and thus to hasten the arrival of the day "when all people in East Asia could be truly happy." The aim of this civilizing and liberating work was to "place the research that originated from East Asia on a legitimate lineage of study."[76] His classification, not the European universal, would be the legitimate classification for all plants related to East Asia.

Nakai's new classification of the genus *Lonicera*, the honeysuckles, and *Camellia*, the tea tree family, revealed that there was nothing fundamentally new in his classification itself. He again displayed his preference for minute details in dried specimens, although by then he could provide a somewhat more developed rationale for his splitter tendency. He said that "East Asian plants, particularly those of the Japanese archipelago, manifested a uniquely wide variation." Applying the "Great Western" method based on "Euro-American or South Pacific plants" was impossible.[77] He had to split genera and species more minutely to fully express this rich variation; he created eight new sections in *Lonicera*.[78] His new classification of the genus *Camellia* revived the discarded genus *Thea* first proposed by Linné and added some sub-sections. His new classification for *Thea* certainly failed to undo the international consensus made in 1887 to combine the genus with *Camellia*, and most of the new names that he proposed were not even adopted by contemporaneous Japanese scholars.[79]

Conclusion: Modern Interactions for a Modern Success

In spite of Nakai's failure at having his clear, yet too minute, differentiation of genera and species accepted widely, he nonetheless had resounding success as a taxonomist of Korean flora representing Japanese botany. His honorary chairmanship at the Seventh International Botanical Congress held in Stockholm in 1950 is no minor proof of that. Although the nomenclatural committee did not show much enthusiasm about Nakai's attempt to increase the clarity of the English phrasing of the nomenclatural principles and rules (most of his proposals were set aside as editorial matters), it obviously made more sense for the committee to have him, again the sole non-Westerner, presenting its solution as a "solution that would be satisfactory from an international point of view."[80] The presence of Nakai, a scientist who had been, from the point of view of most scientists, on the enemy side in the recent war, also sent the message that science transcended the world's political and military conflicts.

Nonetheless, the Japanese path of plant systematics, through which botanists at the Japanese imperial center had paved the way for their European success, is unthinkable outside such militaristic politics. Although it was a fully scholarly strategy formulated by botanists, the Japanese botanical establishment consciously took a path that closely conjoined its scientific practice with the power struggle in the age of imperial competition. This political engagement leading and cheering for Japanese expansion to Asia and the world certainly made it easier for them to claim their place in Japanese society, pre-occupied as it was with Japan's place in the world after its acceptance of unequal treaties. This "foreign-oriented" imperial path was what this botanical establishment chose and carefully guarded for the distinction that it needed in competing with other well-trenched and down-to-earth knowledge practitioners; the substantial and laborious research on domestic plants built on Tokugawa social and intellectual bases is yet to enter Japanese botanical stories of success.

So far, the vociferous victors like Nakai have made this imperial making of Japanese modernity look too natural or inevitable, as if all botanists accepted and worked toward it. It was the path that could be kept only by suppressing the serious questions that the likes of Hayata had raised, and rather more and more detaching himself from the colonial field that produced the specimens for his success. So, did the Japanese center sustaining Nakai's authority produce a European botany or a Japanese one? Can we identify species of knowledge practice so globally connected and negotiated either as Japanese or European?

Nakai outdid European botanists in claiming the legitimacy of his efficient systematics outside of its region of origin. With "yellow faces" looking at him, Nakai never wanted to call his method simply "Korean." He asserted the Oriental, Asian, and East Asian nature of his systematics, eventually claiming a new Japanese universal in lieu of the European one, this time against the European centers that had frustrated him. As a proper imperial scientist, such a resolute claim for his rule's universal authority seemed always necessary to him. He nonetheless never voiced such wish to European centers. He seemed most accepting to these powerful counterparts who judged the properness of his method. It was the strange parity that Japan had achieved in interacting for its success in modern botany. Unbelievably accepting and understanding on the one side, and painfully authoritative despite its own confusion on the other side.

Hayata shows that this was not the only way that Japanese botany could achieve parity in its interactions for botanical pursuits. Hayata, who could not easily accept the one way move of the specimens to the Japanese center, corrected such asymmetry by his open critique on European reliance on inadequate specimens or descriptions, and by immersing himself in various fields unknown to European armchair botanists. Hayata neither pretended his acceptance of European rules in making his open interactions with European botanical communities nor did he seek to be 'White' to Taiwanese who might watch him. He respected the Pescadores living prosperously with their own rules of sharing not enough produce of their land. It was a parity that seems also alert to diversity, creativity, and dynamic adjustment of the plant world in the fields, and faithful to what the European enlightenment said about the critical and progressive spirit of science, ever defying against accepted norms and entrenched authorities.

History within the frame offered by successful moderns did not note on these serious questions from various fields. Yet, Hayata was just one of those who had not obtained full hearing yet. Much more colonial collaborators of Nakai's authoritative botanizing in Korea in fact never ceased to raise their concerns and doubts about Nakai's rather confined choices as an imperial botanist and his way of interacting with them. The rest of this book, unlike Nakai, plunges into colonial fields, where the imperial new names of Japanese botany were variously contested and repurposed.

Notes

1 Hayata Bunzo, "On Taiwania, a New Genus of Coniferae from the Island of Formosa," *The Journal of the Linnean Society. Botany* 37(260) (1906): 330–31; "On the Distribution of the Formosan Conifers," *The Botanical Magazine, Tokyo* (植物學雜誌, *BMT* hereafter) 19 (1905): 43–60; "On Taiwania and its Affinity to Other Genera," *BMT* 21 (1907): 21–28; Ohashi Hiroyoshi, "Bunzo Hayata and His Contributions to the Flora of Taiwan," *Taiwania* 54 (2009): 1–27. Ohashi is a contemporary Japanese botanist.

2 Hayata Bunzo, "What Is it to Have the So-Called Eternal Life (永遠の生命とは如何なるものか)," *Botany and Zoology, Scientific and Applied* (植物及動物) 1(12) (1933): 1743–50, 1747, 1750.

3 It was published in both Japanese and English in the same volume. Nakai Takenoshin, *Flora Sylvatica Koreana*, vol. 18, Seoul: The Government-General of Korea, 1930, 52.

4 Masamune Genkei 正宗嚴敬, "Bibliography of Dr. B. Hayata's Work (早田博士の著書及論文)," *Transactions of the Natural History Society of Formosa* (臺灣博物學會會報) 24(135) (1934): 478–85.

5 Hayata's first sentence in announcing his new genus *Taiwania* was to thank the contribution of this collector. Hayata Bunzo, "On Taiwania, a New Genus of Coniferae from the Island of Formosa," *The Journal of the Linnean Society. Botany* 37(260) (1906): 21, 25.

6 Hayata's first sentence in announcing his new genus *Taiwania* was to thank the contribution of this collector. Hayata Bunzo, "On Taiwania, a New Genus of Coniferae from the Island of Formosa," *The Journal of the Linnean Society. Botany* 37(260) (1906): 21, 2.

7 Hayata was a splitter, who enjoyed finding new plants by differentiating them from existing ones. More on the issue of splitters and lumpers follows in the next section. Ohashi Hiroyoshi, "Bunzo Hayata and His Contributions to the Flora of Taiwan," *Taiwania* 54 (2009): 8–9.

8 Hayata Bunzo, 早田文藏. *Icones plantarum Formosanarum nec non et contributiones ad floram Formosanam: or, Icones of the Plants of Formosa, and Materials for a Flora of the Island, Based on a Study of the Collections of the Botanical Survey of the Government of Formosa,* Taihoku: The Bureau of Productive Industries, Government of Formosa, 1911–1921, vol. 4, ii–v; vol. 5, iii.

9 Hayata Bunzo, 早田文藏. *Icones plantarum Formosanarum nec non et contributiones ad floram Formosanam: or, Icones of the Plants of Formosa, and Materials for a Flora of the Island, Based on a Study of the Collections of the Botanical Survey of the Government of Formosa,* Taihoku: The Bureau of Productive Industries, Government of Formosa, 1911–1921, vol. 4, v–vi; Hayata made one long trip to see specimens at European herbaria in 1910, during which he interacted extensively with European botanists. Hayata Bunzo, 早田文藏. *Icones plantarum Formosanarum nec non et contributiones ad floram Formosanam: or, Icones of the Plants of Formosa, and materials for a Flora of the Island, based on a study of the collections of the Botanical Survey of the Government*

of Formosa, Taihoku: The Bureau of Productive Industries, Government of Formosa, 1911–1921, vol. 1, 2–3.

10 Hayata Bunzo, 早田文藏. *Icones plantarum Formosanarum nec non et contributiones ad floram Formosanam: or, Icones of the Plants of Formosa, and Materials for a Flora of the Island, Based on a Study of the Collections of the Botanical Survey of the Government of Formosa,* Taihoku: The Bureau of Productive Industries, Government of Formosa, 1911–1921, vol. 4, 6, for more on Hayata's concerns about botanical illustrations, see Jung Lee, "Provincializing Global Botany," in Curry et al. eds., *Worlds of Natural History*, Cambridge: Cambridge University Press, 2018, 433-446, 438-440.

11 In 1914, he proposed this as an agenda for the Botanical Congress scheduled the following year, but this conference, postponed on account of the turmoil of the Great War, ended up taking place too late for him to be able to actively participate. The congress at Cornell in 1926 was the next one. Hayata became ill in 1922 and remained weak until his premature death in 1934. Ohashi Hiroyoshi, "Bunzo Hayata and His Contributions to the Flora of Taiwan," *Taiwania* 54 (2009): 3. Hayata did not fully discard dried specimens but he did not take many specimens when he collected. That dried specimens had contributed to the making of the universal botanical standards would not be denied, though the constant uncertainties and negotiations should be noted. Staffan Müller-Wille, "Joining Lapland and the Topinambes in Flourishing Holland: Center and Periphery in Linnaean Botany," *Science in Context* 16 (2003): 461–88.

12 Hayata Bunzo, 早田文藏. *Icones plantarum Formosanarum nec non et contributiones ad floram Formosanam: or, Icones of the Plants of Formosa, and Materials for a Flora of the Island, based on a study of the collections of the Botanical Survey of the Government of Formosa,* Taihoku: The Bureau of Productive Industries, Government of Formosa, 1911–1921, vol. 10, 98, 114, 115, 115–28, 99.

13 Hayata Bunzo, 早田文藏. *Icones plantarum Formosanarum nec non et contributiones ad floram Formosanam: or, Icones of the Plants of Formosa, and Materials for a Flora of the Island, Based on a Study of the Collections of the Botanical Survey of the Government of Formosa,* Taihoku: The Bureau of Productive Industries, Government of Formosa, 1911–1921, 2–3, 75–76.

14 Hayata Bunzo, 早田文藏. *Icones plantarum Formosanarum nec non et contributiones ad floram Formosanam: or, Icones of the Plants of Formosa, and Materials for a Flora of the Island, Based on a Study of the Collections of the Botanical Survey of the Government of Formosa,* Taihoku: The Bureau of Productive Industries, Government of Formosa, 1911–1921, 116, 125. Although this kind of crossing is rare, the history of life is indeed sufficiently long to make it significant enough. According to some estimates, 47 percent of angiosperms and 95 percent of pteridophytes (ferns) are allopolyploid, i.e., hybrids of heterogeneous species. For plant species in general, they estimate that the hybridity would be about 30–40 percent. David L. Hull, *Science as a Process: An Evolutionary Account of the Social and Conceptual Development of Science*, Chicago, IL: University of Chicago Press, 1990, 81–109. 103.

15 Robert J. Richards, *The Romantic Conception of Life: Science and Philosophy in the Age of Goethe*, Chicago, IL: University of Chicago Press, 2002, 407.

16 That he encountered Goethe's work through Ernst Haeckel (1834–1919), "a tireless promoter of Darwin's theory of evolution" and "an assiduous champion of Goethe as an evolutionary precursor" made the irony only deeper. Gordon L. Miller, "Introduction," in Johann Wolfgang von Goethe ed., *The Metamorphosis of Plants*, tr. Miller, Cambridge, MA: MIT Press, 2009, xxii.

17 Johann Wolfgang von Goethe, ed., *The Metamorphosis of Plants*, tr. Miller, Cambridge, MA: MIT Press, 2009, 6.

18 Ferdinand Cohn (1828–1898), a father of microbiology, made the claim. Hayata examined readings of Cohn, Haeckel and the Goethe scholar Albert Bielschowsky (1847–1902). J. Reynolds Green's work, *A History of Botany, 1860–1900* (1909) was also

utilized. Hayata, *Icones*, vol. 10, 90–94. Robert J. Richards, *The Romantic Conception of Life: Science and Philosophy in the Age of Goethe*, Chicago, IL: University of Chicago Press, 2002.

19 Hayata Bunzo, 早田文藏. *Icones plantarum Formosanarum nec non et contributiones ad floram Formosanam: or, Icones of the Plants of Formosa, and Materials for a flora of the Island, Based on a Study of the Collections of the Botanical Survey of the Government of Formosa,* Taihoku: The Bureau of Productive Industries, Government of Formosa, 1911–1921, vol. 10, 76. Originally quoted in German. Sie schafft ewig neue Gestalten; was da ist, war noch nie, was war, kommt nicht wieder; alles ist neu und doch immer das Alte. Es ist ein ewiges Leben, Werden und Bewegen in ihr. Sie verwandelt sich ewig, und ist kein Moment Stillstehen in ihr… Jedem erscheint sie in einer eigenen Gestalt. Sie verbirgt sich in tausend Namen und Termen, und ist immer dieselbe.

20 Johann Wolfgang von Goethe, ed., *The Metamorphosis of Plants*, tr. Miller, Cambridge, MA: MIT Press, 2009, 102.

21 It should be noted that Hayata's gene did not have a fixed blueprint to be unfolded. On the development of molecular biology and for critical assessments of its ascendance, see Lily E. Kay, *The Molecular Vision of Life: Caltech, the Rockefeller Foundation, and the Rise of the New Biology*, New York: Oxford University Press, 1993. Bruno J. Strasser, "Who Cares about the Double Helix?" *Nature* 422 (2003): 803–04. Evelyn Fox Keller, *The Century of the Gene*, Cambridge, MA: Harvard University Press, 2000.

22 Hayata Bunzo, 早田文藏. *Icones plantarum Formosanarum nec non et contributiones ad floram Formosanam: or, Icones of the Plants of Formosa, and Materials for a Flora of the Island, Based on a Study of the Collections of the Botanical Survey of the Government of Formosa,* Taihoku: The Bureau of Productive Industries, Government of Formosa, 1911–1921, vol. 10, 98, 108–9.

23 Hayata Bunzo, "The Succession and Participation Theories and Their Bearings upon the Objects of the Third Pan-Pacific Science Congress," in *Proceedings of the Third Pan-Pacific Science Congress*, Tokyo: Maruzen, 1928, vol. 2, 1869–1875, 1874, 1872, 1870, 1875. For recent critiques on this interpretation of evolution as survival of the fittest, see Daniel S. Milo, *Good Enough: The Tolerance for Mediocrity in Nature and Society*, Cambridge, MA: Harvard University Press, 2019; Brian Hare and Vanessa Woods, *Survival of the Friendliest: Understanding Our Origins and Rediscovering Our Common Humanity*, New York: Random House, 2020.

24 Hayata said, "genealogy is really as intricate as the meshes of the net" and the making of a human race thus must be "a result of interaction, interrelation, interdependence or correlation of many things founded on an intrinsic necessity" of life, denying the validity of racial discussions. Hayata Bunzo, "The Succession and Participation Theories and their Bearings upon the Objects of the Third Pan-Pacific Science Congress," in *Proceedings of the Third Pan-Pacific Science Congress*, Tokyo: Maruzen, 1928, 1874, 1872, 1871–72.

25 Hayata Bunzo, "The Succession and Participation Theories and their Bearings upon the Objects of the Third Pan-Pacific Science Congress," in *Proceedings of the Third Pan-Pacific Science Congress*, Tokyo: Maruzen, 1928, vol. 2, 1873.

26 Hayata Bunzo, "The Succession and Participation Theories and their Bearings upon the Objects of the Third Pan-Pacific Science Congress," in *Proceedings of the Third Pan-Pacific Science Congress*, Tokyo: Maruzen, 1928, vol. 2, 1870.

27 Kurt Krause (1883–1963) introduced Hayata's theory as providing an entirely new perspective on systematics. "Eine Neue Form des natürlichen Systems," *Naturwissenschaften* 4(11) (1923): 60–63. Engler seemed not to make reference to Hayata's work, although they knew each other from Engler's trip to Japan in 1913 when Hayata guided him. Hayata Bunzo, "The Succession and Participation Theories and their Bearings upon the Objects of the Third Pan-Pacific Science Congress," in *Proceedings of the Third Pan-Pacific Science Congress*, Tokyo: Maruzen, 1928, 1883. For more on the

early debate concerning phylogenetic classification, see David L. Hull, *Science as a Process: An Evolutionary Account of the Social and Conceptual Development of Science*, Chicago, IL: University of Chicago Press, 1990, 81–109.

28 Du Rietz, who obviously encountered Hayata through Krause's German introduction, felt that he found what he had been theorizing from his field experience in Hayata's theory. G. Einar Du Rietz, "The Fundamental Units of Biological Taxonomy," *Svensk Botanisk Tidskrift* 24(3) (1930): 333–428, 413, 406. Hayata noted Du Rietz's acknowledgment of his work later. Hayata Bunzo, "On the Dynamic System of Plants (植物の動的分類系に 就きて)," in *Biology* (生物学), Tokyo: Iwanamishoten, 1931, 95–111, 96.

29 Kimura Yojiro 木村陽二郎, Nakai's student and a later historian, left this record of his conversations with both Nakai and Hayata. Although Nakai blamed Hayata for their estrangement, that Nakai's student Kimura asked only Nakai and not Hayata about their obviously strained relationship, despite visiting both of them, can tell us a lot about who appeared as an actor in this relationship. Kimura, "Professor Nakai Takenoshin's Words," in Kimura Yojiro, *Works on History of Biology* (生物学史論集), Tokyo: Yasaka Shobo, 1987, 407–11, 409.

30 On the definition for the "armchair" botanist and the "voyaging" one, see Londa L. Schiebinger, *Plants and Empire*, Cambridge, MA: Harvard University Press, 2004, 24. Most European botanists including Linné were "armchair" botanists. Christophe Bonneuil, "The Manufacture of Species: Kew Gardens, the Empire and the Standardisation of Taxonomic Practices in late 19th Century Botany," in M.-N. Bourguet, C. Licoppe, and O. Sibum eds., *Instruments, Travel and Science: Itineraries of Precision from the 17th to the 20th Century*, London: Routledge, 2002, 189–215. Lisbet Koerner, *Linnaeus: Nature and Nation*, Cambridge, MA: Harvard University Press, 1999.

31 Of course, this was in line with popular sentiment in Meiji Japan since the 1870s concerning Japan's duty to subjugate Korea. Nakai Takenoshin, *A Synoptical Sketch of Korean Flora,* Tokyo: The National Science Museum, 1952, 1.

32 Palibin's *Conspectus Florae Koreae* (1898–1901) recorded only 103 families, 393 genera, 635 species and 20 variations. Nakai Takenoshin, *Flora Koreana*, Tokyo: Imperial University of Tokyo, 1909; J. Palibin, *Conspectus Florae Koreae*, Petropoli, Petropoli: Acta Horti Petropolitani, 1901.

33 George Orwell, *Shooting an Elephant and Other Essays*, San Diego, CA: Harcourt, Brace and Company, 1950, 8.

34 Nakai Takenoshin, "An Observation on Japanese Aconitum-I," *BMT* 22(259) (1908): 127–33, 127.

35 This tension between 'splitters' and 'lumpers' was high and Hooker's agenda for lumping is well discussed in Jim Endersby, *Imperial Nature: Joseph Hooker and the Practices of Victorian Science*, Chicago, IL: University of Chicago Press, 2008, 155–69. The tension was never resolved because no consensus on the concept of species could be made. For a temporarily successful attempt of European botanists in orchestrating nomenclatural stability in the late nineteenth century, see Bonneuil, "The Manufacture of Species." For the conceptual debate on species, Gordon R. McOuat, "Species, Rules and Meaning: The Politics of Language and the Ends of Definitions in 19th Century Natural History," *Studies in History and Philosophy of Science Part A* 27(4) (1996): 473–519; Gordon R. McOuat, "The Origins of 'Natural Kinds': Keeping 'Essentialism' at Bay in the Age of Reform," *Intellectual History Review* 19(2) (2009): 211–30.

36 Gordon R. McOuat, "The Origins of 'Natural Kinds': Keeping 'Essentialism' at Bay in the Age of Reform," *Intellectual History Review* 19(2) (2009): 128; Nakai Takenoshin, "An Observation on Japanese Aconitum-II," *BMT* 22(260) (1908): 133–40.

37 On the difficulties of learning to work with specimens and the methodological difference that the heavy reliance on dried specimens could bring, see Anne Secord, "Pressed Into Service: Specimens, Space, and Seeing in Botanical Practice," in David N. Livingstone

and Charles W. J. Withers eds., *Geographies of Nineteenth-Century Science*, Chicago, IL: Chicago University Press, 2011, 283–310.

38 For the difference of his splitter tendency relying on specimens from those of other non-European botanists in Americas or Oceania, who emphasized their field experience, see Jung Lee, "Between Universalism and Regionalism: Universal Systematics from Imperial Japan," *The British Journal for the History of Science* 48(4) (2015): 661–84, 673–74.

39 Both these influential works consider Korea part of China. Adolf Engler and Karl Prantl, *Die Natürlichen Pflanzenfamilien*, Leipzig: Wilhelm Engelmann, 1900; *Index Kewensis*, Oxford: Clarendon Press, 1895–1905.

40 Nakai Takenoshin, "Research on Korean Flora (朝鮮植物の研究)," *Oriental Art and Science Magazine* (東洋學藝雜誌) 43(534) (1927): 561–71, 564.

41 In comparison, Taiwan, known to the West through its earlier Portuguese visitors as *ilha Formosa*, literally "beautiful island," had drawn and was still drawing Western collectors and botanists to its verdant subtropical forests. Henry F. Hance (1827–1886), Karl Maximowicz (1827–1891), William B. Hemsley (1843–1924), and Augustine Henry (1857–1930), etc., produced works on Taiwanese flora. Ohashi Hiroyoshi, "Bunzo Hayata and His Contributions to the Flora of Taiwan," *Taiwania* 54 (2009): 1–27 5–6.

42 Nakai Takenoshin, "Research on Korean Flora (朝鮮植物の研究)," *Oriental Art and Science Magazine* (東洋學藝雜誌) 43(534) (1927): 561.

43 Nakai Takenoshin, "Research on Korean Flora (朝鮮植物の研究)," *Oriental Art and Science Magazine* (東洋學藝雜誌) 43(534) (1927): 562. Nakai expressed his gratitude to the GGK's support by naming a couple of flowers after the then governor-general. Nakai Takenoshin, "On Naming a New Genus of Korean Native Lilies, Derauchia (Preliminary Report)," *BMT* 27(322) (1913): 441–43.

44 Nakai Takenoshin, "Research on Korean Flora (朝鮮植物の研究)." *Oriental Art and Science Magazine* (東洋學藝雜誌) 43(534) (1927): 562. The colonial government's interest in this ideological element of science is also much more prominent in Korea. Taiwan was given to Japan by Western powers upon Japan's victory in the Sino-Japanese war in 1895, while Korea became its colony through its series of maneuvers, often criticized as unjust by Westerners in the region. Also, the colonial Taiwanese government achieved financial independence from the imperial coffers in 1906, with its lucrative cash crops like sugar cane, camphor trees, tea, opium, and tobacco. It had enough leverage to "tame" the colonized with civilizing measures, or at least to better control them with a better-equipped police system. It did not have to justify its presence to anyone, unlike the financially dependent and ideologically insecure colonial Korean government. Moon Myung-ki 문명기, "The Gap Between the 'Colonial Modernity' of Taiwan and Korea: Comparing the Police Systems (대만·조선의 '식민지근대'의 격차)," *Studies on Chinese Modern History* (중국근현대사연구) 59 (2013): 69–99.

45 Nakai Takenoshin, "Research on Korean Flora (朝鮮植物の研究)," *Oriental Art and Science Magazine* (東洋學藝雜誌) 43(534) (1927): 562. *Flora Sylvatica Koreana* XXI (1936), preface. Other than Nakai's words, we cannot clearly trace the reception of his work.

46 On Nakai's extreme splitter tendency, Chang Chin-Sung장진성, "A Reconsideration of Nomenclatural Problems on Korean Plants and The Korean Woody Plant List (韓國樹木의 目錄과 學名에 대한 再考)," *Korean Journal of Plant Taxonomy* 24(2) (1994): 95–124. For Nakai's disagreement with contemporary Japanese systematists, see this defence by him. Nakai Takenoshin, "Reexamination of Questioned Scientific Names of Plants (問題にされた学名の再検討)," *The Journal of Japanese Botany* (植物研究雜誌, *JJB* hereafter) 26(11) (1951): 321–28.

47 The current Tokyo University Herbarium shows that unlike Hayata, who kept one or two specimens for each plant, Nakai kept many specimens for one plant.

48 Nakai Takenoshin, *Flora Sylvatica Koreana*, vol. 1, Seoul: The Government-General of Korea, 1915. The preface was dated 1913.
49 Nakai Takenoshin, *Flora Sylvatica Koreana*, vol. 1, Seoul: The Government-General of Korea, 1915. The preface was dated 1913.
50 Nakai Takenoshin, "Aconitum of Yeso, Saghaline and the Kuriles," *BMT* 31(368) (1917): 219–31, 219, 222.
51 Nakai Takenoshin, *Flora Sylvatica Koreana*, vol. 2, Seoul: The Government-General of Korea, 1915. The preface is dated 1914.
52 Nakai Takenoshin, "Research on Korean Flora (朝鮮植物の研究)," *Oriental Art and Science Magazine* (東洋學藝雜誌) 43(534) (1927): 569–70.
53 He visited major botanical gardens and herbariums in the US, France, England, Sweden, and the Netherlands from May 9, 1923 to Sep 28, 1925. Committee for Commemoration of Dr. Nakai's Works (中井博士功績記念事業会), *The List of Monographs and Articles by Prof. Nakai and the Index of New Groups, Species and Scientific Names by Him* (中井教授著作論文目録並に教授の研究発表による植物新群名,新植物名及新学名總索引, Nakai Commemoration hereafter), Tokyo: Hokuryukan, 1943.
54 Japan's attempt to reevaluate Japanese civilization while emphasizing China's deterioration had been made since the nineteenth century, especially when it competed with China over Korea. Koyasu says that the earliest usage of *To-a* was the translation for Ernest Francisco Fenollosa's *The Epoch of Chinese and Japanese Art (1911).* For the book that gave "national pride and understanding for the Japanese arts," they chose the Japanese title of *Outline of To-a Art History* (東亞美術史綱). In supporting this new imperial ideology, this term became "a geo-political concept with strong political implications in the 1930s and 1940s" in every discipline, in line with the Japanese claim for the Greater East Asia Co-Prosperity Sphere. Koyasu Nobukuni 子安宣邦, *East Asia · Great East Asia · East Asia: Orientalism in Modern Japan* (동아 · 대동아 · 동아시아: 근대 일본의 오리엔탈리즘), tr. Yi Sŭngyŏn, Seoul: Yŏkbi, 2005, 151.
55 See Tateiwa Iwao 立岩巖, *The Early History of Geological Research in Korea* (韓半島 地質學의 初期硏究史: 朝鮮·日本列島地帶 地質構造論考), tr. Yang Sŭngyŏng, Taegu: Kyŏngbuk University Press, 1996, 589.
56 Nakai Takenoshin, "The Vegetation of Dagelet Island: Its Formation and Floral Relationship with Korea and Japan," in *Proceedings of the Third Pan-Pacific Science Congress, Tokyo, Oct. 30th–Nov. 11th, 1926*, Tokyo: Maruzen, 1928, 911–14, 913.
57 The story of his closest associate Ishidoya Tsutomu is discussed in Chapter 3. It had no mammals except for rats, no reptiles or amphibians and few insects. Ishidoya, "On the Floristic Region of the Ullŭng Island (欝陵島ノ植物區系=閥スル考察 (1))," *Journal of the Natural History Association of Korea* (朝鮮博物學會雜誌) 7 (1928): 21–25. Terada Torahiko (寺田寅彦, 1878–1935) at the Earthquake Research Institute in the Imperial University and Tateiwa Iwao (1894–1982) on the geological survey team in the colonial government pointed out that it was quite unlikely that such a small land mass (72.99㎢) would be left while the whole continent was sinking under. However, Nakai's advocacy of the ancient Japanese Sea Continent appealed to some colonial officers to provide him extra funding for another investigation for other islands around Korea. Terata Torahiko, "On the Bathymetric Features of Japan Sea," *The Bulletin of the Earthquake Research Institute* 12(4) (1934): 650–55; Nakai Takenoshin, "The Comparison of Floras of the Isolated Islands in the East and West Part of the Korean peninsula (朝鮮半嶋ノ東西ニ孤立スル鬱陵島ト大黑山島トノ植物帶ノ比較)," *Oriental Art and Science Magazine* 43(4) (1927): 214–27, 220–21. 214.
58 For the discussion on such naturalistic biogeography, especially since Humboldt, see Michael Dettelbach, 'Humboldtian Science,' in Jardine, Nicholas, James A. Secord, and Emma C. Spary, eds., *Cultures of Natural History*. Cambridge: Cambridge University Press, 1996, 287–304; Janet Browne, *The Secular Ark: Studies in the History of Biogeography*, New Haven, CT: Yale University Press, 1983.

59 Notably, it was quite an exaggeration to say that he had the constant contact with East Asian flora given that up to this time had only made one trip to China, in 1933. Nakai Takenoshin, *East Asian Plants* (東亞植物), Tokyo: Iwanamishoten, 1935, 2–3.
60 See Bruno Latour, *We Have Never Been Modern*, Cambridge, MA: Harvard University Press, 1993.
61 Nakai Takenoshin, *East Asian Plants* (東亞植物), Tokyo: Iwanamishoten, 1935, 3–4.
62 The committee consisted of 29 members including four Americans and two Europeans representing Algeria and South Africa. Benjamin Duggar, *Proceedings of the International Congress of Plant Sciences, Ithaca, New York, August 16–23, 1926*, Menasha, WI: George Banta Pub. Co., 1929.
63 *Nakai Commemoration* (note 52); On 1926 Botanical Congress, Benjamin Duggar, *Proceedings of the International Congress of Plant Sciences, Ithaca, New York, August 16–23, 1926*, Menasha, WI: George Banta Pub. Co., 1929, 1777–1782.
64 Benjamin Duggar, *Proceedings of the International Congress of Plant Sciences, Ithaca, New York, August 16–23, 1926*, Menasha, WI: George Banta Pub. Co., 1929, 1777–1782. Japan's scientific achievement was not limited to botany as discussed in the introduction.
65 Nakai did not go to the 1930 congress in Cambridge because the Japanese government did not provide the funding, which he strongly protested. Nakai Takenoshin, *Regarding the Botanical Nomenclature* (植物命名規則に就いて), Tokyo: Iwanamishoten, 1930. As no serious history of modern Japanese botany yet exists, most detailed information on Nakai's career can be traced from his obituary and the memorial work for his sixtieth birthday. Hara Hiroshi, "Takenoshin NAKAI 1882–1952," *BMT* 66 (1953): 1–4; *Nakai Commemoration*. As a historical note, the following work by one later successor of his post at the university is most substantial. Oba, Hideaki 大場秀章. *History of Botanical Research in Japan: 300 Years of Koishikawa Botanical Gardens* 日本植物研究の歴史：小石川植物園三〇〇年の歩み. 東京大学コレクション. Vol. 4, Tokyo: Tokyo Daigaku Sogo Kenkyu Hakubutsukan, 1996, 101–106.
66 See Sean Hsiang-lin Lei, *Neither Donkey nor Horse: Medicine in the Struggle over China's Modernity*, Chicago, IL: University of Chicago Press, 2014, 12. On the difficulty of finding another reason to challenge Western reason, see Gyan Prakash, *Another Reason: Science and the Imagination of Modern India*, Princeton, NJ: Princeton University Press, 1999.
67 For the challenges made by Americans, see Sharon E. Kingsland, *The Evolution of American Ecology, 1890–2000*, Baltimore, MD: Johns Hopkins University Press, 2005.
68 Sharon E. Kingsland, *The Evolution of American Ecology, 1890–2000*, Baltimore, MD: Johns Hopkins University Press, 2005, 40–95. Doubts about the mandatory usage of Latin were raised by some new generation European botanists as well. For more on various proposals on the nomenclatural issue, see G. Perry, "Nomenclatural Stability and the Botanical Code: A Historical Review," in *Improving the Stability of Names: Needs and Options: Proceedings of an International Symposium, Kew, 20–23 February 1991*, Königstein/Taunus Germany: The International Association for Plant Taxonomy, 1991, 79–93.
69 The mandatory usage of Latin was maintained. While the debates on taxonomy were mostly internal affairs among scientists or a specialist topic of historians of science, recent popular works in science show the unresolved problems regarding classifying and naming nature, rather heightened by the development of molecular analyses, some of which asserts "fish don't exist" as a category. If these work distinguish such scientific discussions as too rational an approach against more instinctive and traditional perception of 'fish,' this book doubts whether the distinction between science and instinct is natural. Carol Kaesuk Yoon, *Naming Nature: The Clash between Instinct and Science*, New York: W. W. Norton & Company, 2009.
70 Nakai Takenoshin, *Suggestions for an Amendment to Be Made to the Rules of Botanical Nomenclature*, Tokyo, 1926, 1. It is a typed manuscript, whose copy is at the New York Botanical Gardens.

71 Article 1 of the International Rules of Botanical Nomenclature in 1930 states: "Botany cannot make satisfactory progress without a precise system of nomenclature, which is used by the great majority of botanists in all countries." And these rules hoped to be used by many were divided into *principles*, *rules* and *recommendations* showing its flexible nature. International Botanical Congress, *International Rules of Botanical Nomenclature Adopted by the Fifth International Botanical Congress, Cambridge, 1930*, London: Taylor and Francis, 1934.

72 Nakai Takenoshin, *Suggestions for an Amendment to Be Made to the Rules of Botanical Nomenclature*, Tokyo, 1926, 3.

73 The similar concerns for exactness and standardization existed in other sciences like anthropological biometrics and pedology. For the case of pedology, see Bruno Latour, *Pandora's Hope*, Cambridge, MA: Harvard University Press, 1999, 24–79.

74 Nakai Takenoshin, *Suggestions for an Amendment to Be Made to the Rules of Botanical Nomenclature*, Tokyo, 1926, 4.

75 Seven families discussed are *Aristolochiaceae*, *Lardizabalaceae*, *Berberidacea*, *Pittosporacece*, *Malvaceas*, *Ericaceae*, and *Urticaceae*, Nakai Takenoshin, *Flora Sylvatica Koreana*, vol. 21, Seoul: The Government-General of Korea, 1936.

76 Nakai Takenoshin, *Flora Sylvatica Koreana*, vol. 22, Seoul: The Government-General of Korea, 1940.

77 Nakai Takenoshin, "A New Classification of the Genus *Lonicera* in the Japanese Empire, together with the Diagnoses of New Species and New Varieties," *JJB* 14(6) (1938): 359–76, 359.

78 Nakai defied his usual tendency to create a new genus. Only one of his eight sections, *Monanthae*, has survived as a valid one in the family. Nakai Takenoshin, "A New Classification of the Genus *Lonicera* in the Japanese Empire, together with the Diagnoses of New Species and New Varieties," *JJB* 14(6) (1938): 359–76, 359; Theis, Nina, Michael J. Donoghue, and Jianhua Li, "Phylogenetics of the *Caprifolieae* and *Lonicera* (*Dipsacales*) Based on Nuclear and Chloroplast DNA Sequences," *Systematic Botany* 33 (2008): 776–83.

79 Nakai Takenoshin, "A New Classification of the Sino-Japanese Genera and Species which belong to the tribe Camellieae (I) (日華兩國產 つばき族 植物ノ 新分類法 (其一))," *JJB* 16(11) (1940): 659–67; Nakai Takenoshin , "A New Classification of the Sino-Japanese Genera and Species which belong to the tribe Camellieae (II) (日華兩國產 つばき族 植物ノ 新分類法 (其二))," *JJB* 16(12) (1940): 691–708.

80 Joseph Lanjouw, *Seventh International Botanical Congress, Section Nomenclature: Report*, Uppsala: International Bureau for Plant Taxonomy and Nomenclature of the International Association for Plant Taxonomy, 1953, xi; Joseph Lanjouw, *Synopsis of Proposals Concerning the International Rules of Botanical Nomenclature, Submitted to the Seventh International Botanical Congress, Stockholm, 1950*. Utrecht: A. Oosthoek, 1950.

3 Unsettling Imperial Universality

In 1910, right after the annexation of Korea, Nitobe Inazo greeted the freshmen of the top high school in Japan with these words: "the annexation of Korea is literally a golden opportunity… the Japan of today is entirely different from that of one month ago. In measuring up to such a great nation, we have to throw away our insularity."[1] Japanese youths had to take up their imperial responsibility by assuming the civilizing mission for this new colony. Historians have noted the role of colonial research in shaping the scientific profession in modern societies, mostly in the cases of Western empires. Many individuals without enough social and cultural capital to allow them build scientific careers at home carved out successful scientific careers abroad by venturing out to colonial outposts.[2] In Japan's "settler colony" of Korea, this aspiration to build a scientific career in the colony was more collectively manifested. For Japanese intellectuals like Nitobe imbued the best minds of Japan with the hope of a "civilizing mission," encouraged by Koreans who pressured on for "scientific" imperialism in return for their hopefully temporary submission to Japanese rule.[3]

From the time of Japan's opening to the world, those Japanese looking for new opportunities in life had easily found a home in Korea, its nearest neighbor. It was the country to which the largest number of Japanese had immigrated between 1876, when Korea was opened to the world by Japan, and 1885, when Japanese immigration to Hawaii began. There were already about 170,000 Japanese residents in Korea in 1910, the largest Japanese overseas population then. It rose to 750,000 by 1942 and a million by the end of colonial rule. Unlike the laboring immigrants in Hawaii or America, the Japanese who went to Korea were required to have better plans, so as not to be looked down on by Koreans. Experts and officials began to fill Japan's extensive colonial administration and subsidiary institutions starting in 1910; approximately a fourth of the Japanese population in the colony played official roles in Japan's civilizing mission.[4] Although some Japanese traveled between Korea, Manchuria, Taiwan, and Japan, many stayed for decades in one place, deepening their ties with the country.

Having new ties can loosen old ones. For many Japanese experts in colonial Korea, it meant developing their own voices about the imperial scientific projects that they were implementing in the colony. Although vocal criticism seems to have been uncommon, Japanese experts in Korea at times produced surprisingly strong

DOI: 10.4324/9781003511755-4

and articulate critiques of Japan's imperial scientific projects. Several factors more or less unique to the Japanese empire were at work: the solid presence of long-staying experts, the vulnerable authority of Japanese scientific centers about Western science, and shared knowledge traditions.

For example, Ishidoya Tsutomu (石戸谷勉, 1891–1958), Nakai's most important colonial associate, who came to Korea in 1911 as a resident forester and stayed until 1939, made the following remark in 1934 about Japanese forestation in colonial Korea.

> Russia had sent scholars long before it gained control of Manchuria to make them thoroughly study its animals and plants. Russian scholars then devoted themselves to this task. Unlike Japanese scholars who would finish their investigation in just one or two months, they would continue their research over a year or so passing the winter in the field with the occupation forces. Therefore, they produced good reports, on which good policies could stand. For example, while Japanese brought Japanese cypresses and cedars from Japan to decorate Korean streets, Russians had brought everything from the local mountains in decorating Manchuria… What they did gives the feeling that humans made adjustments with what was most fitting for the land.[5]

Ishidoya apparently had a certain expectation about how Japan should conduct its colonial research and Japan's current way of conducting scientific investigation in the colonial field was not it. Comparing Japan's practice with the Russian approach in Manchuria before Japan's arrival there, he praised Russia, against whom Japan was competing over the continent. Neither did the resultant policy that ruined the "local color" with Japanese trees receive his approval. Ishidoya showed his disillusion about Japanese scientific projects not just in writing. He had revealed his seriousness by abandoning his 15-year career as a forester in 1925 and shifting to a study of Korean traditional medicinal plants.

What disillusioned this Japanese colonial expert so thoroughly? What did he expect in assuming his civilizing mission and what aspect of Japanese imperial policies failed his earnest mission to civilize the colony with his science? What was wrong with Japanese cypresses and cedars? This chapter follows Ishidoya's knowledge practice as a colonial forester in the first part of his career. It views Japanese policies and the concurrently shaping metropolitan scientific practice in Tokyo through the questions and concerns of ordinary scientific practitioners like him in colonial outposts.

Scientific Forestry to Display Japan's Modernity, Made in Korea

Ishidoya was one of those enthusiastic Meiji youths who wanted to learn modern science. For his education, he chose the first scientific agricultural school established in Japan, the Sapporo Agricultural College. Founded by the Japanese government with American teachers at its head in 1876, it produced influential graduates like Nitobe.[6] Then Ishidoya chose Korea for his scientific career in 1911, evidently answering the call to do away with Japanese insularity and to take up imperial responsibilities; there was a scientific career to be made in Korea while enlightening backward Korea with Japan's advanced science. Ishidoya, enamored of science and invigorated by Japanese expansion, accepted this civilizing mission by launching his serious scientific career in Korea.

However, Ishidoya immediately found that his work in the colony was not what he had dreamt. Above all, carrying out civilizing measures through investment in scientific research was not a priority for the fledgling colonial government. The Government General of Korea (GGK hereafter) had other more serious issues to concern itself with, like suppressing Korean insurrections and finding resources for expensive military operations. Also, in forestry, its priority in all practicality for the revenue tight GGK was to make Korean forest resources pay for these operations. Although forestry was an important area in which it wanted to display the enlightened Japanese governance over the benighted people and land, it did not have to be grand, or expensive. It seemed good enough to display Japan's enlightened "forest love (愛林, airin in J, aerim in K)" by declaring Arbor day with ceremonial tree planting.[7]

Japan's interest in Korean forest resources was shaped early on, as it had exhausted its own rich forests by rapidly building its modern infrastructure.[8] It twice sent forest survey teams to Korea, in 1901 and 1905. In 1906, it made a contract with the Korean government, virtually under its control, to develop deep forests in the mountainous areas in the North. Then it made the consistent move to develop Korean forests throughout the land. Even though Korean forests were seriously denuded near cities and towns, their surveys confirmed that this mountainous land had enough promising sites for development. It enacted the first Forestry Law in 1908, which above all secured the legal foundation to develop forestry by protecting forests from unauthorized access. Upon securing dominance in 1910, it pushed forward this process that ultimately aimed to claim state ownership for the unclaimed mountain areas. However, it had to make constant adjustments to the original plan since such an attempt based on a hasty land survey caused fierce resistance and land disputes, especially involving ancestral burial places and access to firewood. Several amendments to the Forestry Law in 1911, 1912, 1915, and 1918 and a more systematic land survey and repressive actions were necessary.[9] By 1918, the GGK could boast that it made Korea an inviting site for Japanese foresters to come and develop the forests.[10]

Another important factor that dampened Ishidoya's enthusiasm for his colonial post was the status of Japanese forestry, in large part owing to this colonial boon that Japan came to enjoy. Despite the call for scientific measures to deal with the rapid deforestation in Japan since the 1880s, the agricultural faculty at the Tokyo Imperial University did not have the proper research facilities in forestry. As Japan was gaining access to rich forests in Hokkaido, Karafuto, Korea, and Taiwan, it made sense to neglect investment in forestry research aimed at serious reforestation. As in Korea, the most systematic effort that the Japanese government made in forestry was to establish legal measures like property rights to protect state forests and allow for the private development of forest resources, while leaving reforestation to private initiative or nature.[11] If more earnest reforestation efforts in Japan were briefly considered in the late 1910s, even in the 1920s forestry research or afforestation failed to secure necessary support, "reduced [to] a less urgent matter," as the disciplinary history assessed.[12] That the Society of Forestry in Japan (日本林學會) was founded in 1914, decades later than in other fields like botany also shows the slow development of the field. Japanese forestry, when it had colonized Korea, did not have the solid knowledge or experience base to offer the colonial researcher.[13]

This situation had the potential to make the colonies not mere laboratories to test imperial science but the very means to build up Japanese forestry. Colonial posts in Hokkaido, Karafuto, Taiwan, and Korea indeed inspired enough interest among those who were serious in doing research in forestry.[14] In Korea, that Japan had some ideas about using scientific forestry as an exemplar of its "scientific" governance had been an obvious attraction to young forestry researchers who had flocked to Korea before its colonization. Under the protectorate regime, during which Japan advised various scientific measures to the Korean government, the best trained foresters in Japan, graduates of the Sapporo Agricultural School, came to Korea. In 1906, these foresters established model nurseries in five major cities and carried out several reforestation projects in Seoul and Pyŏngyang, to widely demonstrate scientific forestry. In 1908, with legislation, it established a research lab and forestry offices in those five cities.[15]

Such illustrious development in research foundations in Korea did not exist when Ishidoya, another Sapporo graduate, came to Korea in 1911. He was the only forest researcher in the colony and all the model tree nurseries apart from the one in Seoul were closed.[16] The role that the colonial government expected from Ishidoya was to "directly apply the experience and research results from the [Japanese] mainland." The GGK believed that separate forestry research in Korea was unnecessary, as it saw that trees and climates were quite similar in Japan and Korea.[17]

Apparently, simply applying Japanese forestry to Korea was not the civilizing mission that Ishidoya had in mind. Revealing why any civilizing mission was unthinkable without the strong involvement of ground-level agents, he assiduously moved to make the GGK pay some attention to the civilizing mission that he had come for. In 1912, the GGK agreed that it would retry the model nurseries if this was done on provincial initiatives, i.e., if it did not have to pay for it. It could be a way to kindle Japanese entrepreneurs' development interest in colonial forestry. This policy seemed a failure to Ishidoya. The provincial governments "induced much sweat in vain by either selecting the wrong species, or dabbling with too many species, or planting species already known to be unfit, or producing simple garden trees." He highlighted these bad cases to create useful failures for himself. In 1913, the GGK accepted his appeal that a laboratory to guide the "selection of proper species and cultivating and nursing of saplings" was necessary. The Forestry Research Laboratory (林業試驗所) was re-instituted.[18]

However the support of the GGK was not very significant. Ishidoya was still the only forest researcher in Korea. Initially, the only assistant that he could claim for his research was a Korean laborer, Chung Tyaihyon (鄭台鉉, 1883–1971), who had received a one-year training in forestry during the protectorate period. Ishidoya was not completely out of luck. That year, Nakai Takenoshin met the Governor-General of Korea to secure the sponsorship for his botanizing in Korea. The GGK empowered Ishidoya's small lab by appointing Nakai as a guest government botanist. The following year, Asakawa Takumi (浅川巧, 1891–1931), a graduate of a three-year agricultural school in Yamanashi who had happened to come to the colony with his brother and mother for a new life, joined their small workforce.[19]

In this modest lab, Ishidoya could finally launch and shape the civilizing mission that he had dreamed of. He began to produce such impressive research results that in 1922 the GGK approved the expansion of his one-forester lab into the Forestry Research Institute (林業試驗場), with four senior foresters and 16 foresters, by helping him secure Imperial Congress funding. A scientific forestry to display Japan's modernity was ready to be born in the colony through the initiatives of a ground-level researcher who seemed determined to make his colonial post rewarding and meaningful.

Superior Foreign Trees Struggling in a Colonial Land

The crucial researches undergirding his institution building for scientific forestry in colonial Korea stemmed from his small lab's deeper engagement with the colonial land. This neighboring peninsula that appeared so similar to Japan in climate, soil, and flora displayed enough ecological power to unsettle such assumptions. Above all, it failed in a crucial area of forestry, the transplantation of trees for reforestation. Given these failures of various transplantations, Ishidoya made it his task to scale up colonial forestry and to launch better informed forestry policies in colonial Korea.

To Ishidoya's eyes, the unexamined creed of foresters in Tokyo, trained in various Western traditions, was that Western forestry had a ready solution for a problem as universal as deforestation. Japan's forests would be quickly restored if they faithfully copied what had succeeded in the West, which duly included importing 'superior' Western trees: a universal solution for a universal problem. They applied this understanding also to the colony while showing some originality in their applications by analogy. That is, while planting Western trees like North American poplar in the streets of Seoul to display the modernity of Japanese rule, and some fast-growing Western trees like false acacia in the mountains, they also recommended Japanese trees to reforest the colony. Like Japanese people, Japanese trees were, of course, superior to Korean ones, they believed.[20]

Transplanting these foreign trees to Korea was the task that frustrated Ishidoya. Notwithstanding the confidence of his coordinators in Japan, these superior foreign trees, including the Japanese ones, did not grow well in the colony in general, though the fast growing false acacia showed some immediate results. To explain these failures, he and his team embarked on various comparative studies on these trees. By 1917, his lab in Korea had accumulated enough data to present its result to Japan. In their first report sent to Japan, signed by him and Asakawa, the colonial researchers singled out one famous example, American catalpa, to criticize Tokyo's blind reliance on foreign trees. American catalpa, having obtained such grandiose names like "Golden Tree (黃金樹)" and "Spiritual Tree (靈木)," was promoted as the magic bullet needed to rapidly reforest denuded Japanese forests. However, Ishidoya and Asakawa did not witness its promised vigor in Korean forests. They found that its impressive growth was limited to well-maintained nurseries. The researchers contended that foresters ought to reconsider the hasty introduction of foreign trees based on nursery data alone.[21]

The report by Ishidoya team setting out how superior Japanese species performed badly in comparison to similar Korean species when transplanted into Korea soon followed. In the colonial soil, the robust Japanese larch, ambitiously imported from Nagano and Karafuto, fared poorly in every respect—growth, survival rates, and seed and fruit production. The Korean larch that Ishidoya and Asakawa succeeded in cultivating gave decisively better results (See Figure 3.1). This comparative failure of Japanese species was hardly limited to larch; for many trees common to these adjacent lands—chestnut, pine, mountain hazel, and cedars of Lebanon—Japanese ones performed worse in colonial soil in every regard.[22]

The devastating failure in the colony of two representative Japanese trees, Japanese cedar and Japanese cypress, gave Ishidoya some clues. Both trees failed to

Figure 3.1 A photo sent by Ishidoya's team to show successful seedlings and the growth of Korean larch. *The Journal of the Forestry Association of Great Japan* 414 (1917), Front matter (The Courtesy of the Japan Forestry Association).

endure Korean winters in spite of the fact that they had grown well in the Northern temperate zone in Japan, which had similarly cold winters. Ishidoya realized that "the standardized weather table" that had initially suggested the similarities in climate was inadequate. By adding many conditions other than temperature (the daily minimum and maximum, the duration of the cold snap, changes of temperature after the cold snap, the strength of wind, rainfall, and the number of clear days during the winter, etc.), he could see Korean winters were quite distinct from Japanese winters. They were much drier with a much lower minimum temperature. He conjectured that Japanese cedars and cypresses, accustomed to humid and less severely cold winters, could not have developed the mechanism for preserving moisture in that dry and cold winter, which Korean ones seemed to have developed. Consequently, they just withered away by the end of winter, while trees of the Korean species began to renew themselves. This invisible physiological difference between Japanese and Korean trees and the subtle differences in climate could not be ignored.[23]

Ishidoya's consideration for local conditions did not stop at climate. By the early 1920s, he came to reflect on the emerging concept of "forest ecology."[24] He observed that Japanese foresters were not so careful about "the importance of ecological interrelationships" among trees, other living organisms, and the soil sustaining them. Ishidoya felt that when Japanese foresters mentioned ecology and said that the forest was more than an aggregation of trees, they "merely repeat from the Western textbook" without fully appreciating what it meant in the field. Their assertion was devoid of a serious appreciation of the complex interrelationships among forest components. He asserted that forests were "living creatures (生物)," whose sort of "personhood (人格)" should be recognized. Foresters thus should be aware of this ecological complexity before they tried to make the forests yield what people needed. It was another reason why indigenous trees, which had been part of the "collective life" of Korean forests, would fare better in Korea.[25]

Ishidoya, with his carefully thought-out comparisons and clear statistics, made a strong argument that the transplantation of Japanese forestry to the colony was not the solution; to understand these local conditions and to find solutions for the colonial field, much more research in the colonial field was necessary. In 1922, his years of efforts were finally acknowledged, as mentioned above. The imperial congress authorized the budget for his small lab's expansion into the full-blown Forestry Research Institute. Nodding to Ishidoya's initiative, the stipulation for its establishment endorsed by the imperial congress specifically says:

> Given the records provided so far, Korea, due to the influence of the continental climate, was different from Japanese mainland and other provinces of Japan in the progress of deforestation, the kinds and the distribution of flora, and the changes in the forest floor.[26]

A serious colonial lab to conduct Korean forestry was inevitable, it asserted. Ishidoya and his small team, after the struggles of transplanted trees in the colonial land, had successfully transformed their colonial lab into an exciting outpost to build up scientific forestry.[27]

Confined Universality

To Ishidoya, the struggles of those transplanted foreign trees were not an isolated phenomenon; it strongly reflected what went awry in Japan's transplantation of Western science. If mainland Japanese forestry, relying on an easy "universal" solution copied from the West was not enough to push him to this conclusion, there was another case that solidified his mistrust about Japanese scientific practice. It was the internationally recognized plant systematics of Nakai Takenoshin that came to further evidence this problem of ill-transplanted scientific knowledge.

As discussed in the previous chapter, Nakai's success was based on imperial Japanese botanists' fastidious re-enactment of European botany. Just as the European botanist Linné based his universal classification system on dried specimens collected from non-European peripheries including Japan, Japanese botanists searched for less explored territories to build up the Japanese center of botany. Although these "armchair botanists" often did not leave their comfortable centers, the specimens for their studies mostly came from the colonial field. Ishidoya began a systematic collection of colonial plants with his Korean assistant Chung Tyaihyon as a guide as soon as he arrived in Korea, accumulating specimens and notes. He seemed quite happy when the GGK appointed Nakai, an imperial university researcher with an impressive credential, for a more systematic study. Ishidoya assigned Chung as an assistant and translator for Nakai's summer collection trips. Yet, what enabled Nakai's successful botanizing from his Tokyo lab was the collections of numerous colonial workers who assembled Korean plants throughout the year for him. Ishidoya and Chung were most reliable and knowledgeable collectors, while coordinating other foresters and amateur naturalists in colonial Korea to help in Nakai's remote botanizing.

Nakai appreciated these colonial collectors' contributions to his systematics but quite perfunctorily, although he could not entirely dismiss them.[28] As stipulated in the botanical rules, he duly recorded the names of the collectors; he noted Ishidoya's name frequently along with other colonial Japanese experts' names as collectors of the specimens that he was describing; yet he only began to record the name of Chung, his most assiduous Korean collector, in 1921. Nakai also had four plants named after Ishidoya.[29] Nonetheless, Ishidoya seemed unimpressed by these acknowledgments recorded in Nakai's works.

Nakai's discovery of a new Korean willow, *Chosenia*, especially disillusioned Ishidoya and colonial collectors. *Chosenia* was a kind of willow tree in northern Korea that Nakai at first thought just a known species common in Manchuria, *Salix rorida* Laksch. Yet the foresters who saw the trees in early spring when it bloomed found its small catkins quite distinctive. Since it was very difficult to collect and make dried specimens with those delicate catkins intact, it took two years for them to obtain satisfactory specimens. They sent these carefully collected specimens to Nakai in 1918 with a note of their finding. Alerted thus, Nakai gladly registered it as a new species, *Salix splendida* Nakai. Yet his collectors suspected that there was something more to this splendid looking willow and realized that, unlike those of other willows, these flowers had no scent. This suggested that it could not be pollinated by bees but only by wind. As the method of pollination was an important character distinguishing the genus *Salix* from *Populus*, two genera in family *Salicaceae*, a new genus might be more proper. It was only after learning this from his collectors that Nakai decided that this tree should be a new genus in the family, renaming it *Chosenia splendida* Nakai. Nakai, however, did not fully acknowledge the expertise and active input that his collectors had been providing. He mentioned their names only in his second report in 1920 where he treated them as passive collectors working on his instructions.[30]

The following year, Ishidoya chose to publish an unusual paper for a forester regarding the classification of Korean willows. In this paper, published in a colonial journal, he drew attention to Nakai's silence about collectors' active and substantial input in detail by recording this long process of knowledge transfer from the colony to the metropole.[31] By 1928, Ishidoya had produced a systematic report on the role of collectors in the development of plant systematics in Korea, writing a history of plant collectors in Korea since the opening of the country. The much celebrated achievement in Korean botanizing owed much to these colonial "people other than professional [systematists]," Ishidoya concluded.[32]

Ishidoya's reflection on plant systematics did not stop at underlining this disparity between the metropolitan botanist and colonial collectors. Nakai's repeatedly mistaken identifications suggested a certain arbitrariness in the practice of plant systematics.[33] By way of excusing himself for discussing such a fundamental issue outside his field, he gave the reason for his serious engagement:

> For example, a scholar named "A" describes a plant and reports as new species. Then, a scholar named "B" says that the plant is just a variation of a known species. A third scholar "C" yet again asserts that it is a form of a known variety of a known species. Therefore, even a novice in plant systematics could not help but critically engage with these various theories in finding one's own way.[34]

Yet in thus engaging with the fundamental issue of the confusing species concept, Ishidoya seemed emboldened by his position as an outsider. He flatly challenged the authority of "armchair botanists" on several levels. They were not only unable to produce a satisfactory consensus about what constituted meaningful characteristics for genus, species, and forms. They were also slipshod in dealing with specimens collected by others. These botanists made obvious human mistakes, such as mixing up specimens and misrecognizing traits, as he had witnessed first-hand from Nakai.[35] Furthermore, they were at times so unfamiliar with the life cycles of plants in the field as to mistake a seasonal characteristic for a species identifier. The descriptions of plants that they gave, as a result, were only meaningful on dried specimens and, he claimed, of little use in identifying plants in the field.[36]

Ishidoya thought that the most serious problem with these armchair botanists was that they did not fully see the relationship between the plant and the "local soil." To him, the species concept relying on characteristics of uprooted plants combined with their tendency to, in his words "classify genus and species by one or two characteristics that they considered important" seemed the main reason for their production of many unnecessary new taxa. Ishidoya did not go so far as to allow the dynamic re-association of species by multiple comparative criteria, as Hayata did, but he sided with Komarov, who accepted Hayata's point in suggesting a more relational concept of species in association with the "local soil." According to Ishidoya, this concept also better reflected the evolutionary nature of plant species shaped in interaction with local environments.

Ishidoya showed the aptness and the necessity of this species concept bound by local soil by examining claims made on the plants of Cheju Island. This sizeable and mountainous volcanic island south of the Korean peninsula with a high mountain claimed almost half of Korean plant species, due to its diverse soils and climates. Furthermore, due to many plant specimens sent to European armchair botanists by collectors including the French priest, Emile Taquet (1873–1952), who sent the famous specimen of *Prunus yedoensis* to Germany, more and more species had been controversially claimed.[37] Ishidoya claimed that most of them could not be considered new species by Komarov's species concept, because many new plants proved to develop traits of known species when they were transplanted into his Seoul nursery. To him, they were the same plants, which happened to develop different characteristics in different soil and climate. Nonetheless, many armchair botanists were competing to give new scientific names when 'Cheju pine' or 'Cheju mountainous or seaside pine' would have been sufficient identification.

Ishidoya thought that the visibility of many Japanese botanists including Nakai in these competitions was due to a "Japanese attitude that usually over-evaluated anything from outside." Japanese botanists not only accepted all these new designations of "scientific names" but also tried hard to compete with Western botanists by producing more unnecessary names. Ishidoya even made the accusation that this uncritical conversion to the Linnaean system was in effect a repetition of Tokugawa naturalists' blind following of the *Systematic Materia Medica* of the great Chinese systematizer Li Shizhen. Ishidoya urged his colleagues in the colonial field to stop relying on "those taxonomists who recklessly claim new species based on the shades of the bark or the shapes of the fruit." Instead of following these metropolitan workers who produced information quite meaningless for fieldwork, simply in order to make a name for themselves, colonial researchers had to work for "the public good" by delving into local soil.[38]

Conclusion

Ishidoya came to Korea with great hopes for the civilizing mission of Japan. However, he became very much disillusioned by his mission and caustically criticized Japanese imperial science. He took a unique stance as a settler expert utilizing his regional expertise about the colonial soil, climate, and ecology, which revealed the vacuity of many hastily claimed new names. In particular, what made Japan's imperial science most vulnerable to the criticism of this colonial expert was the claim that modern science, together with the tools used to create it, such as dried specimens, was universal. 'Superior' trees from the West and Japan simply did not stand up well in the colonial land and the supposedly objective facts captured in dried specimens turned out to be rather unreliable in the field. Although the powerful promise of universality brought modern science into this non-European contact zone, its weakness in the field seemed to have had the potential to upend the imperial enterprise itself.

In 1925, Ishidoya left the recently expanded Forestry Research Institute, where he had carved out a well-regarded career as a forester for 15 years.[39] What he decided to do next was made clear in his first appearance at the meeting of the Natural History Association of Korea (朝鮮博物學會) that year.[40] He began to study Korean medicinal plants through the *Treasured Mirror of Eastern Medicine*, the text which had inspired many Japanese scholars before him.

We see here a striking transformation of a Meiji youth who wanted to enlighten Korea with his modern science. His long sojourn in Korea with day-to-day interaction with the colonial soil and climate, and the unique dynamics of the Japanese colonial project in this neighboring colony drove this change. In his last analysis, Japanese sciences and European botanical systematics did not seem to have much to offer, a disillusioning realization for this assiduous researcher. Since the unyielding land had much to say about Japan's 'enlightening' knowledge practice and the European assumption of universality, he found himself having more to offer to the metropolitan sciences than they had given him. The problem was that the Japanese metropolitan center, that had proudly proclaimed many new names such as *Salix splendida* and *Chosenia splendida* with the help of his expertise, was much less willing to recognize his invaluable contribution. While many more colonial experts like him seemed to have passed over those questions about the unreasonable asymmetry within transnational botanical interactions, Ishidoya, a powerful institute builder who created his own colonial center of forestry gradually raised his voice. Yet this transformation did not end there. Now not just the colonial land, but the colonial knowledge tradition and the people who appropriated it would unsettle his ever transforming civilizing mission, which remained only in name, as we shall see in Chapter 6.

Notes

1 Quoted from Wŏn Chiyŏn, "Nitobe Inazo and Orientalism," *Tongamunhwa* (동아문화) 40 (2002): 47–72, 57. The colonial ideologues maintained such a call imbuing the civilizing mission as in Western empires to staff the expanding Japanese empire. Aaron S. Moore says, "Under the Banner of 'Constructing East Asia' (tōa kensetsu), Several Thousand Idealistic Engineers Flocked to Korea, Taiwan, Manchukuo, and China Primarily during the 1930s to Construct Roads, Canals, Ports, Dams, Cities, Irrigation, Sewage and Water Works, and Electrical and Communications Networks." Aaron S. Moore, *Constructing East Asia: Technology, Ideology, and Empire in Japan's Wartime Era, 1931–1945*, Stanford, CA: Stanford University Press, 2013, 14.

2 Much has been written about the relationship between scientific professionalization and colonial expansion. Fan, Raj, and Cook's works mentioned in the introduction all provide examples. See also Warwick Anderson, *Colonial Pathologies: American Tropical Medicine, Race, and Hygiene in the Philippines*, Durham, NC: Duke University Press, 2006, etc.

3 Kwon Tae-eok 권태억, *Japanese 'Civilizing Mission' in Korea (1904–1919)* (일제의 한국 식민지화와 문명화), Seoul: Seoul Nat'l University Ch'ulp'anmunhwawŏn, 2014, 53.
4 Studies on Japanese immigration to Korea or Japanese bureaucrats in colonies are expanding. See Uchida Jun's exhaustive bibliography. In comparison, Korean immigrants to Japan outnumbered Japanese immigrants to Korea but they were mainly laborers who found menial jobs in Japanese industrial cities and mining towns. There were more than a million Koreans in Japan by 1938 and there were about 2.1 million after Japan's forced conscription of Koreans in 1945. Park Keong-Suk 박경숙, "Population Dynamics of Korea during the Colonization Period (1910–1945) (조선의 인구 동태와 구조)," *Korea Population Studies* (한국인구학) 32 (2009): 29–58.
5 Ishidoya Tsutomu 石戸谷勉, "Miscellany about Manchurian and Korean Landscapes (滿鮮植物風景雜觀)," *Korea and Manchuria* (朝鮮及滿洲) 325 (1934): 65–68, 66.
6 The school was founded to help the Hokkaido development. It became the Faculty of Agriculture at Hokkaido Imperial University in 1918. Its famous first president was the enthusiastic American educator, William Smith Clark (1814–1886). The History of Science Society of Japan 日本科学史学会 ed., *A Compendium of History of Science in Japan* (日本科學技術史大系 , *Compendium* hereafter), vol. 15, Tokyo: Daiichi Hoki Shuppan, 1965, 14–57.
7 For the complex makings, negotiations, and ideas surrounding Japanese forestry in colonial Korea and how the fresh tradition of Japanese "'forest love' became a centerpiece of its civilizing mission abroad," see David Fedman, *Seeds of Control: Japan's Empire of Forestry in Colonial Korea*, Seattle: University of Washington Press, 2020, 25; Takemoto Taro 竹本太郎 and Komeie Taisaku 米家泰作, "Imperial Forestry and Local People in Colonial Korea (朝鮮における帝国林業と地元住民)," in Nakashima Koji (中島弘二) ed., *The Japanese Empire and Forests: Conservation and Development of Forest Resource in Modern East Asia* (帝国日本と森林——近代東アジアにおける環境保護と資源開発), Tokyo: Keiso Shobo, 2023, 247–97, 261.
8 Yamamoto Hikaru 山本光, *History of Forestry and Forestral Geography* (林業史·林業地理), Tokyo: Meibundo, 1961, 141.
9 Oka Eiji 岡衛治, *History of Forestry in Korea* (조선임업사), vol. I Tr. Yim Kyŏngbin, Seoul: Sallimch'ŏng, 2000–2001, 20–28. Takemoto Taro 竹本太郎 and Komeie Taisaku 米家泰作, "Imperial Forestry and Local People in Colonial Korea (朝鮮における帝国林業と地元住民)," in Nakashima Koji (中島弘二) ed., *The Japanese Empire and Forests: Conservation and Development of Forest Resource in Modern East Asia* (帝国日本と森林——近代東アジアにおける環境保護と資源開発), Tokyo: Keiso Shobo, 2023. Oka Eiji was a Japanese forestry officer commissioned to write this history to celebrate the 30 years of forestry administration in Korea. Due to the war, the publication was delayed to 1944.
10 The Forestry Law stipulates the "transfer the ownership of the forest if the developer who leased the national forest succeeded in their work." As the demand for timber was high in Japan, the profits for colonial development were high. The advantageous lease of state forests was mostly made to Japanese developers. The national forests, which had accounted for more than half of all forests at the beginning, became less than a third in 1939. See Oka Eiji岡衛治, *History of Forestry in Korea* (조선임업사), Tr. Yim Kyŏngbin, Seoul: Sallimch'ŏng, 2000–2001, vol. I, 20; Bae Jae-Soo 배재수, *The Changes in the Forest Ownership in Korea* (한국의 근·현대 산림소유권 변천사), Seoul: Imŏp Yŏn'guwŏn, 2001; see David Fedman, *Seeds of Control: Japan's Empire of Forestry in Colonial Korea*, Seattle: University of Washington Press, 2020.
11 The initial boundary between state and private forests was made in 1881. Forestry District Offices were established in 1886 to police illegal logging, etc. Yamamoto Hikaru 山本光, *History of Forestry and Forestral Geography* (林業史·林業地理), Tokyo: Meibundo, 1961.

12 It further says that the need for urgent reforestation was stipulated in the Revised Forestry Law of 1907 but no action was taken due to a lack of funding. The provision for reforestation was first actualized in 1919 through the forestation encouragement fund. Forestry Association of Greater Japan 大日本山林會, *80 Years of Forestry in Japan* (日本林業八十年史, *80 Years of Forestry* hereafter), Tokyo: 1962, 60; Yamamoto Hikaru山本光, *History of Forestry and Forestral Geography* (林業史·林業地理), Tokyo: Meibundo, 1961, 153. While Totman presents a very positive view of traditional forestry practice in Japan by saying that "the transition to regenerative forestry occurred during the Edo period," this partial transition in some regions could be slower in the Japanese mainland in the Meiji transition owing to its colonial expansion. Conrad D. Totman, *The Green Archipelago: Forestry in preindustrial Japan*, Berkeley, CA: University of California Press, 1989, 4. Conrad D. Totman, *The Origins of Japan's Modern Forests: The Case of Akita*, Honolulu: University of Hawaii Press, 1985.

13 This slow development of forestry professionals would have delayed investment in the field. The Forestry Association of Greater Japan, founded in 1882, was more of a business association. *Compendium*, vol. 15, 265–289.

14 For a pioneering study on colonial forestry in Taiwan, see Kuang-chi Hung, "When the Green Archipelago Encountered Formosa: The Making of Modern Forestry in Taiwan under Japan's Colonial Rule," in Bruce Batten and Philip Brown eds., *Environment and Society in the Japanese Islands: From Prehistory to the Present*, Corvallis: Oregon State University Press, 2015, 174–93.

15 Oka Eiji岡衛治, *History of Forestry in Korea* (조선임업사), vol. I Tr. Yim Kyŏngbin, Seoul: Sallimch'ŏng, 2000–2001, vol. I, 80–82, 27.

16 The development of forestry was used to demonstrate the modernization that happened through Japanese colonization, since Korea at the late nineteenth century was also suffering from serious deforestation. Yet one Japanese historian, under the strong influence of Marxist historiography, characterized Japanese colonial forestry in Korea as a typical colonial exploitation; it was focused on exploiting underdeveloped mountainous areas while actual forestation was not a serious undertaking. The final exhaustion of forestry resources during the Pacific War era left such devastating results that it made some Korean scholars also see forestry as proof of colonial exploitation. Hagino Toshio 萩野敏雄, *On the Development of Forestry in Japanese Korea, Manchuria and Taiwan* (朝鮮, 満州, 台湾林業発達史論), Tokyo: Rinya Kosaikai, 1964; Bae Jae-Soo배재수, *The Changes in the Forest Ownership in Korea* (한국의 근·현대 산림소유권 변천사), Seoul: Imŏp Yŏn'guwŏn, 2001. The complex process of forestry development in colonial Korea to be shown in this chapter, and recent works suggest to move beyond the simple question of exploitation versus enlightenment.

17 Oka Eiji岡衛治, *History of Forestry in Korea* (조선임업사), vol. I Tr. Yim Kyŏngbin, Seoul: Sallimch'ŏng, 2000–2001, vol. 1, 313, 307–33.

18 Oka Eiji岡衛治, *History of Forestry in Korea* (조선임업사), vol. I Tr. Yim Kyŏngbin, Seoul: Sallimch'ŏng, 2000–2001, vol. I, 80–82; vol. II, 59.

19 Until the research lab was expanded to Forestry Research Institute in August 1922, this four-man system was maintained. Oka Eiji岡衛治, *History of Forestry in Korea* (조선임업사), vol. I Tr. Yim Kyŏngbin, Seoul: Sallimch'ŏng, 2000–2001, vol. 2, 307–26.

20 The idea that all things Japanese, including its trees, people, and culture, were better than things Korean, in spite of their similarity, was of course common; investigations into the inferiority of Korean people and history were widely conducted. This racial science was a serious scholarly endeavor. Kim, Ock-Joo, "Physical Anthroplogy Studies at Keijo Imperial University Medical School (경성제대 의학부의 체질인류학 연구)," *Korean Journal of Medical History* (의사학) 17(2) (2008): 191–203; Tessa Morris-Suzuki, "Debating Racial Science in Wartime Japan," *Osiris* 2nd Series 13 (1998): 354–75; Jaehwan Hyun, "Racializing Chōsenjin: Science and Biological Speculations in Colonial Korea," *East Asian Science, Technology and Society: An*

International Journal 13 (2019): 489–510; Jaehwan Hyun, "Blood Purity and Scientific Independence: Blood Science and Postcolonial Struggles in Korea, 1926–1975," *Science in Context* 32 (2019): 1–22.

21 Ishidoya Tsutomu and Asakawa Takumi, "Reporting the Cultivation and Forestation Result of the American catalpa in Korea (朝鮮に於けるカタルペ-スペシオサの養苗及造林成績を報す)," *The Journal of the Forestry Association of Great Japan* (大日本山林会報, *JFGJ* hereafter) 414(1917): 36–40. In spite of their criticism, the unexamined reliance on foreign trees in Japan had been continued until the late 1920s. Forestry Association of Greater Japan 大日本山林會, *80 Years of Forestry in Japan* (日本林業八十年史, *80 Years of Forestry* hereafter), Tokyo, 1962, 64–66.

22 Ishidoya Tsutomu and Asakawa Takumi, "Reporting the Successful Cultivation of the Korean Larch (テウセンカラマツの養苗成功を報す)," *JFGJ* 419 (1917): 36–40.

23 Ishidoya Tsutomu, "The Growth and Development of the Japanese Cedar and Cypress Transplanted into Korea (朝鮮に移植せられたる「スギ」「ヒノキ」の生育)," *Korean Repository* (朝鮮彙報) 4(12) (1918): 20–32.

24 The official history says that the ecological concerns regarding forestry in Japan occurred in the mid-1920s due to "the transmission of the then forestry trends of Northern Europe." Though Ishidoya's reading of German works was obvious, his development of the ecological concept, one of the earliest in Japanese forestry, is unthinkable without his colonial transplantation experience. Forestry Association of Greater Japan 大日本山林會, *80 Years of Forestry in Japan* (日本林業八十年史, *80 Years of Forestry* hereafter), Tokyo, 1962, 60. Cittadino notes that ecological concept in Germany itself was quite strongly influenced by the brief colonial experience of German biologists. Eugene Cittadino, *Nature as the Laboratory: Darwinian Plant Ecology in the German Empire, 1880–1900*, Cambridge: Cambridge University Press, 1990; Grove also notes the colonial origin of environmental concerns. Richard Grove, *Green Imperialism: Colonial Expansion, Tropical Island Edens, and the Origins of Environmentalism, 1600–1860*, Cambridge: Cambridge University Press, 1995.

25 Ishidoya Tsutomu, "About Forests, Forest Canopy, Interior, and Floor viewed from the Ecological Perspective (生態上ヨリ觀察シタル森林ト林冠′ 林幹及林床ニ就)," *Journal of the Japanese Forestry Society* (日本林學會誌, *JJF* hereafter) 10 (1921): 26–29; Ishidoya Tsutomu, "Foresters' Attitude towards the Forest and the Chŏnnam Province's Gift of Nature (삼림에 대한 조림자의 태도와 전남삼림의 천혜)," *The Journal of the Forestry Association of Chŏnnam* (全南山林會報) 1 (1922): 18–26, 19, 20. It is notable that it is printed in both Korean and Japanese.

26 Oka Eiji岡衛治, *History of Forestry in Korea* (조선임업사), vol. I Tr. Yim Kyŏngbin, Seoul: Sallimch'ŏng, 2000–2001, vol. I, 34.

27 For the detailed functions of the lab, see Oka Eiji岡衛治, *History of Forestry in Korea* (조선임업사), vol. I Tr. Yim Kyŏngbin, Seoul: Sallimch'ŏng, 2000–2001, vol. II, 307–26; and David Fedman, *Seeds of Control: Japan's Empire of Forestry in Colonial Korea*, Seattle: University of Washington Press, 2020, Chapter 4.

28 Nakai acknowledged his collectors only in his botanical descriptions, where he had to name them. In comparison, Hayata made long acknowledgments for collectors in his prefaces. Nakai did only once in 1932.

29 Plants that Nakai named after Ishidoya were *Acer ishidoyanum*, *Prunus ishidoyana*, *Salix ishidoyana*, and *Viola ishidoyana*. Though he never mentioned the reason for these dedications, it indicates that Ishidoya had been a capable and important collector for Nakai. Nakai Takenoshin, "Notulæ ad Plantas Japoniæ et Coreæ XIX," *BMT* 33(385) (1919): 1–11; Nakai Takenoshin, "Notulæ ad Plantas Japoniæ et Coreæ. XIII," *BMT* 31(361) (1917): 3–30; Nakai Takenoshin, *Korean Plants* (朝鮮植物), Tokyo: Seibido, 1914, etc. Nakai appeared to record himself as a collector when a specimen was collected by Chung or in their collection trips together until 1921.

30 Nakai Takenoshin, "Notulæ ad Plantas Japoniæ et Koreæ XVIII," *BMT* 32(382) (1918): 215–32, 215; Nakai Takenoshin, "CHOSENIA, A New Genus of Salicaceæ," *BMT* 34(401) (1920): 66–69.

31 Ishidoya Tsutomu, "Classification of *Salix* and *Populus* in Korea (I) (朝鮮に於ける柳屬及上天柳屬の分類(上))," *Korea* (朝鮮) (4) (1921): 37–44; "Classification of *Salix* and *Populus* in Korea (II)," *Korea* (7) (1921): 91–104, 91.

32 Ishidoya left a rather systematically studied report to show collectors' role in systematists' classification. Ishidoya Tsutomu, "The Firsts in Korean Plant Collections (朝鮮植物採集一番鎗の記)," *Korea and Manchuria* 242 (1928): 38–40, 40. Nakai had an unusual remark in 1930 appreciating Ishidoya's years of efforts in collecting flowers of willows. It seems that Nakai had to appease this important collector. Nakai Takenoshin, *Flora Sylvatica Koreana* 朝鮮森林植物編, Seoul: GGK, 1915, vol. 18, preface.

33 On the difficulty on producing a universal or natural concept of species, see Lloyd, G. E. R. *Cognitive Variations: Reflections on the Unity and Diversity of the Human Mind*, Oxford: Clarendon Press, 2007, 39–57.

34 Ishidoya Tsutomu, "Plants of Cheju Island and Expected Problems (濟州島の植物と將來の問題)," *Korea in Art and Education* (文教の朝鮮) 10 (1928): 71–92, 77.

35 Nakai combined specimens from two different plants as one. Ishidoya, who sent those specimens, saw that Nakai had only half done his work and finished Nakai's work by, for the first time, naming a new species with the neglected and mixed-up specimens that he had sent. This Corydalis was, as if to console Nakai for his mistake, named after Nakai as *Corydalis nakaii* Ishidoya. Ishidoya Tsutomu, "A New Species in Korean Corydalis (朝鮮産 延胡索 一新種)," *Journal of the Natural History Association of Korea* (朝鮮博物學會雜誌, *JNHK* hereafter) 6 (1928): 87–91.

36 The primary purpose of his articles on Korean willows where he had made clear foresters' contribution to Nakai's discovery was to provide a useful field classification guide for his colleagues. Ishidoya Tsutomu, "Plants of Cheju Island and Expected Problems (濟州島の植物と將來の問題)," *Korea in Art and Education* (文教の朝鮮) 10 (1928): 71–92, 77.

37 Taquet adopted plant collections to help finance his mission. Yi Ch'angbok 이창복 and Yi Munho 이문호. "Korean Plant Research of a French Priest (프랑스 선교사의 한국 식물 연구)," *Studies in Church History* (教會史研究) (5) (1987): 149–207.

38 Ishidoya Tsutomu, "Plants of Cheju Island and Expected Problems (濟州島の植物と將來の問題)," *Korea in Art and Education* (文教の朝鮮) 10 (1928): 81, 86.

39 A reviewer of Ishidoya's work in 1933 said that "many people talked about such an established forester quitting his job as the strangest thing" at the time, which also reveals the social proximity of the Japanese settler community. Z. Y., "T.ISHIDOYA: Chinesische Drogen,Teil I,1933," *Acta Phytotaxonomica Et Geobotanica* (植物分類 地理) 2 (1933): 315.

40 For more on the Natural History Association of Korea, see Chapter 7.

4 Civilizing Ourselves

Among all scientific fields, botanical research was the field most populated with Korean researchers, who came to produce new forms of knowledge about plants, encouraged by the nationalistic applause of the colonial media as depicted in Yu Chin-o's (1906–1987) novel:

> You owe your salvation to natural science. Such things are possible in natural science because it is universal. For those who study philosophy or economy or law, success in life [for Koreans] is unthinkable. One must either become a Mephistopheles by selling one's conscience, or go into the mountain like Paegi and Sukche and live off bracken ferns....
>
> At any rate, our newspaper wants to report your achievement fully because your achievement [in botany] is not just your own honor, but that of all Koreans. Although we lack everything, it is scholars that we most lack in Korea. Why don't we have good scholars despite the many university graduates that we have now?[1]

In the novel, a struggling middle school teacher, whose school is about to be closed due to financial difficulties, becomes a national hero; he produces a botanical monograph highly praised even by "Dr. Nakai at the Tokyo Imperial University."[2] Poignantly, it shows that nationalistic applause for a Korean teacher's scientific achievement still needed the approval of imperial Japanese authority, regardless of Ishidoya's articulate critiques of that authority.

This seemingly wholehearted acceptance of the civilizing efforts of Japanese authority was a product of lengthy colonial negotiations on the part of Koreans who discovered this familiar colonizer anew. Korea was the closest neighbor to Japan across the straits and had long been a geographical and cultural conduit connecting it to China. Many military, material, human, and cultural exchanges had taken place by this route. Korea and Japan were both peripheral barbarians in the civilizational hierarchy set by powerful China, in which both gradually acquiesced, each creating their own Confucian cultures on the Chinese model. While by the sixteenth or seventeenth centuries, rulers of both countries seemed resigned about their lack of ability to compete with a huge Chinese polity, each cultivated a sense of superiority concerning the other 'barbarian,' mobilizing their people for wars

DOI: 10.4324/9781003511755-5

and creating and re-creating sagas about their own bravery and exploits over the other. A telltale clue is that the Koreans of the Chosŏn Dynasty perceived their diplomatic delegation to Tokugawa Japan as the transmission of their superior culture to the barbaric samurai while Japanese bakufu leaders tried to portray this journey as Korea's paying tribute to Japan.[3] While positive interests in each other had also increased since the eighteenth century, given these competitive and narrow frames of mind quite common between closest neighbors, the reconciliation of Koreans to their defeat by their neighbor was challenging.

This chapter examines how various Koreans under the colonial regime, with these conflicting memories of the past, succeeded in building modest institutes for science with a special success in botany. They did it mostly through an acceptance, not rejection, of Japanese authority and guidance, collaborating actively with the Government General of Korea (GGK hereafter) and Japanese settlers. Furthermore, these new knowledge practitioners had their collaborative actions accepted by their colonized compatriots as a patriotic activity for their lost nation. How could these new botanists transform "collaboration" in the name of a persistent spirit of resistance while securing their research foundation?

Betrayed Brotherhood: Civilizing Ourselves

Koreans' attempts at reconciling themselves to accepting Japan as an authority and a model for their civilizing plans had begun as far back as Korea's forced opening by Japan in 1876. Though a growing number of Koreans had noted with surprise the vigor of the Western imperial powers that had incapacitated powerful Qing China and the breathtaking rise of Japan, Japan's maneuver that forced the unequal treaty upon them marked a watershed. The general outlook about the order of the world was fully shaken. They began to study Japan's allegedly all-out Europeanization and look at the "barbaric" cultures of the West more earnestly. Some Korean observers began to consider Japan as a possible source of models for their reform, and for the study of the West, as well as of support for their revolution against the old regime. Nonetheless, entrenched elites found more harmony in China's gradual transformations and were quite successful in maintaining the hold of Confucian-based ideologies of civilization and social authority.

One common response can be seen not just in Korea, but in Japan, China, and many other cultures faced with the rise of the West, or in Western cultures dealing with the rise of new sciences helped. Thus, Korean Confucians divided the material and spiritual realms, and associated science and technology, the alleged essence of Western civilization, with its material prowess while claiming the continued validity and superiority of its culture's moral and spiritual values. By holding onto this important authority, it was easier to create a consensus about the need to accept Western science and technology as a mere tool for material progress. By the 1890s, many Koreans based the promotion of Western science and technology on such an approach.[4]

Yet such a selective attitude about Western civilization increasingly seemed not enough to most Koreans as they were seized by a full sense of crisis at Japan's victories over China in 1895 and over Russia in 1905 that had led to the transformation of Korea into a Japanese protectorate.[5] Not many Koreans could maintain the haughty claim that Japan had transformed itself into another Western barbarian by all-out Westernization and denial of their own culture. As more and more Koreans adopted the perspective of linear historical progress preached by Western imperial powers, they were ready to throw away everything in order to be on the right path of progress like the Japanese.

In fact, those who strongly imbibed this historicism had rather naïve expectation about Japanese rhetoric about Asian brotherhood and sought more active Japanese guidance for Korea's self-strengthening and reform, even before colonization. Many so-called nationalists praised the Japanese promise of the civilizing mission, *bunmeika* for its Asian brothers. In their view, Japan on its fast track to modernity wanted to help its neighbors shift to the right path to join in this forward march of history. It was none other than this faith in the Japanese promise of brotherly help that so enraged and dismayed Koreans when they saw Japan's encroachments into their sovereignty and the final usurpation that Japan called "annexation." Their faith or wishful thinking was so strong for some that even after colonization, some still argued that Japan would return their sovereignty as soon as it had finished its civilizing mission.[6]

These believers in Japanese altruism toward Asian brothers were finally disillusioned at the GGK's changed attitude. It was obvious that Japan was abandoning the 'civilizing measures' taken earlier in favor of more lucrative colonial developments, displacing Koreans and luring Japanese entrepreneurs and settlers. Japan had a ready justification for breaking its promise, and it was one often used by other imperial powers. The level of Korean people was, regretfully, too low to make it possible to continue civilizing measures. As shown in the introduction, Japanese ideologues loudly sanctioned such a view, asserting that Koreans were irrational primitives belonging to the prehistoric era. It was a striking reproduction of European Orientalist logic, applied to the neighbor that Japan had known for centuries and with whom it had only recently discussed Asian brotherhood, exchanging official delegates and reform plans. Korea suddenly became as foreign and remote as the far-away colonies of the West. In this ideal logic, it was these primitive Koreans, not promise-breaking Japanese, who made civilizing efforts useless and unrealistic. Transforming lazy and dirty Koreans into diligent and hygienic producers of rice and cotton or other valuable resources for industrializing Japan was challenging and civilizing enough.[7]

Deprived of their hope and humiliated, Koreans in general did not welcome Japan's assumption of power in their land, although responses varied widely from military uprising to complete collaboration. As if to display their resistance, some Koreans refused the modest "civilizing" measures offered by the GGK. They found those to be a usurpation of their right to civilize themselves by reducing and distorting what they had envisioned. The disappointment about the GGK's education policy was especially serious. The colonial education system was a two-tier system (three-tier if metropolitan Japanese education was considered), with separate schools for Japanese and Koreans, while most resources were given to Japanese schools accommodating the growing settler population. On every level, Korean schools had a lower level of education than that offered in the colonial Japanese schools. Entering even a middle school was quite unthinkable for a Korean child, as the number of schools for Koreans was too small. Koreans responded to this meager and discriminatory measure of the GGK by boycotting the newly established Japanese primary schools. They chose to modernize the village schools known as 書堂 sŏdang (literally 'book halls'), which used to teach Confucian classics. The number of village schools was about 16,000 in 1912 and grew to 25,000 in 1921, almost matching the number of villages, a peak for this age-old and modest institution.[8]

This heyday of village schools was brief, however, as the Korean people found a different way to express their dissatisfaction. Energized by the concessions that they had achieved through the nation-wide independence marches in 1919, Koreans more forcefully demanded that the GGK had to fulfill the civilizing mission. Historians say that the impressive participations of Korean students enrolled at Japanese colonial schools in those independence marches were significant in changing Korean attitudes to colonial education. Seemingly confirming the Japanese worry that "to be educated was to be anti-Japanese" in Korea, these students proved that Japanese education by Japanese teachers in Japanese language would not make Korean kids Japanese.[9] More Koreans found it safe to accept colonial tutelage in educating their children. They realized, however, that there were still too few Japanese-style schools for their children. They began to build private primary and middle schools in the Japanese style, instead of making and maintaining dead-end village schools. By accusing the GGK of neglecting its basic duties through the newly sanctioned Korean-language media, often using GGK statistics, they gradually secured authorization for their self-civilizing measures. Drawing attention to a farcical competition to enter primary and secondary schools, including entrance exams for primary schools, became a good way of exposing the hypocrisy of the regime's civilizing promise. Under this constant pressure, the GGK agreed to increase the number of primary schools by promising "one school in three townships" and then "one school in one township" in the 1930s.[10] The making of secondary education followed the same pattern of Korean pressure and the GGK's grudging endorsement.

In this Korean self-civilizing scheme, securing a proper science and engineering education was a vital goal. With the rise of cultural nationalism in the 1920s, Koreans decided that they would provide this training for themselves if the GGK would not provide for them. They launched a nationwide movement in 1923 to raise funds for their own university with strong science and engineering faculties. The GGK responded very quickly to this Korean initiative by giving Korea the first imperial university outside the Japan's mainland in 1924. This was the Keijo (Seoul's name during the colonial period) Imperial University. Taiwan, colonized 15 years earlier, got its imperial university only in 1928. However, this generosity for Korea was largely intended to tame the inordinate ambition of Koreans for high-level science and engineering education; the new imperial university only had law, philosophy, history, literature, and medicine in its faculties. The Korean people's supposed ineptitude for advanced disciplines like science and engineering was again the reason.[11]

This guardedness of the GGK about science and engineering education for Koreans, well noted by historians, reveals the typical angst and contradiction of imperial regimes regarding the empowerment of the colonized, and the strong association of science and engineering with power in this imperial era.[12] Belying their confident condemnation of the colonized as innately inferior and irrational primitive, they seemed to understand that the only way to maintain the inferiority of a subject population was through discriminatory restrictions on their empowerment. The GGK authorized a college-level course in mathematics and physics only for the Yŏnhŭi College in 1917, to mitigate criticism of Japanese rule by the Western missionary group that founded the school.[13] The Keijo Imperial University, which admitted just small numbers of Korean students anyway, only gained a science and engineering faculty in 1939, for wartime mobilization. The GGK made a serious study of science impossible in the colony, while also carefully controlling Koreans' study abroad in the field.[14]

In fact, it was hard to get any kind of higher education in Korea. For example, upon colonizing Korea in 1910 the GGK decided that Korea did not even need a normal school for training teachers and closed an already existing one.[15] The GGK had decided to recruit teachers from Japan.[16] However, the changing attitude of Koreans about colonial education in the 1920s created an explosive demand for teachers. The GGK had to meet this demand in order not to be ridiculed by the colonized for ineptitude. It established public normal schools in Seoul and in all eight provinces. While these public teachers' schools favored Japanese students, talented Koreans began to flock to them. If they overcame the tough odds against entry, they could enjoy a free education like the Japanese. And those with good grades (the top 30 percent) could have a stipend that was more than half of the average Korean household income. Furthermore, it guaranteed employment. With the Korean employment rate for graduates from Keijo Imperial University or other public higher schools at only about 70–80 percent, and those from private Korean colleges lower, the normal school became highly popular.[17]

The expansion of Korean-funded middle schools gave the possibility of developing science teaching in such schools, although the route to become middle school science teachers was even more convoluted than it was to become primary school teachers. The GGK refused until the end to build higher normal schools to train middle school teachers, while limiting the increase of public middle schools. Yet, it was hard not to authorize privately funded middle schools in the face of Korean criticism on this obvious lack, and some measures to authorize teachers to teach at those Korean middle schools were necessary. For science teachers, the option chosen was to give teaching permits to graduates from Yŏnhŭi College or other trade schools in agriculture and pharmaceutics that had some science education, sometimes with ad-hoc training courses. To become middle school teachers, some Koreans even went to Japan to attend higher normal schools and took the qualifying exams in Japan, when those became open to Koreans in 1932.[18]

A career in teaching seems to have been especially popular among those who studied science and engineering. Colonial Korea produced about 400 bachelors in science and engineering throughout the whole period, mostly from Japanese universities and some, usually with missionary connections, from North American universities. These bachelors were the best trained in a field that had few Ph.D. and master's degree holders. Among these, about 65 percent found a home in colonial middle and elementary schools. Yet, saying that teaching was popular among Korean scientific talents is only a half-truth. The popularity of teaching among scientific talents rather suggests how little chance that they had to develop a scientific career. Many young Korean youths, even with financial means, hesitated to go into the field as there was so little chance for Koreans of getting any serious academic or research positions, especially back in Korea. For the 400 degree holders carefully studied by Kim Geun-Bae, fewer than 10 of them secured careers in academic/research institutes, and those were mostly in Japan.[19] Studying science and engineering for colonial youths must have meant their accepting the lot of becoming teachers. A tradition of valuing learning and teaching, reinforced by the limited chance for colonized talents, appeared to have made such a modest career an acceptable option for those interested in science and engineering. And some of the most highly motivated could turn their modest teaching post into a research career, as shown in Yu's novel.

Unexpected Collaboration for Colonial Science of Botany

Teaching in elementary or middle schools does not normally involve scientific research. The desire of some to escape from such modest teaching posts into a coveted research career reflected the strong wish of some Koreans to pursue science. Yet it was mostly in botany, rather than in such fields as physics and chemistry, that such career transformations took place. For one thing, Meiji Japan's strong bias toward natural history in its primary and secondary science education for conforming citizens, strengthened in the colony, provided a wide foundation for the field.[20] Also, in comparison to research facilities for physics and chemistry, the tools and facilities for natural history were relatively affordable, allowing the institution building by modest initiatives of the colonized—so long as expensive endeavors like overseas collection trips were not ventured on.

Yet there is a factor that needs a more serious examination: various parties, including the GGK but hardly limited to it, made their moves toward studies of Korean plants and animals. Botanical studies' relation to "practical sciences" like agriculture, forestry, and pharmaceutics was important in ensuring that the GGK and Japanese entrepreneurs moved into these fields before they were pushed by Koreans who built up modest education and research foundations. Such Japanese initiatives were certainly not intended to raise Korean scientific talents in those fields. It was the job of the GGK to make this unrewarding colony pay for its expensive administration of repression, such as the maintenance of police stations in all villages. Japanese developers constantly sought to find lucrative resources in Korea. As this goal was not achievable without native helpers, some room had to be made to nurture and accommodate them. Also importantly, it was not just Japanese who were interested in those developments, as there were Korean agriculturalists and entrepreneurs too. They appeared welcoming to Japanese initiative in joining various projects, collaborating and competing with their Japanese counterparts.

Japanese officials had given agricultural education in Korea the highest priority since the protectorate period. The Suwŏn Agricultural and Forestry School was the first higher education facility that received their support, although when they assumed an advisory role in 1905 they recommended the reduction of the four-year school launched by the Korean government in 1904 as unrealistic for the level of Koreans.[21] By introducing a one-year fast-track course in forestry in 1906, they made it clear that education in Korea was designed to staff Japanese development projects with the necessary Korean assistants. The GGK maintained this priority in agricultural education as it more or less maintained the colonial development scheme of having Korea as a provider of agricultural and other natural resources for industrializing Japan. As sign of this commitment, it even made the exception of stepping up education in agriculture. The GGK had authorized at least one agricultural school in each province, nine in total, while it allowed only one public engineering school.

It is notable how even the establishment of this two- or three-year trade school in agriculture, not meant to produce agricultural scientists, caused concern in the GGK about the possibility of empowering the colonized. The GGK revealed its angst in 1918 when it upgraded the Suwŏn School to a Higher School with a three-year course; it instituted a quota for Korean students, limiting the number of Korean students who could obtain such a high education. Until the end of Japanese rule, the school produced only 100 Korean graduates in forestry, i.e., 4 or 5 a year.[22] It was an unexpected development when these agricultural schools, especially the one in Suwŏn, came to be known to colonial youths as the best institute to study natural science in Korea, which also gave them secure teaching permits in colonial schools. Also, although not as prestigious as the Forestry Experiment Station of the GGK, there were many other modest experiment stations and nurseries for research. In the scarcity created by colonial restrictions on science education and research, agricultural schools and labs strongly attracted colonial talents who had no other way to study science.

The pharmaceutical field also came to provide similar institutional opportunities for colonized scientific youths, mostly through the actions of entrepreneurs, both Japanese and Koreans. Japanese entrepreneurs appeared to move first. Japanese pharmaceutical entrepreneurs at first just jumped in to fill the gap in the medical market created by the GGK's lack of effective medical policies. Like other colonial regimes, the GGK had an interest in showing its modernity through medicine. The main tools were Western-style hospital buildings and services. Even before annexation, Japan established a modern hospital in treaty ports of Pusan, Wonsan, and Incheon to display Japan's successfully acquired Western medicine.[23] Upon colonization, it established 13 modern hospitals, named "Benevolence Hospitals (慈惠病院)," at least one for each province in the peninsula, displaying its benevolent and Western-style modern rule. However, these impressive hospitals hardly touched the rural population of Korea while Japanese doctors and Korean patients felt alienated from each other. Japanese doctors wanted only enlightened Japanese as patients whereas Koreans were suspicious of Japanese doctors, who were too expensive for them anyway.[24] By 1926, the GGK offered enough doctors for its Japanese population, having 708 Japanese doctors for about 500,000 Japanese residents. Yet, the number of Korean doctors serving 20 million Koreans was 703, leaving a big gap in the colonial medical market.[25] Traditional medical doctors filled this gap, which the GGK had to condone and even encourage, although it showed a clear hierarchy between traditional medicine and modern Japanese medicine by calling Korean traditional doctors "medical pupil (醫生)," instead of doctor (醫師, literally medical teachers).[26]

This state of affairs provided a good opportunity for pharmaceutical entrepreneurs. In this age before antibiotics, it was not just the Japanese actors with modern pharmaceutical expertise who saw this chance. Seasoned in the ginseng-based international trade, Korean medicine dealers had been making the transition to patent medicines (賣藥) since the late nineteenth century. Many had re-organized themselves as modern corporations with new patent medicines and marketing technology.[27] They had even weathered the colonial government's monopoly of ginseng by producing new derivative products.[28] Some of them adopted chemical methods to fully modernize themselves. These Korean dealers were significant players in making pharmaceutics a phenomenally lucrative industry. Some companies even offered promotional airplane rides in their advertisements. Japanese entrepreneurs chose to partner with these successful Koreans, whom they found "fast in responding to the changes of the time."[29]

Creating educational institutions needed to staff this growing industry was an essential task for both Japanese and Korean entrepreneurs. While Koreans had made initial moves just before colonization, Japanese pharmacists eagerly joined in. Yet, in the turmoil of 'annexation' that changed their respective status dramatically, the dialogue between the two groups did not immediately come to fruition. However, in 1914, the Korean-only Pharmaceutics Union (藥業總合所) made a further move by establishing a temporary summer program to train pharmacists. The Pharmaceutical Society of Korea (朝鮮藥學會), the Japanese-centered organization with some Korean members, joined forces by sponsoring it.[30] Inspired by the success of this summer course, they further moved to establish a permanent one-year school in 1915. This school, which had modestly aimed to produce licensed pharmacists, became a two-year school in 1918 and a three-year Higher School of Pharmaceutics in 1930.[31]

Japanese pharmacists paid for a bigger share of these expansions partly as the school was primarily for sons of Japanese pharmacists in Korea. It had 10 Korean students out of 100 students in 1918 and 20 out of 120 in 1933.[32] However, 10–20 Korean graduates a year was a considerable number, given that other public higher schools only admitted about 30 students a year in total. Furthermore, it was the first and only co-ed school, although only until it became a higher school in 1930. It produced 17 female Korean pharmacists. Those Korean students who overcame these tough odds must have been excellent. Koreans were often top-ranked in licensing exams, elevating their standing in the field.

These Korean pharmacists could find a place in pharmaceutical research, a growing area in the field. As the competition for new medicines was high, pharmaceutics became the first industry that adopted the emerging R&D strategy of establishing affiliated laboratories. In addition to the Hygienic Laboratory (衛生試驗所) under the GGK, not just Japanese but also Korean companies established specialized laboratories. Ch'ŏnil Drugs and Kŭmkang Pharmaceutics were Korean companies that had their own laboratories. Furthermore, Korean pharmacists and pharmacologists worked not only in these Korean laboratories but also in other related laboratories. Notably, there were quite a few Korean researchers including four women working at the Hygienic Laboratory. The Herbal Medicine Laboratory (生藥研究所) at the Keijo Imperial University also hired some Korean researchers and assistants.[33]

This research in pharmaceutics might not at first sight have been expected to have a strong relation to botanical studies. But the strength and expansion of patent medicine market based on traditional prescriptions, which increased the trade in traditional herbs, was strongly linked with research with botany. Even until the 1930s, the top two Korean pharmaceutical companies, Ch'ŏnil Drugs and Chosŏn Patent Medicine Co., which had branches in Manchuria, Shanghai, and Japan, earned the majority of their income from the traditional medicine trade.[34] As these medicines were also in high demand in Japan, where traditional medical practice was no longer allowed, Japanese dealers could not ignore this long-term trading partner, they needed the expertise of Koreans in collecting, cultivating, circulating, understanding these mostly plant based medicines.[35] The increased openness to Korean talents, even in the GGK lab and the imperial university, was not an unreasonable choice.

The strength of colonial traditional medicine also gave colonial research facilities a favored position within the imperial research network by constantly attracting researchers from Japan. Japanese pharmaceutical experts visiting Korea never failed to mention the need "to utilize the local advantage by focusing on the studies of medicinal herbs" and emphasized the unfading importance and potential of furthering ginseng cultivation. The specialized pharmacological lab for medicinal herbs at the Keijo Imperial University reflected this high expectation. In this atmosphere, the largest portion of the research published in the *Journal of the Pharmaceutical Association of Korea* (朝鮮藥學會雜誌) was about medicinal herbs.[36]

Pharmaceutics became the most important institutional foundation for botanical studies during the colonial period. Unlike agriculture and forestry, the GGK's interest was not the driver, although both imperial and colonial policies that made traditional medicinal products popular in Japan and Korea aided this development. Japanese settlers joined with big and small Korean dealers in the field, and they became active co-builders of these educational and research foundations. In later reflections, one Japanese pharmacist attributed this collaborative atmosphere to the neutrality of science.[37] Nonetheless, what strongly inspired them to cooperate with Koreans and to accept Korean talents in their labs was their need to do so. They had ventured out to Korea in search of lucrative business opportunities, and to ignore the promising and still vibrant traditions was not good business sense.

They needed the trade networks and expertise of Koreans, just as much as Koreans needed Japanese connections to colonial authority, and willingness to pay for their expertise and products. Thanks to individual entrepreneurs' practicality and the strong Korean tradition in the field, a common ground of botanical research for the colonizer and the colonized had emerged.

Making New Koreans: From Observers to Doers

For those Korean talents who overcame the tough odds to enter specialized schools and labs in agriculture and pharmaceutics, the Korean media showed only approval. They proudly reported outstanding exam results, and welcomed of the entry of Koreans into the august laboratories of the GGK. No one seems to have accused them of helping Japanese exploitation of Korean resources or working for the GGK. They only deplored that there could not be more Koreans in the field, owing to the discrimination that favored less qualified Japanese for desired education and research opportunities. As already mentioned, this was a significant change of attitude among Koreans, as they had previously refused to send their children to Japanese primary schools. The GGK's suppression of Korean efforts toward scientific enlightenment led to the abandonment of this negative attitude. Demanding better access in the face of the GGK's restriction of opportunities was a respectable form of resistance to its rule; it was not a betrayal of their defeated nation. In the field of science, the colonized further revised their viewpoint; they made working with the Japanese in the field of science not just acceptable, but a patriotic attempt that might restore their nation.

A certain amount of soul searching about the reasons for Korea's defeat, leading to the identification of the faults of Korea's past intellectual practice, was crucial. They came to accept that it was Koreans themselves, especially the Confucian ruling elites, who had allowed the Japanese to deprive Koreans of their right to civilize themselves. As also shown in the Chinese case, the dichotomous view of Confucian thought as metaphysics full of empty talk and lacking hands-on empiricism emerged, aided by Western views that promoted science as its opposite—practical and hands-on.[38] Of course, most Western natural philosophers would have abhorred such characterization of their intellectual endeavor, while many Confucians who had emphasized "practical studies 實學" would have been perplexed by such criticism.[39] Nonetheless, this was a genuine act of self-reflection that gave a new purpose for some Koreans, shattering habits that might have led them to prefer the comfort of a quiet desk.

Field sciences like botanical systematics especially well answered this need to change the old habits. Trudging up and down the hills of their lost peninsula with specimen bundles might redeem Korean's souls.[40] Also, it was an intellectual practice that best embodied the new scientific attitude. One earnestly claimed that collecting nature from the field and systematically storing those "specimens" to "observe, classify, and describe" their "objective" characteristics was *the* science.[41] That such specimens had almost never been systematically collected in Korea or China in spite of long traditions of studying nature effectively guaranteed this study's modernity.[42] Botanical systematics could be the exemplary science for new Koreans, those doers and makers, not something chosen only because more cutting-edge options like physics were not available to them.

Furthermore, as some astute Korean observers came to note, doing and making by scientific fieldwork was a move *against* a certain civilizing program that the GGK had been exceptionally serious about. Year after year the GGK had held big and small exhibitions celebrating its enlightened rule, and the kindness of the Japanese imperial family, which occasionally sponsored those exhibitions. The GGK widely publicized these exhibitions as evidence of its generous and sincere will to civilize the benighted Koreans. They gave Koreans the chance to see modern material culture, including scientific specimens, for themselves. The GGK strongly urged local Korean leaders to mobilize rural Koreans for trips to bigger cities for those exhibitions, using sales vouchers and other lures. Koreans had to take part in those civilizing exhibitions to overcome their backwardness.[43] However, some Korean elites came to suspect both the intention and the effect of such exhibitions in enlightening Koreans. In their view, by pushing Koreans to gawk at those endless exhibitions, what the GGK sought was not the enlightenment of their Korean subjects, but to create awe-struck spectators who would then submit themselves to Japan's grandeur. The Japanese also used these exhibitions to humiliate Koreans, by "collecting all the shameful things of Koreans" in the name of Korean culture, in contrast with all the brilliant things of Japan. These critical Korean elites urged Koreans not to waste their time and money in showing "idiotic wonder" by following such exhibitions.[44] Instead, Koreans should go out to the field, since the best way not to remain awe-struck observers was to enlighten themselves by adopting enlightening practices. Collecting nature in the field was an excellent way to make themselves doers, thus defying the GGK's scheme.

The Korean colonial media joined this move to transform Koreans into doers by criticizing colonial science education and suggesting collecting for the study of nature as an antidote to it. Colonial science education, limited to natural history and some basic math, was inadequate not just in its makeup but in its method, they claimed. They said any science education would be useless if "students stow away in the closet the scientific principles and mathematical theorems that they memorize for exams as soon as they leave the school with their textbooks." To nurture true empirical attitudes and critical spirits, students had to "learn about nature through nature instead of relying on books and lectures." It was a campaign to preach the virtue of natural history fieldwork for Korean youths. Korean-language newspapers and magazines regularly gave guidelines and tips for collection trips and specimen making, and recommended such collection trips as hobbies and weekend activities for families.[45] Natural history became a highly visible and popular science during the colonial period through such calls to become new Koreans, hands-on doers.[46]

Some attempted to add even more meaning to this natural history practice, beyond the realm of science. They thought that such work would be a good way to imbue patriotism in Korean youth. They worried that it was difficult for young people to have much pride in being Korean. "At present, the Korean people have

Figure 4.1 A photo taken on a field trip of the Dongdeok Girls High School to Mount Kŭmkang, dated May 1934 (Courtesy of the National Mountain Museum).

nothing to take pride in," they lamented. Yet, one thing that they could still boast was their land known for its beauty, with its "silk embroidered rivers and mountains (錦繡江山)."[47] Notably, this Korean attempt to find pride in things Korean interacted with Japan's Orientalizing gaze.[48] Ishidoya worried about colonial policies ruining Korean local color, because he found Korean nature beautiful. Abe Yoshishige (安倍能成, 1883–1966), professor of philosophy at Keijo Imperial University, also did not stint his praise for the unique beauty of Korean nature.[49] Encouraged by such Japanese attitudes, some Koreans expressed boundless pride about the beauty of nature in Korea.

> They say that Italy has the Alps bordering with Switzerland. Korea does not lack mountains as praiseworthy as that! Without even mentioning Mount Kŭmkang, town after town in Korea is endlessly endowed with beautiful mountains and mystical rocks.[50]

Mount Kŭmgang and other famous scenic spots became sacred places to which the elite and young people of the colonized population had to make pilgrimages on foot during their summer vacations. Korean schools organized field trips to these places (See Figure 4.1).[51] Collecting natural objects in Korean forests and mountains was a good way to strengthen this proud attachment to what is Korean.[52]

By struggling to make collecting nature a superbly patriotic endeavor, teachers at colonial schools could also build a research foundation in their schools. For

Figure 4.2 Specimen rooms at the Whimoon Higher Elementary School, *70 Years of Whimoon*, 1976 (Courtesy of the Whimoon High School).

the Korean philanthropists who emerged as part of the cultural nationalism of the 1920s, funding "specimen rooms" at schools was quite rewarding, as it was likely to be followed by positive media accounts of their forward looking investment in science education. By providing specimen rooms and labs, equipped with specimens, diagrams, scale models, and various tools, philanthropists expressed their wish to inspire Korean students to develop scientific curiosity and to do science for themselves. Although many Korean schools struggled with financial problems, they prioritized the provision of good specimen rooms and a science lab. In the two-tier system of education maintained throughout most of the colonial period, Korean elementary school students could not enter public middle schools directly, because their education in Korean schools (essentially specified to be a five-year course) did not match up with the education in public elementary schools (specified as a six-year course). Adding specimen rooms became one element through which Korean schools could demand the upgrade to the six-year curriculum. Therefore, at times students themselves demanded the provision of well-equipped specimen rooms to improve their education. Students at one school even went on strike for that demand.[53] If the GGK had viewed such a requirement as a barrier preventing poor Korean schools from being upgraded, it actually made it a very rewarding cause for various Koreans to support (See Figure 4.2).

Collecting nature thus came to be at the center of the scientific self-enlightenment of colonial Koreans, producing some avid collectors. They often revealed their existence by donating the fruits of their years of labor. Chang Hyŏngdu (張亨斗, 1906–1949), who went to Japan to study plant systematics and learned from the popular taxonomist Makino Tomitaro at the Tokyo Horticultural School, decided to donate 7,000 Korean plant specimens that he had been collecting for ten years to Yŏnhŭi College in 1933. He proposed to build "the best herbarium in the orient" with his "precious Korean treasures which were rare not only in Korea but in the whole orient."[54] It was an ambitious proposal intended to surpass herbariums in Japan. Sungŭi College received 858 insect specimens, which were claimed to be "pure Korean indigenous insects," for research and exhibition. This was an anonymous donation by "a young scientist."[55] Colonized Koreans succeeded in transforming themselves into doers, instead of spectators of Japanese modernity, by taking up this unfamiliar practice and walking miles after miles over the peninsula in pursuit of it.

Conclusion: Botanical Nationalism?

One may use the term "botanical nationalism" for the motivation that led to the creation of small but effective foundations for botanical research in various colonial institutions. The foundation that they built was modest indeed, and was in most cases not able to produce serious botanists. Yet, it was a foundation built by assiduous political action by the colonized and aided, at times, not only by their own traditions and but also by Japanese who found it in their own interests to do so. Although the GGK and various Japanese parties helped or at least tolerated some of this institution building by Koreans, it was Korean efforts with an obvious patriotic tinge that reshaped these colonial institutions to create the beginnings of Korean scientific enlightenment.

However, it is important to keep in mind that like Japanese imperialism, this nationalism of Koreans, created by many people with different ideas and backgrounds, was not a static or a monolithic thing, and had many shades. Furthermore, this botanical nationalism of Korean researchers was hardly anti-Japanese or anti-imperial in that it showed a certain thorough acceptance of Japanese imperial logic. Above all, Koreans acquiesced in the Japanese perspective about the unworthy and inferior nature of Korean intellectual tradition, and strongly and often uncritically sought to follow Japanese ways in order to be on the right track of civilization. In a sense, the famous resistance of Koreans against Japanese rule was mostly a struggle for their right to follow the Japanese path and to be like the Japanese. It was a nationalism of a colonized mind that was ashamed of its past and present and envied Japanese style empowerment.

This search for power, and the wholehearted acceptance of Japanese logic, worked especially well for Koreans during colonial times, thanks to the imperial contradiction that could not be at ease about its ostensible aim of 'helping' the inferior Koreans to become enlightened like Japanese. Koreans made the civilizing mission a liability to the colonizer, simply by accepting it. The limitations and anti-Korean discrimination imposed by the GGK broke their promise of a civilizing mission, and made the Korean adoption of that mission into a challenge to colonial rule and policies. Koreans could thus expose the hypocrisy of Japanese policy, while defying its desire to produce awe-struck but passive spectators of Japanese modernity. They were not successful in every area, but they took pride in making advances in botany, the field that they came to see as the essence of the Western scientific spirit, while deepening their understanding and love of their country's natural riches.

While their nationalism could not be neither pure nor simple, this building of a research foundation by colonized Koreans is important. It showed that colonial politics could not be completely controlled by the colonial regime, despite the GGK's impressive monopoly of power in all areas of governance including the police and armed forces, and despite the disenfranchised status of the colonized people. This basic institution building enabled a people who never stopped thinking and acting for themselves to dictate many of the terms of their relationship with the colonizing power.

Notes

1 Yu Chin-o 유진오, "A Genealogy of Brilliant Wishes (華想譜)," in Yi Hyosŏk 이효석 and Yu Chin-o, eds., *The Complete Collection of Korean Literature* (한국문학전집) vol. 8, Seoul: Minchungsŏkwan, 1959, 229–460, 460, 459. I thank Prof. Kwon Boduerae for letting me know about the novel. Some readers remember that the one who sells his soul to the devil Mephistopheles in the Goethe's novel is Dr. Faust.

2 Yu Chin-o 유진오, "A Genealogy of Brilliant Wishes (華想譜)," in Yi Hyosŏk이효석 and Yu Chin-o eds., *The Complete Collection of Korean Literature* (한국문학전집) vol. 8, Seoul: Minchungsŏkwan, 1959, 458. It shows that colonial intellectuals actually knew the name of Nakai as an important authority in botany, even though details are wrong as this is not a history but a novel.

3 Ha U-Bong 하우봉, *Koreans' Understanding of Japan in Joseon Dynasty* (조선시대 한국인의 일본인식), Seoul: Hyean, 2006; Ronald P. Toby, *State and Diplomacy in Early Modern Japan: Asia in the Development of the Tokugawa Bakufu*, Stanford, CA: Stanford University Press, 1991.

4 The consensus was quite broadly made that Western science and technology should become essential tools for their survival and progress, even though not many Koreans actually did embark on such an unfamiliar task of studying them on their country's opening to the imperial powers in 1876. The ruling monarch and the reform-minded elites both put an emphasis on science and technology, as shown in the first modern state gazette *Hansŏngsunbo* (漢城旬報, 1884–1885) and a pioneering magazine for youth, *Youth* (少年, 1908–1911). Even those who denied the superiority of Western civilization, the alleged product of science and technology, embraced them as useful tools in developing a unique Korean civilization, expressed in a catchphrase like "Eastern Way and Western Means," which had slightly different matching phrases in China and Japan. Park Seong-Rae 박성래, "Western Science Introduced in the Late Chosŏn Period (개화기의 서양과학)," *Journal of the Korean History of Science Society* (한국과학사학회지, *JKHSS* hereafter) 10(1) (1988): 169–72; Lim Jongtae 임종태, "'도리'의 형이상학과 '형기'의 기술: 19세기 중반 한 주자학자의 눈에 비친 서양 과학 기술과 세계: 이항로 (1792–1868) (Yi Hang - No's Neo-Confucian Critique of Western Science and Technology and His Natural Philosophy of Civilizations in the Mid-19th Century Korea)," *JKHSS* 21(1) (1999): 58–91. Cho Hyŏngrae 조형래, "The Science of *Youth* 소년의 과학," *SAI* (사이) 6 (2009): 273–302. For complex variations of the "Eastern Way

and Western Means" idea, see Noh Dae-Hwan 노대환, *Studies on the Making of the Eastern Way and Western Means Idea* (동도서기론 형성 과정 연구), Seoul: Ilchisa, 2005.

5 These Japanese victories over China and Russia also changed Western and other non-Western countries' views on Japan and its civilization. Michael Adas, *Machines as the Measure of Men: Science, Technology, and Ideologies of Western Dominance*, Ithaca, NY: Cornell University Press, 1990, 357–64.

6 Kwon, Tae-Eok 권태억, *Japanese 'Civilizing Mission' in Korea (1904–1919)* (일제의 한국 식민지화와 문명화, 1904–1919), Seoul: Seoul National University Press, 2014.

7 Chung Yountae 정연태, *Everyday Colonial Discrimination in Japanese Korea* (식민지 차별의 일상사), Seoul: Purŭnyŏksa, 2021.

8 This emergence of the traditional school makes for a good contrast with the situation in colonial Taiwan, where the number of traditional schools was reduced. For a comparative study of the colonial education system in terms of cultural assimilation, see Komagome Takeshi 駒込武, *The Cultural Integration of the Colonial Empire of Japan* (식민지제국 일본의 문화통합), tr. Oh Sŏngchŏl 오성철 et al., Seoul: Yŏkbi, 2008. For the comparison of the colonial attitude about colonial education in Korea and Taiwan, see E. Patricia Tsurumi, "Colonial Education in Korea and Taiwan," in Ramon H. Myers and Mark R. Peattie eds., *The Japanese Colonial Empire, 1895–1945*, Princeton, NJ: Princeton University Press, 1984, 275–311. For Korean involvement in making the colonial education system, see Oh Sŏngchŏl 오성철, *The Formation of Colonial Elementary Education* (식민지 초등교육의 형성), Seoul: Kyoyukkwahaksa, 2000. Oh points out that these demands for more education involved various other motives like upward social mobility.

9 Yoshino Sakuzo (吉野作造, 1878–1933), an influential political thinker in Japan, famously noted this after his trip to Korea and Manchuria in 1916. From Mark R. Peattie, "Japanese Attitudes toward Colonialism, 1895–1945," Ramon H. Myers and Mark R. Peattie eds., *The Japanese Colonial Empire, 1895–1945*, Princeton, NJ: Princeton University Press, 1984, 80–127, 105.

10 The emerging Korean-language media found education policies the most legitimate way of criticizing the GGK. Oh Sŏngchŏl 오성철, *The Formation of Colonial Elementary Education* (식민지 초등교육의 형성), Seoul: Kyoyukkwahaksa, 2000.

11 The Imperial University began its preliminary courses in 1924. Kim Geun-Bae 김근배, *The Emergence of Modern Techno-Scientific Manpower in Korea* (한국 근대 과학기술 인력의 출현), Seoul: Munji, 2005, 450–52. Kim's work skillfully documents the efforts of the colonized to create the foundation for science and engineering education and the GGK's attempt to limit it.

12 The Imperial University began its preliminary courses in 1924. Kim Geun-Bae 김근배, *The Emergence of Modern Techno-Scientific Manpower in Korea* (한국 근대 과학기술 인력의 출현), Seoul: Munji, 2005, 450–52.

13 The mathematics and physics faculty was abolished by the GGK in 1923 but reinstated in 1924 through further negotiation. Jeon Chan-Mi 전찬미, "Plans for Science Education in Colonial Korea: The Establishment of the Science Department of Chosen Christian College (식민지시기 연희전문학교 수물과의 설립과 과학 교육)," *JKHSS* 32(1) (2010): 43–68.

14 While the GGK did not move, it is notable that various Japanese settler associations made petitions to have the science-related faculty in the colonial university too. Kim Geun-Bae 김근배, *The Emergence of Modern Techno-Scientific Manpower in Korea* (한국 근대 과학기술 인력의 출현), Seoul: Munji, 2005, 456–66.

15 By comparison, the Japanese government made it mandatory to have at least one normal school per prefecture in 1880 to provide "the foundation of national education" and recruited talent with scholarships. The Korean government in 1895 followed this Japanese normal school system by establishing a first state normal school with full scholarship. The History of Science Society of Japan 日本科学史学会 ed., *A Compendium of History of Science in Japan* (日本科學技術史大系 , *Compendium* hereafter), vol. 9, 9–18; Chung Kyu-Young 정규영, "A History of the Normal School of Korea during the

Japanese Colonial Ruling (日帝時代 師範學校의 歷史)," *Theses Collection (*論文集) 41 (2003): 29–56.

16 As it was difficult to recruit enough teachers from Japan, the GGK soon added special courses in Japanese schools in Korea to train teachers. Chŏng Chaech'ŏl 정재철, *The History of Education Policy of the Japanese Colonial Regime* (日帝의 對韓國 植民地 教育政策史), Seoul: Ilchisa, 1985, 147. Chung Kyu-Young 정규영, "A History of the Normal School of Korea during the Japanese Colonial Ruling (日帝時代 師範學校의 歷史)," *Theses Collection (*論文集) 41 (2003): 38–39.

17 Chung Kyu-Young 정규영, "A History of the Normal School of Korea during the Japanese Colonial Ruling (日帝時代 師範學校의 歷史)," *Theses Collection (*論文集) 41 (2003): 38–39. Oh Sŏngchŏl 오성철, *The Formation of Colonial Elementary Education* (식민지 초등교육의 형성), Seoul: Kyoyukkwahaksa, 2000, See Chapter 3.

18 The middle-school teacher qualifying exam was very difficult to pass for Koreans as it screened "anti-Japanese" or otherwise improper candidates. Lee Won-pil 이원필, "A Study on Teacher Training System in Korea under Japanese Imperialism (日帝强占期의 中等教員養成法制에 관한 研究)," *The Journal of Educational Idea* (부산교육대학 논문집) 5 (1996): 7–41.

19 Studying abroad was a highly controlled matter during the colonial period. Those who studied abroad did so in Japanese higher normal schools or in universities mostly in Japan or in the US. Kim Geun-Bae 김근배, *The Emergence of Modern Techno-Scientific Manpower in Korea* (한국 근대 과학기술 인력의 출현), Seoul: Munji, 2005, 244–78, 269, 274.

20 See Chapter 1.

21 Although not so successful and beset by financial difficulties, the Korean government made attempts to modernize Korean agriculture before Japan's entrance to Korea. It opened an experimental farm in 1884, equipped with machines brought from the US after an initial visit to the US in 1883 and established modern schools on various levels and areas after abolishing the civil service exam. It sought to promote "education for practical values not vain titles." College of Agriculture and Life Sciences 농업생명과학대학, *100 Years of Agricultural Education: 1906–2006* (농학교육 100년: 1906–2006), Seoul: College of Agriculture and Life Sciences, Seoul National University, 2006, 30; *History of Korean Agricultural Education* (한국농업교육사), Seoul: Taehan Kyogwasŏ Chusik Hoesa, 1994, 46.

22 The Governor-General of Korea emphasized the GGK's "civilizing" of Korea through the "progress" of agriculture and forestry while celebrating the upgrade of the school in 1918. "Current Affairs in the Agricultural Field (農界時事)," *News of the Agricultural Society of Korea* (朝鮮農會報) 13(5) (1918): 52–54. For the ratio of Japanese and Korean graduates, see Department of Forest Sciences 산림과학부, *One Hundred Years of Forestry Science: 1906–2006* (산림과학 백년사: 1906–2006), Seoul: Department of Forest Sciences, Seoul National University, 2006, 35–36. *History of Korean Agricultural Education* (한국농업교육사), Seoul: Taehan Kyogwasŏ Chusik Hoesa, 1994.

23 Japan adopted "Western medicine" as the only medical practice in its all-out Westernization after the Meiji Restoration. Traditional practice is still not acknowledged in its medical licensing system. Otsuka Yasuo 大塚恭男, *Oriental Medicine in Japan* (일본의 동양 의학). Yi Kwangchun tr., Seoul: Sohwa, 2000, 29–50.

24 Even those 13 hospitals did not become practical providers of benevolent care for many reasons. Most visitors to those Benevolent Hospitals were Japanese. Park Yun-Jae 박윤재, "Joseon Government-General's Policy of Local Medicine and Local Consumptions of Medical Service (조선총독부의 지방 의료정책과 의료 소비)," *Critical Studies on Modern Korean History* (역사문제연구) 21 (2009): 161–83. Dojinkai, a private organization for the Japanese medical mission in Asia, eventually sponsored by Congress, also established modern hospitals in Korea. Osato Hiroaki 大里浩秋, "Dojinkai and Dojin (同仁会と『同仁』)," *Institute for Research in Humanities Brief* (人文学研究所報) 40 (2007): 47–105.

25 Shin Dong-won 신동원, "The Medicine and Medical Practice in Japanese Colonial Period in Korea (일제 강점기의 의학과 의료)," in Park, Seong-Rae 박성래, Shin Dong-won 신동원, O Tonghun 오동훈 eds., *One Hundred Years of Our Science* (우리 과학 100년), Seoul: Hyŏnamsa, 2001, 120–41.
26 Due to the continued shortage of modern doctors, the GGK had to mobilize traditional practitioners even for its public health campaigns in the 1930s. For more about the development of modern medicine in Korea, seeShin, Dong-won, *A History of Modern Public Health and Medicine in Korea* (한국 근대 보건 의료사), Seoul: Hanul, 1997; Park, Yun-Jae, *The Origin of Korean Modern Medicine* (한국 근대의학의 기원), Seoul: Hyean, 2005.
27 Hong Munhwa 홍문화, *Scattered Thoughts on History of Pharmaceutics* (藥史散攷), Seoul: Dongmyŏngsa, 1980, 20–21. Hong Hyŏno 홍현오, *A History of Korean Pharmaceutics* (韓國藥業史), Seoul: Handokyakpum Co., 1972; Chŏng Minsŏng 정민성, *History of our Medicine and Pharmaceutics* (우리 醫藥의 역사), Seoul: Hakminsa, 1990.
28 Donghwa Drugs (1897), Chesaengdang Pharmacist (1903), Chosŏn Patent Medicine Co. (1913), Ch'ŏnil Drugs (1913), Yuhan Co, Ltd. (1935), and Kŭmkang Pharmaceutics (1935) were all very successful Korean pharmaceutical companies. On the transition of Korean medicine dealers from the late-nineteenth century, see Yang, Jeongpil 양정필. "Modern Medicine Environment and Adaptation of Korean Trader for Medicinal Herbs (한말-일제초 근대적 약업 환경과 한약업자의 대응)," *Korean Journal of Medical History* (의사학, *KJMH* hereafter) 15(2) (2006): 189–209; Yang, Jeongpil, "The Commercialization and Promotion of White Ginseng in the 1910s and 1920s (1910–20년대 개성상인의 백삼(白蔘) 상품화와 판매 확대 활동)," *KJMH* 20(1) (2011): 83–118, etc.
29 On the profits of the industry and various promotional campaigns, see Hong Hyŏno 홍현오, *A History of Korean Pharmaceutics* (韓國藥業史), Seoul: Handokyakpum Co., 1972, 106–14. These Korean entrepreneurs show much of "vernacular industrialism." Eugenia Lean, *Vernacular Industrialism in China: Local Innovation and Translated Technologies in the Making of a Cosmetics Empire, 1900–1940,* New York: Columbia University Press, 2020.
30 The Pharmaceutical Association of Korea was quite open to Koreans. The influence of the pioneering Korean pharmacist Yu Sehwan (劉世煥, 1876–1917), who studied at Tokyo Pharmaceutical School (1900) and Tokyo Medical School (1902) before the annexation and revered by Japanese pharmacists, was significant. Hong Munhwa홍문화, *Scattered Thoughts on History of Pharmaceutics* (藥史散攷), Seoul: Dongmyŏngsa, 1980, 19–28, 26.
31 The big sum of money for the school's operation and upgrade mostly came from Japanese pharmacists in Korea. Hong Hyŏno 홍현오, *A History of Korean Pharmaceutics* (韓國藥業史), Seoul: Handokyakpum Co., 1972, 159–68;Sim, Ch'anggu 심창구 et al., "History of Korean Pharmaceutics (한국약학사)," *Journal of the Pharmaceutical Society of Korea* (약학회지) 51(6) (2007): 361–82, 367.
32 There was an obvious divide between Korean and Japanese students, as seen by the student magazine, *The Student Magazine of Keijo Higher School of Pharmaceutics* (京城藥學專門學校校友會誌). However, the school was unique in that it hired Koreans as full professors, not just as assistants. Hong Munhwa홍문화, *Scattered Thoughts on History of Pharmaceutics* (藥史散攷), Seoul: Dongmyŏngsa, 1980, 19–28, 26.
33 Hong Hyŏno홍현오, *A History of Korean Pharmaceutics* (韓國藥業史), Seoul: Handokyakpum Co., 1972, 93–95. Hong Munhwa says that Koreans did their research "under the disabling discrimination," which would be the usual discrimination in salary and promotion. Hong Munhwa 홍문화, *Scattered Thoughts on History of Pharmaceutics* (藥史散攷), Seoul: Dongmyŏngsa, 1980, 33. On general discrimination in the Japanese imperial system, see Ramon H. Myers and Mark R. Peattie, eds., *The Japanese Colonial Empire, 1895–1945*, Princeton, NJ: Princeton University Press, 1984, 37. That Koreans organized a separate organization for Korean pharmacists in 1928 may

show their response to discrimination. Sim, Ch'anggu 심창구 et al., "History of Korean Pharmaceutics (한국약학사)," *Journal of the Pharmaceutical Society of Korea* (약학회지) 51(6) (2007): 368.

34 Hong Soon-Kwon 홍순권, "The Status of Ginseng Cultivation in the Kaesŏng Area before the Colonial Period (한말시기 開城地方 蔘圃農業의 전개 양상 (上))," *Journal of Korean Studies* (韓國學報) 13(4) (1987): 33–67, 34.

35 It is important to note that the tradition was not the only force that reshaped the medicine market in the colony. The introduction of Western medicines provided an important spur in this development as shown by the fact that the first advertisement of medicine was for an anti-malarial drug from the West called Kŭmkyerab. Yeo In-sok 여인석 et al., *History of Korean Medicine* (한국의학사), Seoul: Research Institute for Healthcare Policy, 2012, 206–15.

36 Hong Hyŏno 홍현오, *A History of Korean Pharmaceutics* (韓國藥業史), Seoul: Handokyakpum Co., 1972, 183. For an analysis of articles by subfield within pharmaceutics, see Hong Munhwa홍문화, *Scattered Thoughts on History of Pharmaceutics* (藥史散攷), Seoul: Dongmyŏngsa, 1980, 21–24.

37 If collaboration with Japanese was frowned upon as "selling out" in other fields, in science and technology, "value-neutrality" was evoked as an acceptable excuse for collaboration. On the shaping of this kind of "value-neutrality" of science, Hong Sungook, "Korean Scientists Look at Japanese Colonization: The Emergence of an Idea of the Value-Neutrality and Objectivity of Science," *Historia Scientiarum* 15 (2006): 268–75.

38 For similar framing about "science" vs. traditions, see Fa-ti Fan, ""Mr. Science", May Fourth, and the Global History of Science," *East Asian Science, Technology and Society: An International Journal* 16(3) (2022): 279–304.

39 A Victorian botanist like Joseph Hooker still portrayed himself as a natural philosopher, not a mere expert about plants. Jim Endersby, *Imperial Nature: Joseph Hooker and the Practices of Victorian Science*, Chicago: University of Chicago Press, 2008. For Confucian concerns to make knowledge useful, seeKim, Yung-Sik, *The Natural Philosophy of Chu Hsi (1130–1200)*, Philadelphia, PA: American Philosophical Society, 2000; William T. Rowe, *Saving the World: Chen Hongmou and Elite Consciousness in Eighteenth-Century China*, Stanford: Stanford University Press, 2001.

40 For a similar line of thoughts that moved Chinese geologists for fieldworks, see Grace Yen Shen, "Taking to the Field: Geological Fieldwork and National Identity in Republican China," *Osiris* 24 (2009): 231–52.

41 Chang Ŭngchin 장응진, "On Science (과학론)," *T'aekŭk Journal* (태극학보) 5 (1906): 9–12, 10.

42 Benjamin A. Elman, *On Their Own Terms: Science in China, 1550–1900*, Cambridge, MA: Harvard University Press, 2005, 58.

43 Ch'a Munsŏng 차문성, *The Formation and the Transformation of Modern Museums* (근대 박물관 그 형성과 변천 과정), Seoul: Hankukhaksulchŏngbo, 2008.

44 Ch'unp'a 春坡, "What You Have Forgotten in Touring Seoul (서울구경 왓다가 니저버리고 가는 것)," *Pyŏlkŏnkon* (별건곤) (9) (1929): 129–31, 129, 130; On various Korean critiques even concerning display methods and the ineptitude of Japanese guides to exhibitions, see Todd A. Henry, *Assimilating Seoul: Japanese Rule and the Politics of Public Space in Colonial Korea 1910–1945*, Berkeley: University of California Press, 2014, Chapter 3.

45 "The Need to Reform Science Education (科學敎育改善의 必要, 敎育者 諸氏에게)," *Dong-a Daily*, January 9, 1924. Such calls became more frequent in the 1930s. For example, see "Educational Values of Science Education for Children (兒童科學敎授의 敎育的價値)," *Dong-a Daily*, February 4, 1932; "The Delights and Pride in Collecting Living Things of My Hometown (내 고장의 생물을 모아보는 기쁨과 자랑)," *Dong-a Daily*, May 28, 1933; "Collecting Beautiful Plants from Spring Mountains and Fields (봄의 산과 들에 나는 아름다운 식물의 채집)," *Dong-a Daily*, April 21, 1935.

46 On the enthusiasm for natural history exhibitions and activities in colonial Korea, Kim Sungwon 김성원, "The Context of a Korean Naturalist's Career-building in Colonial Korea (식민지시기 조선인 박물학자 성장의 맥락)," *Journal of the Korean History of Science Society* 30 (2008): 353–82, 370–75.
47 Kwŏn Tŏkgyu 권덕규, "Finally, Korean People Should Become Their Own Pride (마침내 조선 사람이 자랑이여야 한다)," *Kaebyŏk* (July 1, 1925): 18–21, 19.
48 Like everything else under colonial rule, this 'nationalistic' love for things Korean, which shaped the "Korean studies" movement in the 1930s, was shaped in interaction with Japanese imperialism. Yi Chiwŏn 이지원, *Modern Cultural Thoughts in Korea* (한국 근대 문화사상사 연구), Seoul: Hyean, 2007; Ku Chaejin 구재진 et al., *The Formation of 'Things Korean' and Modern Cultural Discourse* ('조선적인 것' 의 형성과 근대문화 담론), Seoul: Somyŏng, 2007.
49 Abe Yoshishige also expressed much displeasure at colonial policies ruining such beauty of Korean nature. Ch'oe Chaech'ŏl 최재철, "Abe Yoshishige, Keijo Imperial University, and Intelletuals in the Colonized Seoul (경성제국대학과 아베 요시시게, 그리고 식민지 도시 경성의 지식인)," *Japan Research* (日本研究) 42 (2009): 261–79.
50 Yi Kilyong 이길용, "Mountain Climbing and Swimming: Self-Cultivation in Clear Sky (炎天의 修養 登山과 水泳, 夏期休暇特輯)," *Tongkwang* (July 5, 1927): 20–22, 21.
51 Yi Ch'ung-u 이충우, *Keijo Imperial University* (경성제국대학), Seoul: Tarakwŏn, 1980, 93.
52 Lee Deok-Bong 이덕봉, "Plants in the School Ground (학교 구내식물 이야기)," *Paehwa* (배화) 2 (1926): 95–110.
53 Oh Sŏngchŏl 오성철, *The Formation of Colonial Elementary Education* (식민지 초등교육의 형성), Seoul: Kyoyukkwahaksa, 2000. "Expansion of Whimoon Higher Elementary School, 70,000 Yen for Classrooms and 10,000 Yen for Specimens, the Founder's Generosity (徽文高普大擴張, 칠만원을 드려 교실을 짓고 일만원으로 표본을 설비해, 校主 閔泳徽氏의 大奮發)," *Dong-a Daily*, June 13, 1921; "The Library Children's Room, Equipped with Fun and Educational Specimens including Telegraph (無線電信을 爲始하야 各種 標本을 備置하고 자미잇는 중에서, 유익한 지식을 엇도록 힘쓰는 아동실)," *Dong-a Daily*, September 2, 1923; "New Owner for Taesŏng Middle School, First Provided 5,000 Yen Worth of Specimens (大成中學校를 閔庭植氏가 經營; 만주에 큰땅을 사셔 농장 경영, 위션 오천원어치 표본을 작만)," *Dong-a Daily*, March 30, 1924; "Students at Ŭ March Schools on Strike, Unqualified Teachers and Poor Instruments and Specimens (義明學生盟休, 教師中에는 缺点이 만코 器具標本은 不備하다고)," *Dong-a Daily*, June 25, 1924, etc.
54 Chang's career showed how hard it was for a Korean to find a place to do scientific research. He worked with To Pong Sup and Ishidoya for a while and stayed in Yŏnhŭi College cataloging the specimens but could not find a permanent place to work. He became a professor at Seoul Normal School after liberation but died in 1949 during a police investigation amid the political turmoil of the time. Park Mankyu 박만규, "Overview of the Flora of South Chŏlla Province (전라남도 식물상의 개관)," *Korean Journal of Plant Taxonomy* (식물분류학회지) 1(2) (1969): 9–12, 9. "Accumulated Labour of 10 years and 20,000 Yen's Expense, 7,000 Plant Specimens (二萬圓의 巨費와 十個年의 積功, 植物標本 七千餘種 光州)," *Dong-a Daily*, January 21, 1933; "Cataloguing of the Yŏnhŭi College Specimens Done (延專植物標本 整理完了)," *Dong-a Daily*, November 28, 1933.
55 "An Unknown Young Scientist Donating 858 Insect Specimens to Sungŭi as Research Material (숨은 青年科學者 崇專에 昆蟲標本 寄贈, 八백五十八점을 연구재료로)," *Dong-a Daily*, April 10, 1934. There was another Korean who searched for a place to donate his tens of thousands plant specimens. Park Mankyu 박만규, "Overview of the Flora of South Chŏlla Province (전라남도 식물상의 개관)," *Korean Journal of Plant Taxonomy* (식물분류학회지) 1(2) (1969): 9.

5 Becoming Japanese through Collaboration

Figure 5.1 shows the students and teachers of the Keijo Higher School of Pharmaceutics, the biggest professional college in colonial Korea, founded by the pharmaceutics industry. They had gathered for one of their regular field trips to a Korean mountain (See Figure 5.1). Among the many Japanese students, there were a few Korean students and even one Korean professor who organized this botanical field trip. Toh Bong-Syup (1904–?), a Tokyo Imperial University graduate in pharmaceutics, was the lucky Korean who secured this rare professorship for a colonized talent. In the photo, Toh is the one with crossed arms in the second row, bottom right, and the one sitting beside him with a stick is his Japanese colleague. However,

Figure 5.1 A photo taken from a field trip of the Botanical Club of the Keijo Higher School of Pharmaceutics (Courtesy of the Toh family, Toh's second daughter Toh Chung-Ae).

DOI: 10.4324/9781003511755-6

it is impossible to tell from this figure how many Korean students there were in this group: thus, there were no visual characteristics that distinguished Toh from his Japanese colleague. Koreans were born assimilated to their imperial rulers.

In addition, the Japanese colonial regime often professed the aim of assimilating uncivilized Koreans.[1] In this cheerfully harmonious photo taken in this non-Western empire's adjacent colony, could Toh then forget his position and image as a Korean, unlike the "racially" different colonial subjects in Western colonies? Or, was it any Koreans' wish to pass and live as Japanese in pursuing the enlightened life of a botanist? Or were they allowed to have their own wishes in such matters? This chapter closely examines the 'civilizing paths' followed by two Korean botanical researchers, in order to delineate the complexity of the botanical enlightenment of Koreans through collaboration with Japanese colonizers. The relatively well-documented paths of Toh and another Korean, Chung Tyaihyon (1883–1971), the subordinate of Ishidoya Tsutomu (1891–1958) and assistant in forming Nakai Takenoshin's (1882–1952) collection, will be traced. What did it mean for Korean researchers to dare to work with the Japanese to become a botanist and what was entailed in pursuing and maintaining that collaboration? As the collaboration under the asymmetry of power as a colonized subject was challenging at every step, Toh and Chung must often have to ask themselves what it meant to be a Korean and a botanist. Their stories reveal the complex textures of colonial subjectivity and nationalism in botanical collaboration.

Helplessly Korean

In discussions of Chung Tyaihyon's successful life as "the father of Korean plant systematics," no one has ever noted his relationship with Ishidoya. It was not that Korean scholars have tried to hide his collaboration with a Japanese. They simply assumed that his success was owed to his "apprenticeship" to Nakai, the international expert on Korean flora at Tokyo Imperial University, who allegedly had only Chung near him during his collecting trips in Korea while avoiding all other amateur naturalists in the colony, including Japanese. Chung was able to become a systematist, it is said, because he had this unusually direct access to the metropolitan, i.e., valid, knowledge practitioner in spite of his meager education and modest work as a forest technician for the Government-General of Korea.[2] Although Chung did not deny his significant association with Nakai, the shaping of his botanical research and his own words reveal a different story; he appreciated Nakai for "years of on-site lessons" and Ishidoya for "the guidance that I could never forget until I die."[3] By closely examining his interactions with both Nakai and Ishidoya, this section reevaluates the overestimated importance of his remote relationship with Nakai, while showing what this entailed for Chung's collaboration with these different Japanese counterparts.

Chung seemed to take pride in being a descendant of *yangban*, members of the literati class whose members once had the almost exclusive privilege of becoming government officials if they passed the civil service exams. This privilege, however, disappeared in 1895 when the reformist Korean government abolished the exam system and introduced a modern education system. Enacted under the sense of crisis brought about by Western expansion toward Asia, this was one of the modernizing measures largely inspired by the Japanese model. Finding himself without a purpose in life, Chung stopped his study of Confucian classics and took over the family farms, although he still maintained the literati practice of holding poetry gatherings in classical Chinese.[4]

In 1905, when Korea became a protectorate of Japan, Chung went to Seoul to begin his study of Japanese at the age of 23. He said that he became "weary of *pre-modern* farm management" and decided to learn some "practical knowledge." This decision was not only taken in order to modernize his own farm management but also, he reflected, for his "poor and powerless country."[5] This move was prescient, as it led to his being awarded a one-year government scholarship to study at the Suwŏn Agricultural and Forestry School in 1907.[6] Taught by Japanese teachers, the course was designed to train staff for the Japanese development project for Korean forestry resources, as mentioned in Chapter 3.

In accepting, if not eagerly seeking, Japanese tutelage in the name of his "poor and powerless country," Chung's reasoning seems not unlike that of the famous nationalist 'defector' Yi Kwang-su (李光洙, 1892–1950). Yi worshipped Fukuzawa Yukichi as "a hero given to Japan as if the heaven willed to bless Japan." Such praise was given after Fukuzawa's call for "Escaping Asia towards Europe" which justified the subjugation of benighted Korea, as simply following "the custom of Westerners."[7] Yi fully accepted Fukuzawa's logic that such an imperial aggression was a natural exercise of Japan's enlightened and civilized identity, equal to the Western one. While Chung said that he felt "humiliation and fury" at seeing Japan's victories over Qing China and Russia, he did not waste his time in expressing moral indignation but chose to cut off his topknot and learn Japanese. What mattered was might, not right. Envying rather than criticizing Japan's exercise of power, these reformist nationalists tried to become Korean Fukuzawas, by aggressively denying the worth of their own past and emulating Japanese, professedly for the benefit of their poor and powerless country.

This obvious envy, and feeling of lost chances at the alacrity and efficiency of Japan's emergence as an imperial power, was an acknowledgment of Korean failure in comparison with this shrewd neighbor. Chung showed that this sense of failure or inferiority was only deepened by accepting Japanese tutelage in Western science and technology. Chung reflected about his forest school days: "Koreans did not perform well in natural science like physics, chemistry, biology, and mathematics—subjects based on rationality."[8] This may not be so surprising. Such subjects were too foreign to the first group of students who were ill prepared for the one-year intensive course they were following; they had little elementary education and inadequate Japanese. Nonetheless, Chung had internalized this as Korean people's lack of rationality, as his Japanese teachers told him. He also seemed not sympathetic to his classmates who, in tears, refused to cut out their topknots. Saying that cutting one's hair given by parents was against filial duty no longer seemed to him to be rational. People who felt that seemed to him not to be serious about learning the rational and empowering discipline of forestry science.

Chung passed this challenging course with only six other classmates, becoming a forester in 1908 at the Suwŏn forestry office, the principal one of the five that Japan had established in major cities the year before. There he participated in afforestation projects specifically designed to "demonstrate the effect of the new [Japan-guided] administration." They chose "the mountains, the riversides or the railway sites where the public would easily see" for such demonstrative afforestations.[9] As a serious believer in the Japanese civilizing mission, Chung seemed not to mind this showy strategy or the effective demolition of rich forests that was simultaneously occurring in the North. To him, this was the scientific approach that Koreans had never before witnessed, "the beginning of forestry in Korea."[10] He apparently made learning and practicing this Japanese knowledge his vocation; he did not leave the forestry department after the GGK took control of Korea in 1910. It was a difficult decision, as the colonial government demoted him to a manual worker and severely reduced his salary. But Chung's acceptance of the Japanese civilizing mission in Korea, not questioning the Japanese assessment of Korean inferiority and its unfair condition, remained resolute.

It is worth noting that Chung had some fortunate encounters leading him not to be entirely disillusioned about the Japanese civilizing mission. Ishidoya was not the only Japanese who came to Korea believing in his nation's civilizing rhetoric, no matter how patronizing and limited that vision could be. First of all, among the teachers who indoctrinated Chung with Korean irrationality, there was one exception, a forestry teacher named Ueki Homiki (植木秀幹, 1882–1976). Ueki worked at the Suwŏn Agricultural and Forestry School from 1907 until 1945, and held a position at the Forestry Research Institute starting in 1925. Korean students adored him because he did not discriminate against them and often intervened on their behalf. It seems that he did not condemn Korean resistance to Japanese measures as irrationality. He experienced this resistance in various forms throughout his long sojourn: the Korean students' refusal to have their topknots cut off in 1907, the independence marches in 1919, the Korean students' strike against discrimination in 1926, etc. When he left Korea in 1945, he congratulated his Korean students on obtaining independence from Japan.[11]

It is hard to concretely measure Ueki's contribution to the development of Korean scientific talents. Many researchers-cum-teachers in Korean private middle schools noted their collection trips with Ueki as first serious scientific experiences. Some of them, who became professors at their alma mater after further studying in Japan, although at lower salaries and subsidiary ranks, acknowledged Ueki's encouragement and help in going through those experiences.[12] Ueki's support for Chung's scholarly development is shown in his favorable preface for Chung's first monograph in 1943.[13] He was the first Japanese who made Chung believe that the civilizing mission could be more than rhetoric, and that he could empower himself by working with Japanese.

Chung apparently also had a high expectation that he could learn proper Western botany from Nakai. This great scientist at Tokyo Imperial University had already produced authoritative looking monographs on Korean plants in languages that Chung could not yet access. It was a great opportunity for him to have direct contact with Nakai for his mission of learning Western botany through Japanese guidance. Chung did his best to maintain a cordial relationship with Nakai. Despite Ishidoya's increasingly open critique of Nakai's classification, Chung never directly commented on it. To some younger Korean botanists, he was noted for his insistence on keeping Nakai's classification, even after independence from Japan. But while it might have seemed from this that Chung's close relationship with Nakai had made him the heir of Nakai's systematics, there are many instances of words and actions by both Chung and Nakai that suggest otherwise.

Nakai and Chung were close. Yet two facts used as evidence for this closeness both leave some room for interpretation. The first is that Nakai stayed at Chung's house during his visit to Korea. Although true, this seems to be a practical arrangement perhaps made by Ishidoya, who noted that there was no good lodging option near the suburban Forestry Research Institute. Nakai also stayed at Asakawa Takumi's (1891–1931) house. Second is the already mentioned claim that Nakai was closer to Chung than to Japanese colonials. In fact, it was not rare for some Japanese colonials to accompany Nakai on his collection trips, as shown in Figure 5.2. In the photo on the left of Nakai's collection trip to the Ŭllŭng Island, also shown in Chapter 2, the figure standing behind Nakai, seated at the center, is Chung, although the caption misspells Chung's name. Two others sitting beside Nakai are Japanese personnel. There are ten figures in the photo on the right taken during a collection trip to Mt. Paekdu, but only eight, including Nakai, marked with a C in the center, have markings and names, and all of them are Japanese names. As Nakai later mentioned that he and Chung "together escaped from the attack of rioters and the clutch of bandits in the Paekdu Mountain area," Chung must be one of the far right figures unidentified.[14]

Figure 5.2 Photos from Nakai Takenoshin's Collection Trips to Ullŭng Island and Mount Paekdu. *Report on the Vegetation of the Island Ooryongto or Dagelet Island, Corea*, Seoul: GGK, 1919, 10; *Florula of* M't Paik-Tu-San. Seoul: GGK, 1918, 41 (The Biodiversity Heritage Library and Seoul National University Library).

If Nakai usually had Chung near him throughout the collection trip, this may reflect rather necessity than personal affection, as he seemed friendlier to those Japanese companions that he named individually. Nakai just seriously needed this Korean who handled the complex logistics of the collection trip. Chung was in full charge of food, tools, porters and other laborers as well as specimen preparation, often working overnight without much sleep. He also prepared and managed itineraries, dealing with both Japanese officials and local Korean communities.

If Chung was at least within sight during Nakai's collection trips, his presence was almost unnoted in the works Nakai produced from those trips. While Nakai duly provided names of collectors for each new plant that he described, as noted in Chapter 3, he never mentioned Chung's name during these years. Only Japanese names, including that of Ishidoya, appeared. When Chung's name first appeared in 1921, it was not as a collector. Nakai noted, in naming a new species of abelia after Chung, as *Abelia Tyaihyoni* Nakai:

> [Chung] followed me as an interpreter and assistant since the year after I became the government botanist for the GGK, i.e. 1914. Until 1919, we had travelled about 1,100 kilometers of the most dangerous terrains and had several near-death experiences together. We together escaped from the attack of rioters and the clutch of bandits in the Paekdu Mountain area; he saved me from falling from a cliff in the Kŭmkang Mountain. While dealing with ignorant laborers, he was meticulous in making specimens and taking care of all provisions. As the study of Korean flora owed much to him, I named this plant after him as *Abelia Tyaihyoni* to memorialize his contribution to my collection, although the International Botanical Congress resolved to name after only botanists or collectors in 1905. It was collected with Chung Tyaihyon on July 16, 1919.[15]

This looks like an affectionate acknowledgment of Chung's help, and it has often been quoted to show how Nakai cherished Chung. Yet, it is also notable that Nakai provided such a detailed explanation to justify his dedication to a mere interpreter, unrelated to botany. Although Chung was at his side learning botanical practice and thought of himself at least as a collector, Nakai did not see it that way. Nakai put his own name as collector for all those specimens, as for this abelia, apparently because he saw himself as having guided Chung's collecting.

Chung's name was placed on the list of collectors only after ten years of service to Nakai, in 1922, when Nakai reported *Corylus hallaisanensis*.[16] In 1926, Nakai helped Chung report *Abelia mosanensis* Chung, by giving the Latin translation as Chung requested. In 1929, he noted that it was Chung who first identified the new species *Fraxinus chiisanensis* Nakai. In 1932, Chung first appeared in the preface of *Flora Sylvatica Koreana*, for providing Korean names for the plants.[17]

Nakai's positive acknowledgment of Chung, however, seems to end here. In 1937, Nakai suddenly said that it was due to Chung's "begging" that he reported *Abelia mosanensis* under Chung's name, as though it were something he would rather not have done. It was in that year that Chung and other Korean naturalists published their *Vernacular Name Catalogue of Korean Plants* (朝鮮植物鄉名集 Chosŏn shingmul hyangmyŏngjip), which will be discussed in the last chapter. In 1940, Nakai turned down Chung's request to contribute a preface for his monograph on trees of Korea. According to Chung, when he showed Nakai a preface that he instead received from Ueki, Nakai threw it away.[18] Ueki's preface praised Chung's diligence in publishing this handy illustrated volume with useful information, but Nakai seems not to have approved of Chung's autonomous knowledge practice.

Chung, who obviously did not displease Nakai with his hard work and excellent coordination skills, revealed his distance from Nakai in his autobiography. Among the only two mentions that he made about Nakai, one was to criticize Nakai's Japanese sea continent theory, quoting and concurring with Ishidoya.[19] The second mention was about the discovery of a new fir, *Abies nephrolepis* in 1917, not by Nakai but by an American who accompanied them, Ernest H. Wilson (1876–1930), the cherry expert and collector for the Arnold Arboretum. Wilson, by realizing this fir's unique fruit shape, came to name the new species. Chung said, "I remember how Dr. Nakai, who had some in depth knowledge of Korean flora at the time, was stamping on the ground with chagrin that he had lost the honor." Chung does not appear to have sympathized with Nakai's defeat, which he chose to document vividly in his recollection.[20]

Nakai and Chung worked closely together but it seemed this was mostly out of necessity. They were constantly negotiating to come up with compromises from their conflicting terms of collaboration; Nakai did not want a Korean apprentice to his systematics but a docile hand to enable his remote botanizing of Korea while Chung's goal was to learn some authentic botany. Undoubtedly, both of them obtained what they wanted from their relationship. If Nakai wanted a useful collector, it was necessary for Nakai to train Chung to a certain degree, giving "on-site lessons."[21] However, it appears that there came a point at which Nakai did not want Chung's searching eyes, to say nothing of his intellectual independence in classifying and naming on his own. Starting in the mid-1930s, other Koreans than Chung were enlisted in Nakai's collection trips.

Compared to these tense exchanges with Nakai, Chung's interaction with Ishidoya and Asakawa was more relaxed and transformative for both parties. It seems to be have been somewhat characteristic of Ishidoya that he did not treat this studious Korean as a mere laborer. From the beginning, he chose to train his Korean colleague, who was about ten years older than himself, in order to have his assistance in his own ambitious yet underfunded studies. He decided to study plant classification with Chung, as a necessary part of his job as a forester. He had to be able to identify plants in this unfamiliar land in order to afforest it. They collected specimens and made notes together from 1911, as Chung fondly remembered. In 1913, Ishidoya recruited Chung for his newly organized laboratory and assigned him as an interpreter-cum-assistant for Nakai's collection trips, although it was not as though Ishidoya had other choices.

However, Ishidoya's relationship with Chung seems to have changed when Asakawa, the Japanese subordinate, joined in their small lab in 1914. Not that

Ishidoya was ever seriously unfair to Chung: Ishidoya treated him quite equally with Asakawa in terms of salary and promotion, rather against the custom at the GGK.[22] However, Ishidoya made only Asakawa his coauthor for the important papers that he presented to Japanese experts, in spite of Chung's active participation in the relevant transplantation experiments.[23]

While he and Asakawa were immersing themselves in figuring out the colonial land and the climate and the new strategies of transplanting trees, Ishidoya set other tasks for Chung. In 1916, the colonial government launched an investigation into traditional medicinal plants, in a scheme intended to make this economically unrewarding colony pay for its colonial rule.[24] Ishidoya mobilized Chung for this project. Chung may have been disappointed that he had to spend time working on the tradition that he had rejected as impractical, if not irrational. Yet, he was ready to reciprocate Ishidoya's teaching, and owing to his study of Confucian classics up to his teens and his continued enjoyment of poetry in classical Chinese, he proved to be a capable informant. Chung appears to have made great efforts to locate useful Korean texts. He not only brought out rather widely-circulated medical books including *Treasured Mirror of Eastern Medicine* but also presented the then-unknown text of *Farm Management* (山林經濟 Sallimgyŏngje, ca. 1700), based on which the 1916 report on 169 commonly traded herbs in Korea was produced.[25]

Ishidoya did not acknowledge Chung's contribution in his report.[26] One could note Ishidoya's inconsistency, remembering his criticism of Nakai's under-acknowledgment. It is more likely, however, that Ishidoya found nothing to acknowledge in what Chung had to offer, which he found to be shocking than enlightening. He deplored the fact that Korean people were still "searching for roots of grass and bark of trees," and that these roots and barks were even traded in specialized herbal markets for medicinal purposes. In Japan, such practices had safely relegated to the past. Ishidoya assessed that "in Korea, the benefits of advanced medicine provided by the colonial government were not yet widespread enough for average people to enjoy." For this civilizer of modern science, the task that he identified on seeing Chung's laboriously compiled list of traditional medicines in Korea was to help develop new medicines from these sources, through "the advancement of chemical research."[27]

Yet even as he deplored the conditions of Korean herbalism, Ishidoya found something quite intriguing in his first encounter with traditional Korean texts through Chung. The academic lingua franca of premodern Japan, Korea, China, etc.—classical Chinese—had strikingly different usages regarding plant names, depending on which country the text originated from. In each, they used different Chinese characters for the same plants. It could be a serious issue for his forestry work. He conducted a systematic investigation in 1917 by comparing Chinese character names for trees found in all three countries. For common trees such as Japanese cypress, silver magnolia, fir, cinnamon tree, and camphor tree, the Chinese characters used in the three countries were all different. For example, "檜" mainly referred to spindle tree or *Euonymus sachalinensis* in Korea, cypress tree or *Chamaecyparis obtusa* in Japan, and fir tree or *Abies holophylla* in China. Furthermore, in Korean texts matters were further complicated by the transliteration of vernacular names into Chinese characters. Ishidoya warned other Japanese experts that they could not obtain exact information from Korean texts without confirming with someone familiar with the local names of plants.[28] He realized that the mediation of Koreans like Chung, who knew the meanings of those Chinese characters in their knowledge tradition, was crucial.

In 1918, when Ishidoya made a follow-up report on herbal medicine, he again relied on Chung's assistance but not for a further investigation of old texts. Ishidoya, with Chung as an interpreter, decided to interview contemporaneous doctors and medicinal dealers. Curiously, these Korean doctors, almost regardless of region or social standing, said more or less the same thing about any given medicine. The information was said to be derived from the seventeenth-century text *Treasured Mirror* to which Chung had drawn attention in 1916. Ishidoya made much of this unanimity, but saw it as "Korean doctors' lack of empiricism and slavish reliance on the old authority." The existence of a standardized traditional knowledge practice did not impress this missionary of modern science. Chung's help was again unacknowledged.[29]

Yet, the nature of Ishidoya's interaction with Chung was changing, as he came to rely on Chung's knowledgeable help in areas with which he was not familiar. Furthermore, Ishidoya's trusted Japanese subordinate, Asakawa, showed him different ways to see the colonized people's traditions. Asakawa seems to have been a person of exceptional sensitivity and sympathy. After being relocated to Korea, he considered immediately leaving the colony, because he felt so sorry for Koreans, witnessing the everyday atrocities inflicted on them by the colonial regime. While he could not just abandon the colonial life that he had begun with his brother and mother, he decided to express his sympathy for Koreans in various ways, like studying Korean language and arts.[30] Furthermore, after Korean people expressed their discontent about the Japanese regime in the nation-wide movement of 1919, Asakawa began to wear Korean clothes and a Korean style beard to demonstrate how Japanese police pushed around anybody who appeared to be "Korean" for no good reason. Ishidoya was fully aware of Asakawa's conduct as Asakawa almost failed to deliver his lectures to local foresters because the police tried to prevent this daring "Korean" from approaching the podium. Apparently, Ishidoya did not reproach Asakawa for such behavior, which would have been frowned upon by other Japanese settlers.[31]

The report that he and Asakawa produced in 1919 signaled Ishidoya's changing views of colonial institutions. As in Japan, the GGK decided to designate natural monuments in Korea, professedly to cultivate the love of nature among the colonized.[32] This project might have been the occasion for a report that revealed the colonized people's inadequate appreciation of nature, and of forests in particular, and stressed the contrast with the long tradition of Japanese appreciation of such things. Yet from all over the country reports poured into Ishidoya's department, showing that Korean people had also long nurtured love for 'nature,' leading to the compilation of a list of 5,300 especially loved and protected trees by region. The co-authored report by Ishidoya and Asakawa listed these monumental trees by province with the reason of their longevity or fame. They found that these deeply-rooted trees survived not simply due to their adaptation to Korean soil but by becoming part of Korean people's lives. These trees had received special communal protection because they offered shade, gathering places, spiritual guidance as well as aesthetic enjoyment and collective memories to the regional people.[33] The report did not dismiss these reasons as superstition or irrationality but appreciated them as part of Korean culture that had ensured the longevity of so many trees. With this new understanding of colonial culture, Ishidoya came to see that soil, climate, and ecology were not the only foundation for trees.

This report's appreciative tone on Korean tradition was probably not just due to Asakawa's influence. In 1922, Ishidoya left an intriguing remark, published both in Japanese and in Korean, that appears to reveal his sympathy with Asakawa's concerns. He said, "Things have different meanings depending on from where and how they are seen." While forestry was a legitimate effort to "produce the maximum yield with the minimum investment," trees and forests, living things with their autonomous life purposes, would not "find such intervention very amusing although they may appear thankful for such efforts for cultivation."[34] Before making his more all-out critiques on colonial forestry policies for its disregard of local nature and culture in the 1930s, he was developing this kind of perspective by reflecting on the benefit of intervention as seen by the intervened. By analogy, he seemed to realize that his sincere civilizing efforts, and needless to say also the overbearing attitude of other colonizers that saddened Asakawa, would be seen very differently by the recipients like Chung. Ishidoya did not come to Korea to be a "vulgar" colonizer of the speechless colonized, who like Chung just wanted to learn and did his best for reciprocation.[35]

Ishidoya's changing perspective on Korean tradition and his soul-searching about colonial relations became apparent in his changing relationship with his Korean protégé. In his last work as a forester, *The Identification Keys of Korean Forest Trees*, published in 1923, Ishidoya made Chung his co-author for the first time. *The Identification Keys* further reveals that his disillusionment about Nakai's slights on colonial experts played a significant role in this change. The preface presented the book as a guide for "fieldworkers" like themselves, to "complement what was missing for fieldworkers and collectors" from Nakai's *Flora Sylvatica Koreana*. Since it was written in "foreign languages," Nakai's *Flora* was quite inaccessible and, since it was full of sumptuous illustrations, it was too expensive. Furthermore, it was an inadequate guide for identifying plants in the field. They needed plant identification keys based not on dried specimens but on living plants. In their guide, they made the distinction between shrub and tree, and used a tree-specific common category (consisting of needle-leaved, broad-leaved, and bamboos) to indicate the overall shapes of trees. While the morphological characteristics of leaves, flowers, and fruits constituted an important part of their system, as it did for Nakai, they did not care about the details except those "permanent characteristics discernible to the naked eye in the field." They provided a tree's usual habitat, soil type, and other plants often observed together as more helpful tips for field identification.[36]

Ishidoya's changing attitude to his work with Chung was further evidenced when he gave Chung a chance to work for what Chung remembered as the most important among his projects at the forestry department.[37] This was an ambitious project that Ishidoya had launched at the expanded institute in 1922. Known as "The Right Tree for the Right Land (適地適樹)" investigation, it aimed to find the most suitable trees for each region by considering the climate, soil, and ecological condition of the land together with physiological characteristics of trees, thus neatly addressing all of Ishidoya's concerns. Although Ishidoya left the forestry department before he had finished this dream project, it seems that he decided to make Chung a full-fledged researcher before his departure. He gave Chung, instead of Asakawa, who was also involved, responsibility for the department's interim report in 1925. This was a mark of recognition and trust by Ishidoya for which Chung expressed his deepest gratitude. It was the outcome of a long-term and mutually transformative interaction.

Chung's first and only independent report at the department, *Distribution and the Right Land for Important Korean Trees*, came out in 1925. This independent work clearly illustrates Chung's approach to forestry. The defining characteristic of Chung's report was its complete absorption of Japanese sources, including data accumulated at the expanded Institute. The report provided suitable soil characteristics and the geographical distribution of each tree, accompanied by a map that showed the vertical as well as horizontal limits of a tree's habitat. Notably, for the use of trees, he utilized various Japanese references to enumerate the specific Japanese uses of 77 representative Korean trees—a decision that stood in marked contrast with the approach taken by Ishidoya in the *Identification Keys*. The approach of that work on the use of trees was special, as it focused on specifically Korean use of trees obtained through Chung's fieldwork with vernacular names, for which Ishidoya credited him. It discussed how Korean people made cookies out of pine pollen and valued nut pine trees for coffin-making, and how "Korean doctors" used *acanthopanax* plants for tuberculosis and stroke, and hawthorn fruits for indigestion, etc.[38] Chung did not utilize any of this information that he himself had laboriously gained. He researched to find specifically Japanese uses for the trees he discussed, such as furniture or tools that could be made from those trees, with the clear aim of showing that Korean forest products could find a promising market in Japan. In fact, the audience for the text was a group of Japanese foresters visiting Korea for the annual meeting of the Forestry Association of Greater Japan, being held for the first time in Korea. Chung, by corroborating the various Japanese uses with the "objective" data of the wood's strength, elasticity, and so on, just fully displayed his ability to use the data obtained through "modern" investigations.[39] With no tinge of Korean-ness, Chung presented himself as a reliable practitioner of Japanese science.

Unfortunately, this comprehensive and clear 50-page report signaled not the beginning of Chung's successful career as an independent researcher but its end. The exceptionally fair treatment that Chung had enjoyed rapidly evaporated after Ishidoya left the department in 1925 and Asakawa died prematurely in 1933. That same year, although Chung had chosen to stay as a contract worker at a much-lowered salary, he lost his position in the department.[40]

In keeping his contract position, Chung resumed his engagement with the Korean tradition. The colonial government re-launched a serious investigation and cultivation project of Korean medicinal plants in 1933, at a time when Chung desperately needed a job assignment. Chung succeed in being assigned to work on this project, apparently having convinced his new boss of his expertise in Korean medicine by producing an article that was essentially rehashed from Ishidoya's 1916 and 1918 reports on the topic.[41] Chung had to accept being assigned to work on a knowledge tradition that he regarded as inferior, even irrational, as it was almost the only thing that the colonial government wanted from him. It did not recognize him as a capable practitioner of Japanese science but relegated him to the status of a reliable native informant, mainly useful for investigating colonial resources for development opportunities.

In pursuing his scientific life under the colonial regime, Chung seems to have believed that absolute submission to Japanese needs was the only attitude that a colonized person could safely assume; his two dismissals, several demotions, and many more abuses must have been a fearful reminder of Japan's power over a Korean like him. When he showed that he fully internalized the Japanese imperial ideology that asserted Japan's superiority over irrational Koreans, his acceptance was induced by this fear and helplessness, which had little chance of escaping owing to the immensity of his Japanese counterparts' power over his fate. In this drastic asymmetry of power, he seemed ready to work under any conditions. Yet the last condition for collaboration at the GGK must have been the most unbearable to him, since he had to study the tradition he despised instead of the modern Japanese science that he wanted so much. It was clear that he had to stay as a Korean.

As Good as a Japanese

In comparison with Chung, who grew up in benighted Korea and entered civilized society under the Japanese rule as a benighted colonial being, the next generation of Koreans like Toh Bong-Syup had the chance of a better lot. That is, they might even dream of being really like a Japanese. They had the 'privilege' of starting their lives in a world enlightened by Japanese rule. They could learn Japanese as a child and assume an immaculate Japanese look—if, of course, they could afford the education and proper modern clothes. However, the possibility of assimilation existed mostly in rhetoric. Fearing that it would be difficult to distinguish between Japanese and Koreans, Koreans were not allowed to Japanize their names, until the GGK forced it on them as part of the war mobilization in 1940. If the colonized chose to civilize themselves by studying science like a Japanese, the path became very challenging. The difficult path that Toh, one of the most fortunately situated Koreans of the day, had to follow in order to pursue his scientific education and research exemplifies how much struggle was necessary for any Korean who might mistakenly believe that scientific enlightenment was a possible option for them.

In pursuing his scientific training at the metropolitan Tokyo Imperial University, Toh like Chung appeared to have embraced the validity of Japan's civilizing mission. Toh, born in 1904 into a wealthy merchant family in Hamhŭng, the central city of the South Hamkyŏng province in northern Korea, was of the first colonial generation that could receive a full modern education. Although it was not at all impossible to buy the privilege of the yangban class even during the Chosŏn era, the abolition of the state examination system that unsettled Chung's life path meant different things for Toh's merchant family. Toh's father allowed his children to choose the lives that they wanted and Toh, instead of inheriting the family business, chose an expensive education in Tokyo.[42] He wanted the best education that the empire boasted.

This was a daring initiative, apparently unexpected by the Japanese authorities. As discussed earlier, the Japanese empire maintained a three-tier system, with separate provision for educating colonized Koreans, Japanese settlers in Korea, and the domestic Japanese, almost to the end of its existence. One may say that it was fortunate for Toh that the original system was somewhat loosened with the onset of the so-called "cultural rule" in the 1920s after the setback of the March First movement. From that time on, Koreans had at least a theoretical possibility of taking the Japanese path in the system, so long as they could manage the laborious task of assembling all the extra bits required for Koreans to follow that. Toh was one of the few Koreans who made this possibility offered by the new system into a reality, through his steadfast determination and family wealth.

Toh, after graduating from a higher elementary school in his hometown in 1921, went to Seoul to study for two more years for secondary level education at a public higher elementary school intended for Japanese settler children. Yet, this education in a colonial Japanese school was not enough for him to qualify for studying science at one of the imperial universities in Japan; science curricula at the colonial secondary schools, even in schools intended for Japanese, were insufficient for that. He had to go to Japan to attend a proper high school with enough science courses. After all this effort, he was qualified for the competitive entrance exam, a new possibility given to Koreans just the year before, in 1924. Toh was one of few persons who went through this entire process in order to be allowed to take the exams for official admission to the Imperial University; most other Koreans who aimed at this goal chose the less costly method of using the foreign student transfer system, although it did not allow for a specific major or a degree. Due to his tremendous effort, he was able to major in pharmaceutics and become one of the five Koreans who graduated from the Tokyo Imperial University in 1930.[43]

However, it was even tougher for a Korean science and engineering graduate to obtain a research career. Furthermore, Toh graduated at the height of the Great Depression, thus lessening his chances. For a graduate from physics that year, it took five years to find a job as a teacher in a colonial middle school.[44] Nevertheless, Toh, a disciple of Asahina Yasuhiko (朝比奈泰彦, 1881–1975), known for his chemical analysis of traditional medicinal plants, secured a professorship in a pharmaceutical college in Seoul, the Keijo Higher School of Pharmaceutics. Although higher schools in Korea were mostly staffed with Japanese professors, this school, recently upgraded owing to a generous donation by a successful Japanese medicine dealer, named Wakamoto, found this Korean with the best education in the empire fit for the job.[45] Toh's choice of this booming field involving Korean traditions seems not unwise since it gave him a respectable research career back home.

Toh's employment shows such an advanced education was not seen as threatening by most Japanese when it was accomplished by such an individual effort. They seemed, rather, to celebrate Toh's achievement as a successful fruit of Japan's civilizing effort, although the reality of Toh's laborious ascent belies any such a self-serving notion. Whatever the case may have been, they were certainly ready to use his talent. Toh was thus re-located in colonial society, working independently in the newly equipped school lab, as if he was like any other Japanese scientist. Like Japanese researchers, he would be able to set his own research agenda if it satisfied his sponsors. However, what his powerful sponsors expected from Toh on the basis of his hard earned Japanese education cannot be described as expansive. Despite his much better credentials, the expectation for him seems to have been the same as it had been for Chung: a native informant, the only role conceivable for him from the viewpoint of colonially shaped relations.

The first Japanese who led Toh to this obvious task of studying Korean plants for Japanese development was none other than Ishidoya, who, as we have seen, had developed a favorable interest in Korean nature and culture. Ishidoya, then in the pharmacological section at Keijo Imperial University studying medicinal herbs, initiated the relationship in May 1930, soon after Toh's arrival. Ishidoya invited Toh to join him for a rare chance to collect plants in a former palace garden in Seoul, which, unlike the sacred Japanese equivalent, had been partially transformed into a public park by the GGK. Toh readily accepted Ishidoya's invitation, recording this first meeting as the origin of his botanizing in Korea, which he came to conduct more vigorously as of 1932 by organizing a research group in his school, the Botanical Club of the Keijo Higher School of Pharmaceutics (Botanical Club hereafter). In the meantime, it was just the two of them with "no other kindred spirits." They met almost every month, collecting several times more at the restricted part of the palace garden and then moving out to the fields and mountains around Seoul.[46] They obviously found each other good company. For Ishidoya, Toh must have been a useful interpreter as well as a connection to the metropolitan pharmaceutics network. In well-established Ishidoya, Toh found an experienced teacher, who could guide him to his own country's nature and culture.

The result of their collaboration was the "Florula Seoulensis," a regional flora for Seoul published in 1932, which was full of praise for the beauty of Korean plants. It praised native Korean species in comparison with Japanese ones, or Japanese transplanted Western species. For example, it contended that whereas there was no interesting lilac in Japan, Korea had many native lilacs "comparable to the best lilacs in Europe in their beauty and fragrance." Also, "the native Korean silver poplar adds colors to the suburban landscape by sprouting its emerald-green leaves, when the cherry blossoms have had their time." In contrast, the North American poplar, transplanted as a street tree by the GGK, "sticks out like an upended broom to frighten Japanese who first visit Korea."[47] Korea, if not ruined by Japanese policies, was floristically more beautiful and richer than the Japanese mainland.

It could not have been difficult for Toh to work with a Japanese who was so openly critical of Japanese policies, and so attached to Korean nature. Yet, if Ishidoya's attachment to Korea was a choice, it was less so for Toh. When Asahina, Toh's teacher at Tokyo Imperial University visited Korea in 1934, the expectation that his Japanese counterparts had of him became clearer. Asahina, on a sweeping journey to investigate Chinese and Korean medicinal plants during his summer vacation, asked his Korean disciple to guide his investigation in Korea. Toh accompanied Asahina's survey of the South Hamgyŏng Province, "the alpine region known for the production of medicinal plants since the Chosŏn period."[48] It was Toh's home region. Asahina must have been pleased to have produced such a reliable informant rooted in this important region.

Yet, in his report, Toh did not mention that this was his native region. Instead, he emphasized the pre-investigation of the region that he made with the Botanical Club preparing for Asahina's arrival. He presented that experience, not his birth, as his credentials for being a guide. This apparent attempt not to reveal his link by birth to Korea appeared in all his reports on Korean medicinal plants that he sent to a journal of the pharmaceutical association in Japan.[49] First, if he asserted his familiarity with the regional plants, it was attributed to his wide collection experiences; he emphasized how he had explored almost every weekend various mountains and many forests since his return to Korea. He stressed these close contacts with the regional flora especially when he asserted his authority over less well-based reports from the colony, often written by Japanese bounty hunters. Many of them claimed to have discovered valuable medicinal plants from the colony. Yet, on investigation, he found most of them to be of dubious validity. They mistook a white sandalwood for a juniper, a lilac for a clove, and a claim to have found tons of guaiacum, used for in the treatment of syphilis, was a mirage. He could reveal these mistakes not because he was a Korean but thanks to his meticulous examinations in the field. Also, while it was rather common even for Japanese experts or collectors to mention Korean custom or traditional texts in making their claims, Toh did not do so. He instead distinguished himself by his connection to the metropolitan scholarly network and development. He could confer with the best experts in the field, like Asahina and Ishidoya, before cautiously presenting his latest findings.[50] Toh played his role as a "native" informant, but he seemed to reject any claim for qualification based on his simply being Korean.

Toh's strategy of establishing himself as a diligent and exact researcher seemed to have worked. Another Japanese dealer, the founder of the Arai Pharmacy (新井藥房), further supported Toh's Botanical Club starting in 1935. In line with the expansionist rhetoric, the dealer asked Toh's Botanical Club "to contribute to the progress and improvement of pharmaceutics in Korea and Manchuria by moving beyond the Seoul area."[51] Toh's club expanded their expeditions to all parts of Korea, although they never moved beyond Korea, as Manchuria was seen as only open to Japanese like Ishidoya or Asahina. Toh seemed to have not disappointed his sponsors despite his confinement to Korea, by continuously publishing reports on promising medicinal plants in region after region. In 1937, he reported on medicinal plants of the North Hamgyŏng province including the one called fraxinella which "grows wild all over Korea but not in Japan" and in 1938 on those of a volcanic island with unique flora, Ullŭng Island. He was skillful in exhaustively amassing various useful and important data such as chemical composition, production quantity, current use, market demand, and the development potential of each medicinal plant.[52]

Although Toh's collaboration with Ishidoya or Asahina was not frowned upon, there was no doubt that his investigation of "Korean" plants was still mostly for the benefit of his Japanese sponsors.[53] Also, there was certainly nothing "Korean" in his perspective and method. Not only that, but also Ishidoya's encouragement of Toh's finding of *Viola Websteri* in central Korea in 1934 clearly illustrates where these Koreans were. Ishidoya said:

> This is the most authentic [*Viola Websteri*]. Mr. Hemsley named this one, collected by Mr. Webster in the riverside of Aprok. The specimen is in the British Museum. In 1897, Mr. Komarov reported that he found it also in Manchuria but this is the first time that it has been found in Japan.[54]

Unlike the case of lilacs in Seoul Mountains, Ishidoya did not highlight the regional specificity of central Korea within the Japanese empire. This discovery of the "biggest violet from the Manchurian region" in central Korea was a fully Japanese discovery. Toh had not emphasized his Korean birth, and presented his reports as something that the best trained Japanese scientist could produce. So this recognition of his discovery, as made in Japan, seems not entirely different from how he presented his work. It could even be an even-handed recognition of his discovery, the same one to be given to any Japanese, and one that Nakai would not give to Chung. Yet, was Toh delighted by these words of congratulations given for his "Japanese" discovery?

Conclusion: Contingent Nationalism in Colonial Collaboration

Chung and Toh were born in different times and circumstances, and made different choices in the course of their efforts to become botanists during the colonial period. Their stories show the remarkable determination needed by the colonized under the Japanese colonial regime if they were to realize their simple dream of becoming botanists. Historians have often attributed their success to their fortunate relationships with the imperial center, namely Chung's direct contact with metropolitan authorities on Korean flora such as Nakai, and Toh's study in Tokyo. What this close examination of their paths and interactions reveal is that such factors played only minor roles, compared with their lengthy and complex everyday interactions and struggles in colonial laboratories as their scientific expertise matured. The vital factor was their persistent efforts within the colonial institutional settings that they helped build and sustain, and their assiduous search for ways to overcome the constant barriers that confronted them as Koreans. Chung persevered in practicing science against Nakai's or the GGK's efforts to dismiss and limit his inordinate ambition. Toh also had to persist in the struggle to find a role as a Korean researcher in an environment made for Japanese, undeterred by the discrimination he suffered as a Korean pursuing a Japanese education.

Despite their commendable perseverance and efforts and the visible progress that they made as serious botanical researchers, their stories further reveal the weight of the imperial power in every step of their paths. Although exceptional circumstances might offer Koreans the chance to carry out botanical research, the asymmetric relations and skewed thinking generated by imperial subjugation inflicted deep ideological and practical violence on these botanists at times. Under such circumstances, it is not surprising that Chung and Toh may seem to have been docile native informants who fully accepted Japan's civilizing mission and their own constrained positions. They could show no sign of thinking for themselves, and even could certainly not follow a Japanese contemporary like Ishidoya in doubting the validity of Japanese scientific practice. Their common tactic was to hide their "Korean-ness," and to make their research look as Japanese as possible.[55] For most Japanese working with them, except the apparently insecure Nakai, it does not seem to have been no problem to accept their efforts to practice 'Japanese' science. Whatever Chung's and Toh's dreams of achieving some kind of "Korean" enlightenment might have been, it was easy to see their successes as "Japanese" achievements manifesting the civilizing effects of imperial guidance.

The Korean post-independence narrative on these endeavors of colonized Koreans did not pay enough attention to these vulnerabilities. Instead, their achievements were hastily rebranded as Korean, as part of an overstated and homogenizing Korean patriotism. However, what the complex paths followed by Chung and Toh reveal is not their committed patriotism, which would have been a rather unlikely thing for them to have openly expressed given the harsh realities of a colonial rule that could dismiss and demote Koreans at any time. Their achievement, though it ran counter to the GGK's original intention, was not cast by them as 'Korean,' but, as shown in Ishidoya's references to their work, mostly aided and advanced imperial development. Simply applauding their cautiously achieved successes as admirable patriotic exercises elides the complexity of colonial experience, and the depth of imperial violence that existed in many forms in the lives of Japan's colonial subjects. Without overcoming the fear-ridden status of mind imposed on them, they could not pursue alternative ways of thinking that might have enable them to follow right beyond might.

Notes

1 Assimilation was a policy practiced on a rhetorical level only until the Second Sino-Japanese War in 1937, when Japan forced Koreans to adopt names in the Japanese style, which had been forbidden in the previous period in order to make it possible to distinguish Koreans from Japanese. Kwon Tae-Eok 권태억, "Japan's Assimilation Policy (동화정책론)," *The Journal of the Korean Historical Association* (역사학보) 172 (2001): 335–65; KwonTae-Eok , "Japan's Assimilation Policy of the 1920s and 30s (1920, 30년대 일제의 동화정책론)," Han'guk saron (韓國史論) 53 (2007): 403–42; Lee Jeong-seon 이정선, *Assimilation and Exclusion: Japanese Assimilation Policies and the Korean-Japanese Marriage* (동화와 배제: 일제의 동화정책과 내선결혼), Koyang: Yŏksabipyŏngsa, 2017.

2 The most detailed information on his life is in the following biography by his student, which emphasized Chung's close relationship with Nakai, and his autobiographical notes, originally published in 1964. Yi Uch'ŏl 이우철, "A Biography of Haŭn Chung Tyaihyon (하은 정태현 박사 전기, A Biography of Chung hereafter)," in Committee on Haŭn Biology Award, *Haŭn Biology Award: The 25th Anniversary* (霞隱生物學賞: 二十五周年), Seoul: Committee on Haŭn Biology Award, 1994, 53–100; Chung Tyaihyon 정태현, "50 Years of Carrying Field Press (야책을 메고 50년, Fifty Years hereafter)," *Forest and Culture* (숲과 문화) 11(3) (2002): 52–62.
3 Chung Tyaihyon, *The Illustrated List of Korean Forest Trees* (朝鮮森林植物圖說), Seoul: Chosŏn Pangmul Yŏn'guhoe, 1943, Preface.
4 Chung's father also passed away that year, so that he had to manage the family estate and farms. Yi Uch'ŏl 이우철, "A Biography of Haŭn Chung Tyaihyon (하은 정태현 박사 전기, A Biography of Chung hereafter)," in *Haŭn Biology Award: The 25th Anniversary* (霞隱生物學賞: 二十五周年), Seoul: Committee on Haŭn Biology Award, 1994, 53–100.
5 Yi Uch'ŏl 이우철, "A Biography of Haŭn Chung Tyaihyon (하은 정태현 박사 전기, A Biography of Chung hereafter)," in *Haŭn Biology Award: The 25th Anniversary* (霞隱生物學賞: 二十五周年), Seoul: Committee on Haŭn Biology Award, 1994, 53–100.
6 College of Agriculture and Life Sciences 농업생명과학대학, *100 Years of Agricultural Education: 1906–2006* (농학교육 100년: 1906–2006), Seoul: College of Agriculture and Life Sciences, SNU, 2006, 30.
7 Yi Uch'ŏl 이우철, "A Biography of Haŭn Chung Tyaihyon (하은 정태현 박사 전기, A Biography of Chung hereafter)," in *Haŭn Biology Award: The 25th Anniversary* (霞隱生物學賞: 二十五周年), Seoul: Committee on Haŭn Biology Award, 1994, 65; Tsukiashi Tatsuhiko月脚達彦, "Korean Intellectuals for Civilization an Enlightenment and Fukuzawa Yukichi (朝鮮開化派와 후쿠자와 유키치)," *Hankukhakyŏnku* (한국학연구) 26 (2012): 307–35.
8 Yi Uch'ŏl 이우철, "A Biography of Haŭn Chung Tyaihyon (하은 정태현 박사 전기, A Biography of Chung hereafter)," in *Haŭn Biology Award: The 25th Anniversary* (霞隱生物學賞: 二十五周年), Seoul: Committee on Haŭn Biology Award, 1994, 63.
9 As its primary purpose was to display scientific forestry to Koreans, the forestry offices were also located in major cities. For forestation, it searched for locations like "the mountains, the riversides or the railway sites where the public would easily see" as explicitly stated in the following 1912 decree by the Government-General of Korea. "An Outline of the Regional Forestation Projects (地方造林事業概要)," *The Monthly of the Government-General of Korea* (朝鮮總督府月報) 2(11) (1912): 1–11, 1.
10 Chung Tyaihyon 정태현, "50 Years of Carrying Field Press (야책을 메고 50년, Fifty Years hereafter)," *Forest and Culture* (숲과 문화) 11(3) (2002): 52.
11 Ueki is known for his study of Korean pines and for the successful introduction of the North American *Pinus rigida* to Korea. Park Seong Rae, "Ueki Homiki, who introduced *Pinus Rigida* to Korea," *The Science & Technology* 252 (2007): 106–07; Yi Kyŏngjun, *Pioneers' Footprints in Korean Agriculture: Hyun Shin-Gyu*, Suwŏn: Hwanong Yŏnhak Chaedan, 16.
12 A serious study on Ueki's work would be very desirable. Chi Yŏnglin (池泳鱗, 1900–1973), Cho Paekhyŏn (趙伯顯, 1900–1994), and Hyun Sin-Gyu (玄信圭, 1912–1986), pioneering figures in Korean forestry and agriculture are examples. Sun You-Jeong 선유정, "The Trajectory of Hyun Sin-Kyu's Forestry Research (현신규의 임학연구 궤적: 과학연구의 사회적 진화)," Ph.D. Thesis, Chŏnbuk University, 2012; Ch'unyŏng Yi 이춘영, *Pioneers' Footprints in Korean Agriculture: Cho Paek-hyŏn* (韓國農學 巨聖의 발자취-조백현), Suwŏn: Hwanong Yŏnhak Chaedan, 2002.
13 Chung Tyaihyon, *The Illustrated List of Korean Forest Trees* (朝鮮森林植物圖說), Seoul: Chosŏn Pangmul Yŏn'guhoe, 1943, Preface.
14 Nakai Takenoshin, *Flora Sylvatica Koreana* 朝鮮森林植物編. Seoul: GGK, 1915, vol. 11, 50. Nakai Takenoshin, *Florula of M't Paik-Tu-San* (白頭山植物調査書),

Seoul: GGK, 1918, 41; *Report on the Vegetation of the Island Ooryongto or Dagelet Island, Corea* (鬱陵島植物調查書, Seoul: GGK, 1919, 10.

15 It is also notable that he used Chung's first name, Tyaihyon. When he dedicated new species to Ishidoya or other Japanese collectors, he used their last names. Although it could show their intimacy, in an East Asian setting where an adult male is rarely called by his first name, it could be a hierarchical infantilization. Nakai Takenoshin, *Flora Sylvatica Koreana* 朝鮮森林植物編. Seoul: GGK, 1915, vol. 11, 50. About Nakai's collection trips and members, see Kim Hui 김휘, et al., "Re-Examination on Foreign Collectors' Sites and Exploration Routes in Korea: Nakai Takenoshin (외국인의 한반도 식물 채집행적과 지명 재고(2)-나카이 다케노신)," *Korean Journal of Plant Taxonomy* 식물분류학회지36(3) (2006): 227–55.

16 Nakai Takenoshin, "Notulæ ad plantas Japoniæ et Koreæ XXVII," *BMT* 36(426) (1922): 61–73.

17 Nakai Takenoshin, "Notulæ ad Plantas Japoniæ & Koreæ XXXI," *BMT* 40(472) (1926): 161–71, 171; "Notulæ ad Plantas Japoniæ & Koreæ XXXVII," *BMT* 43(513) (1929): 439–59, 446–47. Nakai Takenoshin, *Flora Sylvatica Koreana* 朝鮮森林植物編. Seoul: GGK, 1915, vol. 19, 89. Instead of Chung Tyaihyon, Nakai used the Japanese style Romanization of Chung's name, "Tei Dai Gen," here. As Nakai was admonishing other Japanese scholars for not knowing how to correctly use foreign names and made known his displeasure of having his name mispronounced by foreigners, this was not done to flatter Chung. Nakai Takenoshin, "About the Last Names of Western Scholars with 'De,'" *BMT* 411 (1925): 154–55, 155. Lee Ch'angbok 이창복, "New Natural History: Plants (신박물기: 식물)," in Lee Ch'angbok, *The Life of Lee Ch'angbok* (樹友 李昌福 教授의 발자취), Seoul: Chŏngminsa, 1984, 1–10, 7.

18 Lee Ch'angbok 이창복, "New Natural History: Plants (신박물기: 식물)," in Lee Ch'angbok , *The Life of Lee Ch'angbok* (樹友 李昌福 教授의 발자취), Seoul: Chŏngminsa, 1984, 1–10, 7.

19 See Chapter 2.

20 Chung Tyaihyon 정태현, "50 Years of Carrying Field Press (야책을 메고 50년, Fifty Years hereafter)," *Forest and Culture* (숲과 문화) 11(3) (2002): 52–62

21 About the constant efforts made by the armchair botanist to train useful collectors, see Jim Endersby, *Imperial Nature: Joseph Hooker and the Practices of Victorian Science*, Chicago, IL: University of Chicago Press, 2008, Chapter 3.

22 Even though the salary was the same, Asakawa could receive the compensation for settlers. On the systematic discrimination in the Japanese imperial system, see Ramon H. Myers and Mark R. Peattie eds., *The Japanese Colonial Empire, 1895–1945*, Princeton, NJ: Princeton University Press, 1984, 37.

23 See Chapter 3.

24 Shin, Chang-Geon 愼蒼健, "Institutionalization of Research on Traditional Medicine in Keijo Imperial University (경성제국대학에 있어서 한약연구의 성립)," *Society and Culture* (사회와 역사) 76 (2007): 105–39.

25 Ishidoya Tsutomu, "Herbal Plants Produced from Korean Mountains and Fields (朝鮮の山野より生産する藥科植物)," *Korean Repository* (朝鮮彙報) 2 (1916): 75–148; *Farm Management* was not on the traditional Korean book list compiled and updated by the GGK at the time and firstly appeared in the 1921 list. *The Catalogue of Old Books by the Government General of Korea* (朝鮮總督府古圖書目錄), Seoul: The Government General of Korea, 1913; *The Catalogue of Old Books by the Government General of Korea*, Seoul: The Government General of Korea, 1921, 183. The author of this exemplary book in Korean Practical Studies was only identified in 1937. Miki 三木榮, and Togashi Naojiro 富樫直次郎. "On Farm Management 山林經濟考," *Korea* (朝鮮) 262 (1937): 1–19.

26 Chung's role in this was specified in a later report on medicinal plants in 1936. The Forestry Research Institute 林業試驗場, *Wild Medicinal Plants of Korea* (朝鮮産 野生藥用植物), Seoul: The Government-General of Korea, 1936.

27 Ishidoya Tsutomu, "Herbal Plants Produced from Korean Mountains and Fields (朝鮮の山野より生産する藥科植物)," *Korean Repository* (朝鮮彙報) 2 (1916): 75.

28 Ishidoya Tsutomu, "Chinese Characters Used for Tree Names in Korea (朝鮮に於て用いらるる樹木漢字名)," *Korean Repository* 3 (1917): 91–94. See also note 27, Chapter 1.

29 "Report on Korean Traditional Medicinal Herbs (朝鮮漢方藥科植物調査書)," in *Report on Plants of Nobong Mountain in Korea* (朝鮮鷺峯植物調査書), Seoul: GGK, 1918.

30 With Yanagi Muneyoshi, Asakawa founded the Korean folk art museum in Seoul in 1924.

31 Takasaki Soji 다카사키 소지, *Becoming the Part of Korean Soil: Asakawa Takumi* (조선의 흙이 되다: 아사카와 다쿠미 평전). Translated by Kim Sunhi 김순희, Paju: Hyohyŏng, 2005; Kim Sŏkkwŏn, 김석권, "Reflection of Asakawa Takumi from Forest Historical Point of View (아사카와 타쿠미의 임업사적 재평가)," *Forest and Culture* (숲과 문화) 20(5) (2011): 45–56; Chang Hyechŏng 장혜정, "Life of Asakawa Takumi and 'Korea' (아사카와 타쿠미의 삶의 궤적과 '朝鮮')," *Comparative Studies of World Literature* (世界文學比較硏究) 14 (2006): 103–22.

32 See "Imperializing Forestry," in David Fedman ed., *Seeds of Control: Japan's Empire of Forestry in Colonial Korea,* Seattle: University of Washington Press, 2020. Kären Wigen, "Discovering the Japanese Alps: Meiji Mountaineering and the Quest for Geographical Enlightenment," *The Journal of Japanese Studies* 31(1) (2005): 1–26.

33 *Big, Old, and Famous Trees in Korea* (朝鮮巨樹老樹名木誌), Seoul: The Government-General of Korea, 1919. The move to designate "natural monuments" was first proposed in 1911 by Miyoshi Manabu (三好學, 1862–1939), professor in botany at Tokyo Imperial University. Though the Forestry Association of Great Japan first issued their Old Tree list in 1913, most Japanese studies on old trees were after the 1920s, showing the simultaneity of imperial and colonial development again.

34 Ishidoya Tsutomu, "Foresters' Attitude towards the Forest and the Chŏnnam Province's Gift of Nature (삼림에 대한 조림자의 태도와 전남삼림의 천혜)," *The Journal of the Forestry Association of Chŏnnam* (全南山林會報) 1 (1922): 18–26, 21.

35 For the notions on "polite" and "vulgar" imperialism, see Fujitani Takashi, *Race for Empire: Koreans as Japanese and Japanese as Americans during World War II*, Berkeley: University of California Press, 2011; Frantz Fanon, *Towards the African Revolution: Political Essays*. Translated by Haakon Chevalier, New York: Grove Press, 1988, 31–44. I thank late Aaron S. Moore for his advice on this.

36 Omitting names of botanists (mostly Nakai) from Latin scientific names could be a significant gesture in denying credit given wholly to botanists. Ishidoya Tsutomu and Chung Tyaihyon, *The Identification Keys of Korean Forest Trees* (朝鮮森林樹木鑑要), Seoul: The Government-General of Korea, 1923.

37 Yi Uch'ŏl 이우철, "A Biography of Haŭn Chung Tyaihyon (하은 정태현 박사 전기, A Biography of Chung hereafter)," in *Haŭn Biology Award: The 25th Anniversary* (霞隱生物學賞: 二十五周年), Seoul: Committee on Haŭn Biology Award, 1994, 82.

38 Such emphasis on the use of plants also contrasted with Nakai's academic approach not usually mentioning the practical aspect. Ishidoya Tsutomu and Chung Tyaihyon, *The Identification Keys of Korean Forest Trees* (朝鮮森林樹木鑑要), Seoul: The Government-General of Korea, 1923.

39 It was a special standalone issue of the FRI journal. Chung, "Distribution and the Right Land for Important Korean Trees (朝鮮産主要樹種ノ分布及適地)," *Current News of the Forestry Research Institute* (林業試驗場時報, *CNFRI* hereafter) 5 (1925): 1–45.

40 In 1933, Chung received 55 yen, less than the 66 yen that he received as a forester in 1921 and much less than the 97 yen in 1931. It was also less than the average starting salary of an elementary school teacher, which Chung might have become if he had wanted. "The Employee List of the Government-General of Korea (조선총독부직원록)," in Korean History Database. (http://db.history.go.kr/) accessed 1.5.2011.

41 Chung Tyaihyon, "On the Medicinal Plants produced in Korean Mountains and Fields (朝鮮の山野より生産する藥科植物に就て)," *The Journal of the Forestry Association of Korea* (山林會報, *JFAK* hereafter) 97 (1933): 21–28. As if to prove his worth, he published three more articles in that year, all rehashes from Nakai's reports. Chung Tyaihyon, "On Plants of Kanghwado (江華島所産森林植物に就て)," *JFKA* 99 (1933): 40–45; Chung Tyaihyon , "On Plants of Kŭmkang Mountain (金剛山産森林植物に就て)," *JFKA* 102 (1933): 19–27; Chung Tyaihyon , "On Plants of Ŏch'ŏng Island (於青島所産植物に就て)," *CNFRI* 9 (1933): 31–43.

42 Toh Chung-Ae 도정애, *Toh Bong-Syup: Celebrating the 100th Anniversary of His Birth* (都逢涉誕生百週年記念資料集), Seoul: Chayŏnmunhwasa, 2003; Moon Manyong, "Toh Bong-Syup," in *A Biographical Dictionary of Who's Who in Science and Engineering* (과학기술인명사전), Seoul: National Research Foundation of Korea, 2012, 121–127.

43 The entrance exam for Koreans was allowed only in 1925, the year before Toh's entrance. Toh Chung-Ae 도정애, *Toh Bong-Syup: Celebrating the 100th Anniversary of His Birth* (都逢涉誕生百週年記念資料集), Seoul: Chayŏnmunhwasa, 2003, 144. Other students who studied at Imperial Universities mostly took what was called "Elective Courses" offered to transfer students, instead of belonging to a department like Toh. Hong Hyŏn-o 홍현오, *A History of Korean Pharmaceutics* (韓國藥業史), Seoul: Handokyakpum Co., 1972, 101–03. Until 1923, there were 11 Korean graduates out of 16 Koreans who studied at Tokyo Imperial University. Pak Kihwan 朴己煥, "History of Cultural Exchange between Modern Japan and Korea: Koreans' Study Abroad in Japan (近代日韓文化交流史研究: 韓國人の日本留學)," Ph.D. Thesis, Osaka University, 1998. The first Korean Bachelor of Science graduated in 1925, in mathematics. Kim Geun-Bae 김근배, *The Emergence of Modern Techno-Scientific Manpower in Korea* (한국 근대 과학기술 인력의 출현), Seoul: Munji, 2005, 244–77.

44 Im Jong-Hyok 임정혁, "On Works of Physicist Toh Sangrok during the Colonial Era 식민지시기 물리학자 도상록의 연구활동에 대하여," *JKHSS* 27(1) (2005): 109–26.

45 Sim Ch'anggu 심창구 et al., "History of Korean Pharmaceutics (한국약학사)," *Journal of the Pharmaceutical Society of Korea* (약학회지) 51(6) (2007), 361–82, 367.

46 The official launch of the Botanical Club was January 1932. The history of the Botanical Club was detailed in its first monograph. Botanical Club 京城藥專植物同好會, *Catalogue of Korean Plants: Central-Korea* (朝鮮植物目錄 1: 中部朝鮮編), Seoul: The Botanical Club, 1936.

47 Ishidoya Tsutomu and Toh Chung-Ae, "Florula Seoulensis (京城附近植物小誌)," *Journal of the Natural History Association of Korea* (朝鮮博物學會雜誌, *JNHK* hereafter) 14 (1932): 1–48, 37, 38.

48 Toh Bong-Syup, "Alpine and Medicinal Plants in the Alpine Area of the South Hamkyŏng Province (咸鏡南道山岳地帶に於ける 高山植物及び藥用植物)," *Journal of the Pharmaceutical Association of Korea* (朝鮮藥學會雜誌, *JPAK* hereafter) 15 (1935): 212–25.

49 *Pharmaceutical News of Japan* was a semi-monthly tabloid magazine, printed by the Association of Pharmaceutics of Japan (日本藥報社). Toh Bong-Syup, "Wild Medicinal and Indigenous Plants in Central Korea (中部朝鮮の野生植物及び朝鮮特産植物に就て(一))," *Pharmaceutical News of Japan* (日本藥報) 9(16) (1934): 4–6.

50 Toh Bong-Syup, "Wild Medicinal and Indigenous Plants in Central Korea (中部朝鮮の野生植物及び朝鮮特産植物に就て(一))," *Pharmaceutical News of Japan* (日本藥報) 9(16) (1934): 4.

51 Botanical Club 京城藥專植物同好會, *Catalogue of Korean Plants: Central-Korea* (朝鮮植物目錄 1: 中部朝鮮編), Seoul: The Botanical Club, 1936, preface.

52 Toh Bong-Syup, "Plants in Korea," *Pharmaceutical News of Japan* 12 (1937): 24–26, 24. Toh Bong-Syup and Sim Hakchin 沈鶴鎮, "A Comparative Study of the Korean Medicinal Plants from Granite and Limestone Regions (朝鮮に於ける花崗岩地帶と石灰岩地帶とに分布せる藥用植物の比較研究)," *JPAK* 17(3) (1937): 91–98; Toh

Bong-Syup and Sim Hakchin, "Medicinal Plants produced from Ullŭng Island (鬱陵島所産藥用植物-附 島勢一班)," *JPAK* 18(2) (1938): 59–81.

53 Toh's research gained more attention from these sponsors as they tried to cut down the importation of traditional medicines from China to fund Japan's war with China and the West. Toh was fully aware of that. Toh Bong-Syup, "A Researcher's Note: The Future of Traditional Medicine Seen through Medicinal Plants (學究一家言: 藥草를 通해서 본 漢醫學의 將來)," *Dong-a Daily*, June 17, 1938.

54 Ishidoya's exclusion of Manchuria from Japan, in spite of the establishment of Manchukuo, is notable. Toh Bong-Syup, "Wild Medicinal and Indigenous Plants in Central Korea (pt. 2)," *PNJ* 9(17) (1934): 5–7, 6.

55 Members of a discriminated group of people, who had to survive the undue discrimination imposed on them by the dominant group, have commonly responded in this way. Du Bois's discussion of "double-consciousness" and Fanon's discussion of black people's adoption of white masks provide two exemplary discussions. William Edward Burghardt Du Bois, *The Souls of Black Folk: Essays and Sketches*, Chicago, IL: A.C. McClurg & Co., 1903; Frantz Fanon, *Black Skin, White Masks*, New York: Grove Press, 1967.

6 Imperial Transformation of Traditions

Takeuchi Yoshimi (竹内好, 1910–1977), an important critic of modernity in Japan, shared what he felt on his first trip to China.

> I made a trip to China when I was a sophomore in college [1932]. I was moved by the fact that there were the same human beings who had the same ideas… I had never thought that there were actual human beings, just like us, living in the Chinese continent. If it had not been in China but Europe or the US, I would not have had that kind of feeling. You naturally expected to meet human beings even better than you in Europe or in the US. Why were we not aware of the fact that there were people in China who were like us?[1]

His sense of discovery, and the surprise at the impact of that discovery seems to have been shared by many Japanese who visited Korea or China. As for Takeuchi, these feelings of shared 'Asian' or 'oriental' culture often led to critical reflections on Japan's modernizing path. Instead of pity and contempt, these Japanese youths in China and Korea even expressed a certain kind of envy for these supposedly irrational and unenlightened people who had kept their traditional culture, resisting Western or Japanese modernity. Of course, there were misunderstandings in this view, as in the idea that there would be no "actual human beings" or that there would be completely different kinds of people in China or Korea. It ignores the fact that no culture ever stays the same and what made certain traditions traditional was often part of a confused claim for modernity.

This chapter follows the rest of the tumultuous journey that the unique settler expert Ishidoya Tsutomu (1891–1958) made after his turn toward the Korean colonial tradition in 1925. His uneasy relationship with colonial tradition shows how a Japanese researcher like Ishidoya, despite his seeming freedom in his knowledge practice, especially compared with Koreans like Chung Tyaihyon and Toh Bong-Syup, was also much confined by imperial power relations. Like Takeuchi's interest in Chinese literature, Ishidoya's interest in Korean traditional medicines was inseparable from another Westernizing project of Japan that had been

DOI: 10.4324/9781003511755-7

simmering and maturing. Shiratori Kurakichi (白鳥庫吉, 1865–1942), the founder of Oriental studies in Japan, asserted in 1934:

> To prevent the Japanese, the very same orientals, from succumbing to this research by European and American scholars, I advocated the necessity of establishing a major association for the research of oriental history by forming a union of scholars, industrialists, and politicians and conducting fundamental research on the orient.[2]

As "orientals" themselves, Japanese scholars had to lead this field in competition with Western scholars. He presented Manchuria-Korea as an especially promising place to start Japanese Oriental studies.

> The fact is, however, that Westerners had pioneered most of this research, leaving hardly anything for Japanese scholars to take up. The exception to this was Manchuria-Korea, a region over which previous wars had been fought and which Japan was about to conquer. Western scholarship on this region was still scarce. While some research of course existed, such as several useful language studies, issues of geography, history and archaeology yet remained unexamined.

This was a project to use Manchuria-Korea and gradually all of China, as a way of showcasing Japan's ability to study the ancient culture and history of this "Eastern" region using "Western" methods.[3] Echoing precisely the sentiment of the botanists at the Tokyo Imperial University, this Western-educated professor at the university wanted to seize the past of these colonized lands, like botanists did the plants, to display Japan's prowess.

Ishidoya's path in studying traditional medicines, however, demonstrates that those things from the past were not a mere matter for nostalgia and antiquarianism. Especially, things like medicinal plants were so materially present and valid for imperial economic and ideological developments as to confuse any streamlined narrative of modernity. In combination with the globally reshaping market for traditions, the shadow of Japan's "barbarous" past and the glory of China's and even Korea's "civilized" one complicated his task. What would he imagine in the name of Korean, Chinese, Japanese, or Eastern traditions?

The Slow Discovery and Rapid Denial of Colonial Tradition

As described in previous chapters, Ishidoya's concern about the colonial land as the site of knowledge production began with its natural conditions. The singularities of Korea's soil, climate, and ecology made the colonial land not just a place to transmit Japanese knowledge but a place from which its own localized knowledge ought to be generated. If unique natural conditions were important attractions in shaping his civilizing mission, his encounter and deepening contact with colonial tradition was not part of his original project. Korean traditional culture seemed at first to be too foreign and irrelevant to his knowledge practice. He was shocked at Korean people's seeking "roots of grass and bark of trees" as medicines and in his view the fact that Korean doctors still used texts like *Treasured Mirror of Eastern Medicine* seemed to prove Koreans' lack of empiricism and subservience to old authority. It was quite natural that Ishidoya did not show immediate kindred feelings for Koreans like Chung of the kind that Takeuchi described. He mostly treated Chung as a suitable recipient of his civilizing guidance, teaching him everything from collecting plants and making specimens to cultivating and planting trees. His decision to abandon his 15-year career as a forester, and to give a presentation on the *Treasured Mirror* in the gathering of naturalists in colonial Korea in 1925, was a reversal of his previous position, and it was not made without major doubts in the preceding years.

In fact, it turned out to be quite easy for Ishidoya to find new collaborators and secure a sponsor for his new subject, the traditional medicines of Korea. After his talk on 45 notable herbs from *Treasured Mirror* at the meeting of the Natural History Association of Korea in 1925, he got an invitation from Keijo Imperial University. Sugihara Noriyuki (杉原徳行, 1892–1976), then a pharmacology professor in a private medical school, came to think that Ishidoya's research on traditional medicine could be crucial in establishing a distinctive pharmacological section in the new imperial university that he was asked to lead. Sugihara appointed Ishidoya as a lecturer for the section and gave him a handsome budget to purchase old texts. Ishidoya's role was to investigate traditional texts and identify original plants of traditional medicines for his pharmacologist colleagues' further analyses.[4] It was a promising start for Ishidoya's study of traditional herbs in Korea. But it did not last long, as by the early 1930s he had fully shifted his focus to what he saw as Chinese traditions. Until this transition, his constant reassessment of Korean tradition has been the focus of his new career.

Ishidoya's 1925 study of Korean medicinal plants was based on Hŏ Chun's (許浚, 1546–1615) *Treasured Mirror*, the book that he had thought hindered the empirical spirit in Korean traditional medicine. Yet, as a result of this first careful investigation of the book, in the course of which he identified original plants for about 200 medicines prescribed in the book, he came to formulate a different interpretation. The investigation was challenging enough, although getting the medicines described in the book was not a problem; most of them were in wide circulation in herbal shops in Seoul. The problem was that the medicines in the shop, mostly dried parts of plants, did not give good clues as to the plants from which they were produced. He visited provincial herbal markets, where the herbal collectors or growers brought unprocessed plants, and sometimes followed these collectors on trips into the mountains. He managed to identify original plants for about 200 medicines in the book through these field investigations, although he picked only 45 of them for his first report.

The criterion for his selection was that those medicines were uniquely Korean, and not found in other Japanese or Chinese texts. His expectation seems to have been that those plants unique to Hŏ's prescriptions came from Hŏ's attempt to "find locally produced substitutes" for medicines in Chinese texts.[5] And given the longevity of his prescriptions in Korean medicine, Hŏ must have been quite successful in that empirical search in his land. Hŏ's seemed to exemplify the kind of successful localization of foreign knowledge that Ishidoya had always been looking for.

Ishidoya articulated his changed evaluation of the work in the first article published from the pharmacological section in 1926. This 16-part article, "Korean Traditional Medicine," revealed his enthusiasm vividly.[6] It presented Korean traditions as something original that had incorporated traditional Chinese medicine in its own way, particularly in contrast with Japanese texts that appeared to accept the authority of Li Shizhen's *Systematic Materia Medica* until the late nineteenth century. The article characterized a fifteenth century Korean text, the 85-volume *Compilation of Native Korean Prescriptions* (鄉藥集成方 Hyangyak chipsŏngbang, 1433), as a significant initiative, encouraged by the Chosŏn court. Then it depicted Hŏ's *Treasured Mirror*, completed "only 33 years" after Li's work, as an achievement comparable to Li's work. Hŏ's work, he wrote, was an original Korean contribution much ahead of similar Japanese works. "Such a great work was rare even in the West" at the time, he added. He continued his praise by noting that whereas the number of herbal medicines recorded in China had rarely reached 1,000 until Li Shizhen's time, Hŏ contemporaneously described about 1,350 medicines, including many local ones.[7] Ishidoya indicated that the value of the *Treasured Mirror*, comparable to that of the great Chinese and Western traditions, was the result of its own search for what was good in its own land. Thus, portraying it as the wholesome opposite of ill-transplanted Japanese imperial science, Ishidoya appeared to display a thorough appreciation of the Korean knowledge tradition.

Yet, in just five years, Ishidoya made another dramatic reappraisal of the *Treasured Mirror* and the Korean tradition. At the center of this reappraisal was the view that the *Treasured Mirror* had not utilized the great achievement of Li Shizhen's work. He dropped his earlier evaluation of Hŏ's work as an original achievement almost contemporaneous and comparable to Li's work. His new enthusiasm for Li's voluminous work, and his realization that it was introduced to Japan in 1607, seems to have been at work. Ishidoya looked carefully into Li's work, whose impact on the Tokugawa *honzo* (本草) tradition was well known, and found it an impressive empirical approach comparable to Western modernity, as did many before and after him.[8] It appeared to have a superb natural classification system with hierarchical groupings, and also discussed objective details of plants. Li's *Systematic Materia Medica* was completed in 1578 and published in 1596 and the *Treasured Mirror* was completed in 1610 and published in 1613. Ishidoya seemed to believe that if Li's book had come to Japan in 1607, Koreans must have had it earlier and Hŏ must have known it. He specifically criticized Hŏ for creating his own classification system, without following that of Li. This time, he did not see it as Hŏ's originality. Instead he concluded that that Hŏ had relied more on out of date Tang and Song period systems, laden with Confucian cosmology, instead of accepting Li's naturalistic approach. Hŏ's work was, he concluded, a retrogressive and arbitrary copy of an outdated Chinese tradition that failed to learn from the sprouting Chinese modernity.[9]

Despite this sudden reversal, Ishidoya's evaluation criteria had some consistency. They were all qualities as praised as strengths of Western scientific methods, which Japanese metropolitan sciences had failed to live up to, much to his disappointment. He was excited that Hŏ carried out empirical research into local flora to find alternative to Chinese medicines, but Hŏ's empiricism now simply looked pale in comparison to Li's lifework. Li, immersed in a different Confucian cosmology, aimed at a thoroughly philological as well as empirical investigations of all medicinal materials. Li also tried to establish a comprehensive classification system for all medicinal materials that existed, quite like Linné. In comparison, Hŏ failed to conceive such an intellectual agenda on the basis of his medicinal materials, allotting only 3 volumes to the topic in his 25-volume *Treasured Mirror*. He in fact titled his section on *materia medica* as "decoction section," showing the significance of his focus on final clinical products for his broader discussion of medicine.[10] This generalist approach of Hŏ began to be seen as a drawback, as Ishidoya saw more specialized endeavors like Li's as more scientific and promising.

Another familiar criteria for Ishidoya's reassessment of Korean tradition is the all-pervasive historicism, i.e., the idea of linear progress made upon constant improvement of the latest development.[11] It assumes that is how science always progresses, by constantly proving and disproving the latest, best theory. As historians of all culture note, this was hardly a widely held view in any culture until the latter part of the nineteenth century. Even the word "progress" was rarely used. Most cultures rather assumed a moment in the past when all things were ideal. The important task was to find the truth and wisdom that made things closer to that ideal past, and that restoration was the purpose of reform. Yet, under the new reign of historicism, going back to a certain past or seemingly persisting with any successful past achievement looked misdirected and sterile.[12]

Ishidoya's reassessment of Li's approach as comparable to the Western naturalist tradition moreover allowed him to reappraise the Japanese tradition, which he had earlier denigrated as slavishly following Li's work. In his changed view, it was much more praiseworthy than Korean scholars' neglect of the Chinese 'modernity' exemplified by Li. Japanese *honzo* scholars inherited this promising and cutting edge tradition, and made constant progress applying it in the Japanese context. Its culmination, he claimed, was seen in Ono Ranzan's exhaustive and empirically rich discussion of Li's work that included many materials of Japanese origin.

Shin Dong-Won has argued that Hŏ's primary aim was to find a reliable basis for medical knowledge, as "conflicting medical theories abounded, and numerous prescriptions were mixed in various medical books, making it difficult to separate the wheat from the chaff." Hŏ did not view contemporaneous Chinese works as being the most advanced, but saw them as fragmented and trapped by competing lineages. He compiled his work precisely in order to separate the wheat from the chaff. His method was to apply his own criterion of reliable knowledge to help restore and cultivate the healthy self, while providing efficacious decoction prescriptions for illnesses, with available local substitutes for imported medicines. Since Hŏ's critical concerns about the status of medical knowledge, which Li may be said in part to have shared, were not particularly Korean, his judicious selection came to be known outside Korea, through multiple reprints in China and Japan.[13]

Japanese scholars in the Tokugawa period, who showed keen interest in Hŏ's work, seem not to have thought that their or Li's approach was particularly superior to the practice-oriented or generalist approach of Hŏ, eclectically using diverse Korean works as references. As mentioned in Chapter 1, Niwa Seihaku made on-site investigations of medicinal plants in the *Treasured Mirror* and Ono Ranzan utilized it. Ono additionally used Korean works as diverse as *Hawk Husbandry* (鷹鶻方 Ŭnggolbang, 14th C), *Compilation of Native Korean Prescriptions*, *First Aid Prescriptions for a Country Home* (村家救急方 Ch'on'ga gugŭppang, 1538), and *Chibong's Classified Commentaries* (1614), the first work that referred to Li Shizhen's work in Chosŏn, and so on.

Ishidoya's constant reassessments of Korean traditions show how his faith in the certainty of modern science and the linear progress of history confused and narrowed his engagement with Korean, Chinese, and Japanese traditions. All these complex knowledge traditions, in this perspective, existed only in fragmented forms. And those fragments, selected owing to their similarity to Western modernity, were to be ranked against one another, regardless of their meanings in their respective knowledge tradition and in the lives of the collectives that generated and used them. Ishidoya was an articulate critic of Japan's metropolitan science. Nonetheless, it seems that his easy reconciliation with colonial tradition came about because he unquestioningly accepted an idealized view of Western science as the ultimate standard for all intellectual endeavors.

Discovering China, the Continent, in Colonial Korea

In a sense, Ishidoya's shift to the Chinese tradition is not unexpected. To the Japanese, Korea had always been a conduit to things Chinese. However, it was not at all the same with any of the previous moves that aimed at more expansive learning. This time, the vast Chinese tradition was seen as a target for takeover by the Japanese empire, and the right place to initiate this takeover was the colonial outpost of Korea, not the metropolitan centers in Japan itself. Ishidoya, who had strongly criticized certain Japanese colonial policies that disregarded local culture and nature, spearheaded the move, transforming his colonial site into a center of Japan's new civilizing mission based on "Chinese" tradition.

Ishidoya explained his reason why Korea, neither Japan nor China, should become such a site for Japan's new civilizing mission. The long history of exchanges between China and Korea provided a good model. He had drawn attention to a striking example of such exchanges in 1917, while pointing out Japanese metropolitan forestry's failed attempts at transplanting 'superior' foreign trees like Japanese cedar and cypress into Korea. In contrast to this failure, he realized that there were some foreign trees well established in Korea, such as Chinese ginkgo, chestnut, and bamboos, which had been introduced to Korea centuries ago "following the geographical connection and the movement of ancient culture." These plants grew robustly, and were adapted to local soils. Unlike Japanese trees that neither grew well in the colony nor had any practical value, these "Chinese" trees had also displayed "industrial worth."[14]

In this story of positive exchanges between the geographically connected lands of China and Korea, what seized Ishidoya's imagination seems to have been the physical connection itself. It was a connection that did not exist between Japanese and Korean lands, or indeed between the Japanese archipelago and any other lands. This direct connection to the continent (大陸) was what rendered Korea, now Japanese territory, more significant for Japan. Ishidoya had emphasized this connection of Korea to China as a continent in the floral guide to Seoul that he had made with Toh Bong-Syup. Ishidoya's enthusiasm for Korean flora was rooted in the fact that Korea was not a peninsula at the edge of the continent. It was the continent. Japanese people, living and visiting Seoul, could feel "for the first time that they are on the continent," Ishidoya asserted. Japanese could experience such a continental atmosphere "when they catch sight of the Siberian chrysanthemum on Mount Nam," the mountain at the center of Seoul.[15] Explicitly addressing a Japanese audience several times in the text and introducing 'continental beauties' of Seoul unfamiliar to Japanese people, Ishidoya's floral guide was an attempt to impress upon the Japanese people the continental nature of colonial Korea.

Ishidoya was hardly alone in this effort. What he did was in line with the moves of other Japanese experts in Korea, who tried to promote Korea as a base for Japan's scramble for China, especially in competition with the newly acquired Japanese territory of Manchukuo. Korea had previously enjoyed its geopolitical importance in the Japanese empire as a foothold on the continent, but the more advantageously-situated Manchukuo, seized in 1931, inspired a Manchurian dream, or as they called it, the "Manchurian fever" among Japanese.[16] The desire not to lose the unique geopolitical importance of Korea in this changing situation was widely shared by Japanese experts in Korea. Researchers at the colonial imperial university quickly moved, organizing research teams, and promoting their "continental research" projects from Korea to various parties, and in this they were quite successful. Some big Japanese businesses sponsored their initiatives, and even the Manchurian regime welcomed these initiatives since it lacked the built infrastructure that might have made it possible for Manchuria to conduct its own research.[17]

Ishidoya's boss at the pharmacological section, Sugihara, provided an articulate voice for Japanese experts in Korea. He said that now Korea should become the "ideological" as well as the "material" base for Japan's expansion to China. By an "ideological base," he meant that Korea should become the place where a new imperial ideology could be forged. Sugihara was keenly aware of certain derogatory views of Japan's successful "Westernization." He noted that the apparently complimentary assertion that the "Japanese were exceptional at mimicking" implied that the Japanese "lacked originality." He believed that Japan could undermine that criticism by presenting a new civilizational model that creatively combined what Japan had successfully copied from "the West" with what Japan shared with its Asian brothers, the "Eastern" tradition.[18] His pharmacological section was especially well-suited for such an endeavor owing to its expertise in traditional medicines. By creatively fusing those "Eastern" traditions with the "Western" scientific method, his section would shape and propagate the original new civilization of the Japanese empire.

Inviting Japanese to experience the continent through Korea, Ishidoya appears well prepared to join his boss's new civilizing mission. Above all, he knew well that some clarification was necessary for Japan to embark on this so-called "East-West Fused" civilizing mission. Japan's claim for the "Eastern" tradition was quite tricky since it was taken for granted that the center of that "Eastern" tradition was the Middle Kingdom, China, and Japan was just one of its "barbarian" neighbors, no matter how much it claimed to be on a path to modernity that built on the true tradition of China. If relying on borrowed Western modernity for its earlier civilizing mission was not ideal, relying on a borrowed tradition for its new mission might be seen as equally problematic. Japan's leadership for this new civilizing mission in its expanding Asian territories was not consonant with retaining the concept of Japan's "barbarian" past. Therefore, for this "East-West Fused" civilizing mission, it was necessary to enunciate a new historical understanding of what 'Eastern traditions' might be.[19]

Ishidoya took up this task of rewriting history for his pharmacological section's "East-West Fused" medicinal research, joining the wider move in other disciplines. No longer did Ishidoya condemn Japanese traditions as slavish obedience to Chinese authority. His previous conception that the Japanese tradition was a progressive localization of the true Chinese tradition was not good enough either. Ishidoya re-conceptualized the Japanese tradition as a virtual equivalent to that of China. The Japanese search for medicinal plants had a history as long as that of the Chinese search that was famously begun by the mythical figure Shennong. Japan had its own divine figures of Onamuchi and Sukunahikona as "the Ancestors of Herbs." These ancestral deities introduced the practice of finding herbs for people, which was claimed to have been inherited by the Japanese emperors and scholars who created the solid scholarly tradition of *honzo* studies. Although Ishidoya admitted that the Chinese *Systematic Materia Medica* of Li Shizhen played an important role in shaping later Japanese research on medicinal plants, he insisted that the Japanese could boast many original contributions, such as their fine botanical illustrations whose accuracy was, he claimed, recognized by the Chinese themselves.[20]

Ishidoya's new history of Japanese herbal tradition could have been seen as anti-imperialistic, had he also applied it to Korea. But he did not, despite the fact that Korea's long and transregionally shaped tradition had initially helped him dream of a new civilizing mission from his colonial outpost. He now specifically excluded Korea from such considerations, reducing Korea to a nameless region within the Japanese empire. Koreans like Chung and Toh could make "Japanese" discoveries in their own country thanks to his guidance, and it was not necessary to recognize their native land as a peninsula with its unique history. It was just a new continental Japan where Japanese people could venture out from their insular mainland to carry out a new civilizing mission to the wider continent. As such a territory within the Japanese empire, his colonial outpost was all ready to become a new imperial center, where he could conduct a creative new imperial science without the meddling of the inept yet overbearing metropolitan scientific establishment.

Doing Imperial Science through the Market

Ishidoya, however, gradually distanced himself from the collective movement of his colonial university colleagues sponsored by the Manchurian regime and big businesses, etc. His active diversification of sponsors for his research on Chinese medicines showed his serious will to take a different path from his colleagues. Part of this attitude was enabled and dictated by material conditions of his study, herbal medicines, and part of it seems to have reflected his constant concerns about constructing an ideal scientific practice suitable for his civilizing mission. The international market, made vibrant by regenerated material traditions, was a strong element in his thought and knowledge practice.

The market, above all, provided the material base for his study of Chinese tradition. While most of his colleagues in continental research had to go to mountains or villages for their investigation, his study could begin at the market. The international herbal market that he first visited in 1929 impressed him greatly. This single encounter seems to have excited and overwhelmed Ishidoya. He reported with awe that one medicine warehouse (貨棧) was as big as one Korean herbal market, consisting of a huge storehouse, lodging facilities for international dealers, and a trading space. In spite of the turmoil in China, the huge Chinese market was connected to the international market by the dealers' union in this well-established industry. So gigantic were the operations of these international medicine dealers that to them "Japan, Korea, or the South Sea Islands were not foreign nations but simply a region comparable to Yunnan or Sichuan," he noted. Tons of herbal medicines from all around the world, including the so-called American ginseng (mostly from Canada), poured in, changing the market for Korean ginseng; all medicines were classified by origin and quality, and traded constantly by merchants from everywhere according to their exchange value.[21]

Previously he had learned by following Korean herbal collectors to the mountains and fields and identifying the original plants of dozens of Korean medicinal plants. But now an entirely different and immensely promising possibility was open to him. But given the expense of expeditions to the continent and the increasing risk of war, it was not easy to follow up these new opportunities. Ishidoya's exciting first trip in 1929 was sponsored by the *Dojinkai* (同仁会, 1902–1945), an association for Japanese medical missions in Asia partially funded by the Japanese Congress. It appears that the *Dojinkai*, which had emphasized Japan's Western modernity to its Asian brothers so far, wished to add some "Eastern" element to its medical mission in China. This need may have been inspired by the increasing presence of Western medical missions in China, supported by big philanthropic foundations.[22] The Sugihara team's proposal for the "East-West fused" civilizing mission with traditional medicines answered their need for some authentic justification for their competitive presence in China. The continental research activities of their imperial university colleagues added a further impulse to the pharmacological section's research into China. All of Ishidoya's subsequent trips to China, in 1934, 1936, and 1938, became part of the scientific investigations organized by Keijo Imperial University and conducted under the sponsorship and the protection of the Japanese government, the Manchurian military, and Japanese conglomerates.

Ishidoya, however, revealed concerns about this continental research. As mentioned in Chapter 3, he expressed his dissatisfaction with Japanese knowledge practice in Manchuria in 1934, by comparing it with that of its competitor Russia. His Japanese sponsors seem to expect a quick report after a brief trip unlike the Russians who, he claimed, made their scientists stay in the field for a long time, seeking for a policy that fitted the local custom, soil, and climate. Few of his colleagues in continental research shared his concerns. Their reports enthusiastically thanked their sponsors for the short and well-guarded trips that gave them safe continental experiences. They made their confident knowledge claims and proposals based on fleeting experiences, as if they knew all along what it would be best for them to claim and propose. Ishidoya revealed no such enthusiasm or confidence, and in later reports, he even described his work as something "commanded" by the university.[23]

Being "commanded" was not what Ishidoya appreciated. He voiced his distaste about such intervention again in 1936, by invoking the colonial Korean tradition that he had seemed to abnegate long before. His former forester colleagues asked him to write a preface for a book designed to encourage the cultivation of medicinal plant gardens among Koreans. It was a carefully prepared book with modest but charming illustrations of medicinal plants that were not only useful but also beautiful, accompanied by witty haikus and flowery language. Although Japanese colonial foresters had written it, the originator of the book was someone else, the chief of police who came to Korea in 1931. He proposed gardening as a measure to lighten the mood of the colonial people who seemed displeased with just about every activity of the GGK. As he did not see any problem in colonial administration, he located the problem in Koreans who had become depressed as a result of their tasteless lives without proper gardens. Colonial experts made the police chief's proposal look more purposeful by recommending medicinal plants. But while the rose of Sharon, *mugung hwa*, might seem to have been a very suitable plant for inclusion in this work, the book made no mention of it, perhaps following the example of the colonial police who had just imprisoned one Korean principal for planting *mugung hwa* in school gardens and distributing seedlings. The police destroyed the school gardens by uprooting the plants.[24] The book instead restricted itself to plants cherished in Japanese culture, starting with morning glories, whose name in Japanese could be confused with that of the roses of Sharon.[25] Ishidoya's preface took a rather unexpected line. Instead of saying much about the book or morning glories, he chose to write a long note on a Korean text, *Farm Management*, to which Chung had introduced him in 1916. He said that this book was "a perfect manual for the ideal country life of peace and plenty" that he kept it by his pillow to plan his life in retirement. It celebrated a life free from "coercion or command."[26] Did he consider his former colleagues' recommendation of morning glories in the place of roses of Sharon "coercion" for Koreans?

Despite his drastic reappraisal of the *Treasured Mirror*, Ishidoya apparently maintained his doubt about certain coercive imperial impositions or hastily executed

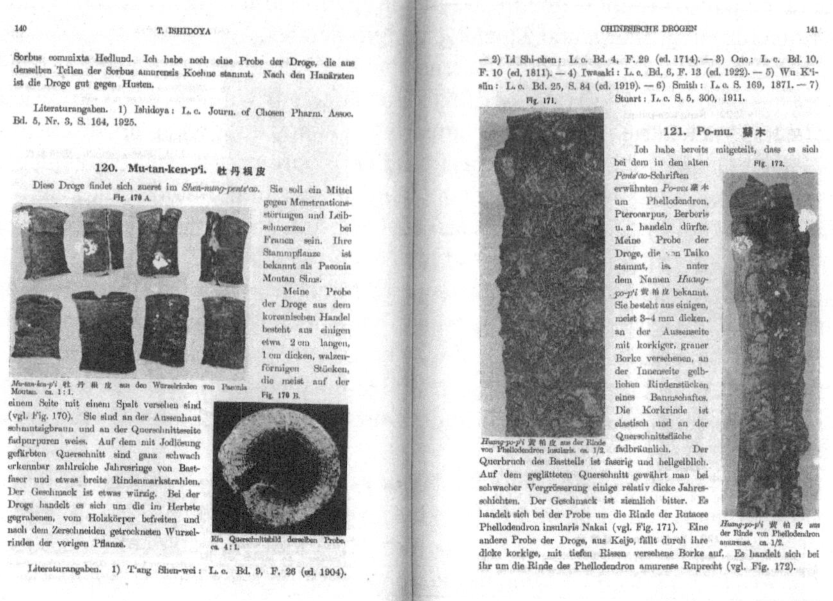

140 T. ISHIDOYA

Sorbus commixta Hedlund. Ich habe noch eine Probe der Droge, die aus denselben Teilen der Sorbus amurensis Koehne stammt. Nach den Hanärzten ist die Droge gut gegen Husten.

Literaturangaben. 1) Ishidoya: L. c. Journ. of Chosen Pharm. Assoc. Bd. 5, Nr. 3, S. 164, 1925.

120. Mu-tan-ken-p'i. 牡丹根皮

Diese Droge findet sich zuerst im *Shen-nung-pents'ao.* Sie soll ein Mittel gegen Menstruationsstörungen und Leibschmerzen bei Frauen sein. Ihre Stammpflanze ist bekannt als Paeonia Moutan Sims.

Fig. 170 A.

Mu-tan-ken-p'i 牡丹根皮 aus den Wurzelrinden von Paeonia Moutan. ca. 1:1.

Meine Probe der Droge aus dem koreanischen Handel besteht aus einigen etwa 2 cm langen, 1 cm dicken, walzenförmigen Stücken, die meist auf der einem Seite mit einem Spalt versehen sind (vgl. Fig. 170). Sie sind an der Aussenhaut schmutzigbraun und an der Querschnitteseite fadpurpuren weiss. Auf dem mit Jodlösung gefärbten Querschnitt sind ganz schwach erkennbar zahlreiche Jahresringe von Bastfaser und etwas breite Rindenmarkstrahlen. Der Geschmack ist etwas würzig. Bei der Droge handelt es sich um die im Herbste gegrabenen, vom Holzkörper befreiten und nach dem Zerschneiden getrockneten Wurzelrinden der vorigen Pflanze.

Fig. 170 B.

Ein Querschnittsbild derselben Probe. ca. 4:1.

Literaturangaben. 1) T'ang Shen-wei: L. c. Bd. 9, F. 26 (ed. 1904).

CHINESISCHE DROGEN 141

— 2) Li Shi-chen: L. c. Bd. 4, F. 29 (ed. 1714). — 3) Ono: L. c. Bd. 10, F. 10 (ed. 1811). — 4) Iwasaki: L. c. Bd. 6, F. 13 (ed. 1922). — 5) Wu K'i-sün: L. c. Bd. 25, S. 84 (ed. 1919). — 6) Smith: L. c. S. 169, 1871. — 7) Stuart: L. c. S. 5, 300, 1911.

Fig. 171.

Huang-po-p'i 黄柏皮 aus der Rinde von Phellodendron insularis. ca. 1/2.

121. Po-mu. 檗木

Ich habe bereits mitgeteilt, dass es sich bei dem in den alten *Pents'ao*-Schriften erwähnten *Po-mu* 檗木 um Phellodendron, Pterocarpus, Berberis u. a. handeln dürfte. Meine Probe der Droge, die von Taiko stammt, ist unter dem Namen *Huang-po-p'i* 黄柏皮 bekannt. Sie besteht aus einigen, meist 3–4 mm dicken, an der Aussenseite mit korkiger, grauer Borke versehenen, an der Innenseite gelblichen Rindenstücken eines Baumschaftes. Die Korkrinde ist elastisch und an der Querschnittsfläche fadbräunlich. Der Querbruch des Bastteils ist faserig und hellgelblich. Auf dem geglätteten Querschnitt gewährt man bei schwacher Vergrösserung einige relativ dicke Jahresschichten. Der Geschmack ist ziemlich bitter. Es handelt sich bei der Probe um die Rinde der Rutacee Phellodendron insularis Nakai (vgl. Fig. 171). Eine andere Probe der Droge, aus Keijo, fällt durch ihre dicke korkige, mit tiefen Rissen versehene Borke auf. Es handelt sich bei ihr um die Rinde des Phellodendron amurense Ruprecht (vgl. Fig. 172).

Fig. 172.

Huang-po-p'i 黄柏皮 aus der Rinde von Phellodendron amurense. ca. 1/2.

Figure 6.1 Pages from Ishidoya's *Chinesische Drogen.*

Source: Ishidoya, *Chinesische Drogen*, vol. I, Keijo: Pharmakologischen Institut der Kaiserliohen, 1934 (Seoul National University Library).

investigations. It seems that he was still searching for a truly enlightening knowledge practice. He needed a concrete way out. The vibrant herbal markets seemed to him to open the way. He found means in the market that allowed him to conduct his research outside imperial research projects "commanded" by imperial authority. Big and small medicinal dealers that he met in the market came to sponsor his research, enabling him to do longer fieldtrips independent of the continental research expeditions. As already mentioned, this field of traditional medicines was very transnational, and included seasoned Korean dealers, whose support Ishidoya welcomed. His section even managed to establish a research laboratory on herbal medicines in Kaesŏng, a city near Seoul that had long been famous for its international medicinal trade based on ginseng. It seems that by making these diverse alliances through the heterogeneous market, he attempted to avoid the homogeneous imperial power that commanded him to do a certain type of research in certain ways.

Chinesische Drogen (1933–1941), the four-volume work in German that became his doctoral dissertation demonstrates a new kind of knowledge practice that he had achieved based on this new sponsorship (See Figure 6.1).[27] First, there was his choice of German for his work on Chinese medicines. German, notably, seems not to have been a language that he could have chosen in writing his official reports. For example, Nakai's *Flora Sylvatica* had all its sections on the use of plants only in Japanese while other academic sections were written largely in Western languages so as to impress Western audiences. More explicitly, Nakai complained in his report for the first scientific expedition to Manchukuo that he could not discuss the uses of plants in his report because the "military authority prohibited their inclusion."[28] Obviously, the Japanese imperial establishment believed that such profitable information should be owned and kept secret by the state. By choosing German for his thoroughly useful investigation of Chinese medicines, Ishidoya in effect begged to differ. As a one-time Esperantist, he seemed to be saying that his knowledge was not for just Japanese audiences, much less for the military. In fact, he presented his research as discovering "unique botanical resources for our human benefits," for "all human beings."[29]

His choice of a Western language, however, reflects the complexities of such cosmopolitan presentation.[30] It above all reflected how the "West" dominated the global studies of traditional Chinese medicines in a way that made German a more marketable choice for his knowledge product. In the global search for lucrative health products, the Chinese herbal market maintained its impressive vibrancy through a strong participation by Western traders as well as scholars. Ishidoya joined this transnational community, shaped by the increased circulation of 'traditional' medicines. He globalized his study of Chinese medicines by engaging with works by Western scholars, which greatly aided the laborious identification of original plants for medicines in the market. He cited a German scholar, a French priest, an American missionary, and even a Russian general and made clear his appreciation of their contributions to his study. Japanese was not a language that could make him part of this community populated by Western scholars from powerful imperial nations, who inherited "the most precious intangible assets that *our* ancestors had bestowed on *us*."[31]

Chinesische Drogen shows another strong influence of the market by its research materials. He focused exclusively on medicinal products from the market. His own investigation began with the medicines that he bought from the market and moved on to their characteristics as dried products processed for sale. He essentially provided visual information on the shapes and colors of those medicines while adding tactile and olfactory information when those were important for their quality evaluation. He carefully included characteristics obtained through commodification, when the plants were changed by various means to prepare them for the market. Leaves, an easy identifier for live plants, were often absent in plants at the market, and sometimes roots had to be tied up or trimmed. Other plants developed corneous tissues or spots during their preparation. He also noted which parts of stems or roots were taken. Based on this information, he produced a unique classification system that might be called "the classification for the market," grouped into seven categories: (1) the whole plant, (2) leaves, (3) flowers, (4) seeds and fruits, (5) roots and rhizome, (6) sprouts, stems and twigs, (7) the timber and bark. Classified by the parts from which the processed medicine was taken, it was a system that helped identify good medicine in the market and helped growers or collectors process their medicines for the best value in the market.[32]

Thus, Ishidoya made his work on Chinese medicines a hands-on empirical investigation of commercially recognized products. His own experience, other similar contemporaneous endeavors, and chemical analyses helped him identify original plants and illuminate medicinal effects and uses. By shaping his research in such a way, he did not have to rely on plants that he had collected under the protection of the Japanese military. He excluded Korean texts that discussed many of those common medicinal herbs from his reference list and "Eastern" traditional knowledge as a whole became subsidiary to his investigation. Neither Li's natural classification nor Hŏ's retrogressive system nor the Linnaean system mattered to his new approach. As a scientist who had criticized an armchair approach detached from the field, he produced a work immersed in his new field of the international herbal market. Nevertheless, his original classification for the market divided by plant parts seems to symbolize how his research was only able to touch on fragments of what came to be called tradition.

Conclusion: Navigating Within or Beyond?

Ishidoya's constant movement did not end. While working on *Chinesische Drogen*, Ishidoya chose to leave colonial Korea in 1939, much to the surprise of his boss, Sugihara.[33] He moved to Beijing, then called Beiping, to join the National Beijing University under Japanese control. There he carried on his new method of research for and through the market, exploring the herbal markets in the old city. He had knowledgeable local collaborators in such explorations, and built meaningful and fruitful relationships with them, including Zhao Yuhuang (趙燏黄, 1883–1960), who was known as a pioneer in modern studies of herbal medicines in China. Ishidoya returned to Japan after the war and finished his career as professor in pharmaceutics at Tohoku University. Although he did not publish any new work after the war, his German work circulated internationally, like Chinese medicines he had studied.[34]

His tumultuous journey offers much to think about. His idealism about proper scientific practice and the Japanese civilizing mission provided uniquely articulate critiques of Japanese scientific practice or imperial policies, while revealing the potency of colonial nature, culture, and people. He and his colleagues in the colony also showed that Japanese settler experts were not mindless agents of the imperial plan created by a few ideologues in metropolitan centers. They, in competition with emerging professionals in all parts of Japan, creatively shaped and promoted their own ideologies for their own rewarding missions.

Ishidoya was a most assiduous navigator in the Japanese colonial field, which he helped shape and expand. Yet, he often veered away from the paths embraced by other Japanese researchers both in mainland Japan and in colonial sites. He did not take the attitude that blatant discrimination or coercive intervention was natural under colonial rule, and gave support to several colonials. He was quite free and brave in expressing his criticism about the powerful metropolitan establishment, while resourcefully carving out his own path in the dynamically changing political situation in which his colonial knowledge practice was set. In fact, his critique was persuasive enough to resonate in locally rooted studies of plants in colonial Korea, as will be shown in the following chapters.

What, then did Ishidoya achieve in the final field that he explored for his research, the traditional medicine market? What distinguished his achievement from other Japanese scientific activity? Was it, as he suggested, a cosmopolitan achievement for all human being? Apart from the Japanese name of the author, people might have thought this was a work written by a Western scholar because of its wide engagement with Western scholarship, and its use of German to facilitate direct communication. His work was quite detached from the cultures that cultivated, collected, named, and used the plants he described and spent no space on "local color." His ideal science came about by uprooting and fragmenting his knowledge practice in a certain Western style, and by changing names of uprooted traditions, just as Nakai changed the names of his willow tree. By doing so, he successfully completed his career as a missionary of imperial science, strengthening the Japanese imperial presence in the market and the global knowledge community but excluding the colonized inheritors of living and transforming traditions. Cosmopolitanism through the market could not be fully cosmopolitan.

Notes

1 Takeuchi Yoshimi竹内好, *Japan and Asia* (일본과 아시아). Translated by Sŏ Kwangtŏk서광덕 and Paek Chiun백지운, Seoul: Somyŏng, 2004, 142–43. For the English translation, see Yoshimi Takeuchi, *What Is Modernity? : Writings of Takeuchi Yoshimi*. Edited and translated by Richard Calichman, New York: Columbia University Press, 2005.

2 Quoted from Stefan Tanaka, *Japan's Orient: Rendering Pasts into History*, Berkeley: University of California Press, 1993, 232.

3 Stefan Tanaka, *Japan's Orient: Rendering Pasts into History*, Berkeley: University of California Press, 1993, 232–33.

4 The pharmacological section had three founding members, including another pharmacologist specializing in the chemical analysis of medicine, Kaku Tenmin (加來天民, 1895–?). Sugihara Noriyuki, "On the Pharmacological Section at Keijo Imperial University (京城帝國大學 藥理學教室)" in The Alumni Association of the Keijo Imperial University京城帝国大学同窓会, *The Deep Blue Yonder: Fifty Years of the Foundation of the Keijo Imperial University* (紺碧遙かに: 京城帝國大學創立五十周年記念誌), Tokyo: The Alumni Association of the Keijo Imperial University, 1974, 212–14.

5 Ishidoya Tsutomu, "On the Original Plants of Korean Traditional Medicine (朝鮮ノ 漢方藥ト其ノ原植物ニ就テ)," *Journal of the Natural History Association of Korea* (朝鮮博物學會雜誌, *JNHK* hereafter) 3 (1925): 1–10, 1.

6 Notably, the original title "朝鮮の漢藥" literally means "Korean Han [i.e. Chinese] Medicine." It shows the great influence of Chinese medicine on Korean traditional medicine, Han being a dynasty in China in antiquity.

7 Ishidoya Tsutomu and Sugihara Noriyuki, "Korean Traditional Medicine (1) (朝鮮の漢藥)," *The Medical World of Manchuria and Korea* (滿鮮之醫界, *MWMK* hereafter) 68 (1926): 17–24, 18, 22. The series was published in the magazine from November 1926 to August 1928. In fact, a high evaluation of the Korean medical tradition was not new in Japan. Even a book published that year in English to celebrate the Japanese scientific past and present, *Scientific Japan: Past and Present*, provided a positive evaluation of the earlier Korean influence on the development of Japanese medicine. The book was published in conjunction with the Third Pan-Pacific Science Congress. Mitsutaro Shirai, "A Brief History of Botany in Old Japan," and Yu Fujikawa, "A Brief Outline of the

History of Medicine in Japan," in The National Research Council of Japan ed., *Scientific Japan: Past and Present*, Kyoto: Maruzen Co., 1926, 213–27, 229–42.

8 For critiques of such presentist readings, see Georges Métailie, "The Bencao gangmu of Li Shizhen: An Innovation in Natural History?" in Elisabeth Hsu ed., *Innovation in Chinese Medicine*, Cambridge: Cambridge University Press, 2007, 221–61; Carla Nappi, *The Monkey and the Inkpot*, Cambridge, MA: Harvard University Press, 2009.

9 He went on to say that all Korean medical texts after Hŏ seemed "simple abstracts of Ming era Chinese texts." Being so certain that Hŏ made had consulted *Systematic Materia Medica*, he further accused Hŏ of making many errors in "just copying" the book. Actually, *Systematic Materia Medica* was introduced to Korea much later than 1610 and Hŏ did not know that book. As full citations were not made in traditional works, Ishidoya apparently mistook one of Hŏ's references as coming from *Systematic Materia Medica*. Ishidoya Tsutomu, "A Story about Herbal Medicines (藥草ノ話)," *JNHK* 10 (1930): 1–7, 6. On classification, see Ishidoya Tsutomu, "The Past and the Future of Korean Medicinal Plants (朝鮮藥用植物の 過去と將來)," *Korea* (1931): 62–74, 70. On the references used for *Treasured Mirror*, see Shin Dong-won신동원, *Hŏ Chun, A Man of Chosŏn* (조선사람 허준), Seoul: Hangyereh Sinmunsa, 2001, 83–87.

10 Oh Chaekun 오재근, "Behind the Naming of Herbal Section as the Decoction Section in Treasured Mirror of Eastern Medicine (조선 의서 『동의보감』은 왜 본초 부문을 「탕액편」이라고 하였을까)," *Korean Journal of Medical History* (의사학) 20 (2011): 263–90. Oh Chae-kun and Kim Yong-jin 김용진, "A Study on the Classification of Materia Medica in Medicinal Part of *Treasured Mirror of Eastern Medicine* (동의보감「탕액편」의 본초 분류에 대한 연구)," *Journal of Oriental Medical Classics* (大韓韓醫學原典學會誌) 23(5) (2010): 55–66, 561; Shin Dong-won신동원, *Hŏ Chun, A Man of Chosŏn* (조선사람 허준), Seoul: Hangyereh Sinmunsa, 2001, 222–32.

11 Prasenjit Duara, *Rescuing History from the Nation: Questioning Narratives of Modern China*, Chicago, IL: University of Chicago Press, 1996.

12 Yet, as many have pointed out, this simplified story that once governed the history of science too did not represent reality. Disenchantment and enlightenment were not polar opposites, and an account of Newton's role in the scientific revolution is much richer when we are given a more realistic picture of Newton as absorbed in alchemical and biblical research and perusing medieval and ancient texts. Steven Shapin, *The Scientific Revolution*, Chicago, IL: University of Chicago Press, 1996; Jed Z. Buchwald and Mordechai Feingold, *Newton and the Origin of Civilization*, Princeton, NJ: Princeton University Press, 2012.

13 Shin Dong-won신동원, *Hŏ Chun, A Man of Chosŏn* (조선사람 허준), Seoul: Hangyereh Sinmunsa, 2001, 82. In Korea, the following works can be mentioned. Kang Myŏngkil's (康命吉, 1737–1801) *New Edition on Universal Relief* (濟衆新編, 1799), and the famous doctor Hwang Doyŏn (黃道淵, 1807–1885) and his son's *Gains and Losses of Medical Orthodoxy* (醫宗損益, 1868), and *Collected Edition of Prescriptions and Medicines* (方藥合編, 1884) are examples. The increasing publication of medical books by doctors reflected changing and more competitive medical markets in late Chosŏn. See Shin Dong-won, "Changes in Late Chosŏn Medical Life: From Gift Economy to Market Economy (조선 후기 의약생활의 변화: 선물경제에서 시장경제로)," *History Review* (역사비평) 75 (2006): 344–91.

14 Ishidoya Tsutomu, "The Growth and Development of the Japanese Cedar and Cypress (朝鮮に移植せられたる「スギ」「ヒノキ」の生育)," *Korean Repository* (朝鮮彙報) 4(12) (1918): 20–32, 32.

15 Ishidoya Tsutomu and Toh Bong-Syup, "Florula Seoulensis (京城附近植物小誌)," *JNHK* 14 (1932): 1–48, 37, 38, 46.

16 Louise Young, *Japan's Total Empire: Manchuria and the Culture of Wartime Imperialism*, Berkeley: University of California Press, 1999.

17 It reveals the unique dynamics of regionalism in Japanese settler colonialism again. On Keijo Imperial University's China research boom, see Chŏng Kyuyŏng 정규영,

"Colonialism and the Politics of Knowledge: The Continent Research by Keijo Imperial University (콜로니얼리즘과 학문의 정치학 -15년전쟁하 경성제국대학의 대륙연구)," *Educational History Research* (교육사학연구) 9 (1999): 21–36.

18 See note 53, Chapter 2, for the similar ideological emergence of "East Asia." Sugihara Noriyuki, "Scientific Examination of Chinese Traditional Medicine (1) (漢方醫學の科學的檢討)," *Kor. & Man* (朝鮮及滿洲), (January 1939): 41–43, 43; "Scientific Examination of Chinese Traditional Medicine (2)," *Kor. & Man* (February 1939): 34–37. As Japan came to be in more direct competition with the West over China in the 1930s, the notion of distinguishing itself from its adjacent colonies with borrowed "Western modernity" became more vulnerable to criticism, too. Japanese imperial ideology, which had been vacillating between brandishing its successfully acquired "Western modernity" and courting its Asiatic "brothers," showed a clear shift around this time toward Asian brotherhood, although the empire's shared tradition with the colony was hardly forgotten by the earliest imperial ideologues like Goto Shinpei (1857–1929). Stefan Tanaka, *Japan's Orient: Rendering Pasts into History*, Berkeley: University of California Press, 1993.

19 Siratori's role was important. Stefan Tanaka, *Japan's Orient: Rendering Pasts into History*, Berkeley: University of California Press, 1993; Koyasu Nobukuni 子安宣邦, *To-a, Great to-a, East Asia: Orientalism in Modern Japan* (동아·대동아·동아시아: 근대 일본의 오리엔탈리즘) Yi Sŭngyŏn 이승연 tr. Seoul: Yŏkbi, 2005.

20 Ishidoya Tsutomu, "The Exploration of Medicinal Plants and Philosophy (藥草の探險と哲學思想)," *Kor.& Man* (1930): 50–55; Ishidoya Tsutomu, "A Story about Herbal Medicines (藥草ノ話)," *JNHK* 10 (1930): 1–7, 5–6. His previous works hardly mentioned Japanese tradition seriously. In addition to Sugihara's concern, his growing contact with Japanese orientalists and philologists must have helped. Okada Nobutoshi (岡田宣捷, 1857–1932) and Fujitsuka Chikashi (藤塚鄰, 1878–1948) were often mentioned in his works.

21 Chinese used to believe that "Western" ginseng from America (in fact Canada) was from France. Ishidoya Tsutomu, "A Story about Herbal Medicines (藥草ノ話)," *JNHK* 10 (1930): 7; Ishidoya Tsutomu, "The Past and the Future;" "An Examination of Original Plants for Ginseng, recorded in *Materia Medica* works (歷代本草ニ所載スル人蔘ノ原植物ニ關スル考察," *JNHK* 9 (1929): 7–16.

22 Though *Dojinkai* was led by a group of politicians and doctors, this organization had tens of thousands of members, showing the Japanese people's support for "civilizing Asia." Osato Hiroaki 大里浩秋, "Dojinkai and Dojin (同仁会と『同仁』)," *Institute for Research in Humanities Brief* (人文学研究所報) 40 (2007): 47–105. For medical missionary works by the West, see Mary B. Bullock, *The Oil Prince's Legacy: Rockefeller Philanthropy in China*, Washington, DC: Stanford University Press, 2011, etc.

23 Ishidoya Tsutomu, "The Flora and its Elements in Jilin Region (吉林地方に於ける植物景と其の要素に就いて)," *Bulletin of a Cultural Survey of Manchu and Mongol by Keijo Imperial University* (京城帝國大學滿蒙文化研究會會報) (1934): 3–9, etc.

24 Lee Jung, "Dreamland for Japanese Bureaucrats: Harmonious Conflicts in Japanese Colonial Korea (관료들의 천국: 일제강점기 약초재배운동의 조화로운 동상이몽)," *The Journal of the Korean Historical Association* (역사학보) 238 (2018): 299–342.

25 Ji Su-Gol 지수걸, "The Militaristic Fascism of the Japanese Empire and the Rural Development Movement in Colonial Korea (일제의 군국주의 파시즘과 '조선농촌진흥운동)," *History Review* (역사비평) 47 (1999): 16–36;

26 It was not so necessary for him to mention the Korean text, as the manual mostly dealt with decorative medicinal plants known to Japanese through Japanese references. The Forestry Association of Kyŏnggi Province 京畿道林業會, *The Beautification of the Countryside and the Cultivation of Medicinal Plants* (農村美化と藥草栽培), Seoul: The Forestry Association of Kyŏnggi Province, 1936, Preface.

27 The research and publication was largely sponsored by a Japanese medicine dealer, Fujisawa Tomikichi (藤澤友吉, 1866–1932). Ishidoya Tsutomu, *Chinesische Drogen*, vol. I–IV, Keijo: Pharmakologischen Institut der Kaiserliohen, 1934–1941.

28 Note that this scientific expedition from Japan was later than the one by the colonial university. Nakai Takenoshin and Honda Masaji 本田正次, *The List of Vascular Plants in Jehol Province* (熱河省ニ自生スル高等植物目録. 第一次滿蒙學術調查研究團報告), vol. 4.4, Tokyo: The First Scientific Expedition Team to Manchukuo 第一次滿蒙學術調查研究團, 1936, Preface.

29 Ishidoya Tsutomu, "Plant Resources of North Asia (北亞細亞の植物資源)," *Research on Continental Culture* (大陸文化研究) (1940): 457–90, 457. Ishidoya tried Esperanto once, a language intended for those who hoped to transcend nationality and foster peace and international understanding, in an abstract for his work. Ishidoya Tsutomu, "Origins of Chinese Ancient Medicines," *Journal of the Pharmaceutical Association of Korea* (朝鮮藥學會雜誌, *JPK* hereafter) 8 (1928): 97–144.

30 In spite of the ascendance of Nazi Germany and Japan's alliance with it just around this time, his choice of German probably did not reflect that political development. German seems to have been the only Western language of which he had a good enough command to write a book, and Ishidoya did not mention anything about the Germany-Japan relationship or Japan's invasion of China, or its attack on Pearl Harbor. Although Ishidoya's work is still found in many libraries and on-line bookstores around the world, have not investigated its reception in Western countries. The effect of Japan's changing relationships with the West on Japanese academia in general and Japan's investigations of China seems a promising topic for further investigation. I thank Mortiz Epple for this point.

31 Although he did not do so in *Chinesische Drogen*, in other works, he was quite specific in acknowledging Western works. Ishidoya specifically noted that he figured out the Chinese names of Mongolian and Tibetan medicines through a German work. Franz Hübotter, *Beitrag Zur Kenntnis Der Chinesischen Sowie Der Tibetisch-Mongolischen Pharmakologie*, Berlin: Urban & Schwarzenberg, 1913. Ishidoya Tsutomu, "Mongolian Medicines and their Original Plants used in Manchukuo Area (蒙疆地方に行る蒙古藥とその原植物)," *JPK* 19 (1939): 97–110.

32 He created his categories for the 515 most-circulated medicines. In the case of medicines for which this kind of division was meaningless, such as those from ferns and fungus, he dealt with them separately. Instead of illustrations, he used high-resolution and rather beautiful photos to accompany his morphological descriptions. Ishidoya Tsutomu, *Chinesische Drogen*, vol. I–IV, Keijo: Pharmakologischen Institut der Kaiserliohen, 1934–1941.

33 Sugihara noted his surprise when Ishidoya suddenly asked for the transfer. Sugihara Noriyuki, "On the Pharmacological Section at Keijo Imperial University (京城帝國大學 藥理學教室)" in The Alumni Association of the Keijo Imperial University京城帝国大学同窓会, *The Deep Blue Yonder: Fifty Years of the Foundation of the Keijo Imperial University* (紺碧遙かに: 京城帝國大學創立五十周年記念誌), Tokyo: The Alumni Association of the Keijo Imperial University, 1974, 212.

34 Ishidoya's work from this period has not been analysed. Hopefully scholars knowledgeable about the Chinese situation can further this research. Notably, Ishidoya continuously published his work on Chinese medicines on the market in Germany, though they were also published locally in Chinese magazines. Zhao Yuhuang and Ishidoya Tsutomu, "Original Plants of the Mongolian Herbal Medicines (蒙古本草之原植物)," *Medical Journal of the Beijing University* (北京大學醫學雜志) 3(2) (1942): 9–19, 115–16; Zhao Yuhuang, Ishidoya Tsutomu and Mi Ching-Shen 米景森, "Research on the Dried Edible Plants in the Beiping Market (北平市場所見食用干菜類之研究)," *Medical Journal of the Beijing University* 4(4) (1942): 213–54, 319–24, etc.

7 Confined to Imperial Privilege

Despite his critical spirit, the case of Ishidoya Tsutomu has shown how difficult it was to work outside of imperial power relations and hierarchies. Furthermore, most Japanese around him in colonial Korea were much more comfortable than he was about the asymmetry of power through which they imposed their rule over the colonized. Their education, as well as what they read in Japanese media, had inculcated in them a form of national pride that saw Koreans as primitive.[1] Novelist Natsume Soseki (夏目漱, 1867–1916) deplored:

> There is no great national emergency. It is not like Japan is disappearing off the face of the earth. There is thus no need to endlessly repeat "nation, nation" all the time… The morality of a nation is of a lower order than the morality of an individual human being. Though diplomatic exchanges look civil and exact, there seems to be no morality between nations. Nations deceive and lie shamelessly to each other. So, if we consider the nation as a unit and make it a standard, we have to be content with a very low degree of morality. To heighten the standard of morality, we have to push to think in terms of individuals.[2]

Despite this call from Natsume for individualism, colonial Japanese generally adopted a nationalistic frame of mind. This chapter looks into how most Japanese maintained such an outlook without actually interacting with the colonized, who they saw as unreasonably proud, despite their subjugated status.

"Science teachers should be educators before they are scientists." It was a solemn admonition for Korean teachers-cum-naturalists made in 1956 by Kamita Tsuneichi (上田常一, in Korea 1924–1945), a Japanese natural history teacher who had taught in Seoul until Japan's defeat. Such an expression of displeasure after so much time had passed suggests that he may have deeply regretted that so many Korean natural history teachers looked more like "scientists before they are educators." Yet the eagerness of Koreans to become botanists or zoologists, and to name their own plants and animals, made Kamita's 20-year service in Korea quite rewarding; the natural history boom among Koreans made the collaboration between Japanese and Korean naturalists on locally rooted (郷土 kyodo in J, hyangt'o in K) research and education possible, although Kamita attributed this

DOI: 10.4324/9781003511755-8

collaboration mostly to the unusual openness of Japanese natural history teachers toward Koreans. "Unlike more [ethnically] exclusive communities in humanities and social sciences," things were "different in the field of natural sciences."[3]

Yet, this allegedly open collaboration between Japanese and Korean naturalists on so-called locally rooted studies of plants needs a closer examination. It was an official policy, since the two most active natural history organizations in colonial Korea, the Natural History Association of Korea (朝鮮博物學會, 1923, NH Association hereafter) and the Association of Natural History Teachers in Seoul (京城博物教員會, 1935, Teachers' Association hereafter) were both open to Korean members. Yet, the open membership policy does not necessarily mean that there was substantial interaction between Japanese and Korean members. Like teachers in other fields, Japanese teachers of natural history also lived in Japanese quarters, separately from Koreans.

Japanese settler naturalists did not say "nation, nation" all the time. Some seem to have found the colony a good escape from the excessive jingoism in mainland Japan. Some seem just interested in the exciting research possibilities that this benighted land might present. In claiming locally rooted research and education, some even seemed to have developed a regional identity attached to Korea. However, being Japanese in colonial Korea could be a great deal only by keeping Koreans under them. By tracing the ups and downs of Japanese natural history teachers in colonial Korea, this chapter reflects on this confined nationalism of the Japanese settler naturalists, and its consequences on their knowledge practice.

Politics for Locally Rooted Natural History Education

As mentioned, the Government General of Korea (GGK) intended to staff colonial schools entirely with Japanese teachers, as it seemed to make no sense that Koreans should be educated by equally uncivilized Korean teachers. The recruitment of teachers from Japan was a serious affair, since colonial schools were the Japanese equivalent of Christian churches in Western colonies, as Komagome Takeshi insightfully notes.[4] Japanese teachers, recruited with scholarships and other benefits, had to become missionaries for the Japanese civilizing mission. In 1929, there were 5,000 Japanese teachers in colonial schools. They constituted only about 30 percent of the teaching staff at the elementary school level, because by then there were so many private Korean schools. Yet they dominated all education above this level, accounting for 85 percent of staff in middle schools, 84 percent in teachers' normal schools, 82 percent in all vocational schools, 91 percent in professional schools and colleges, and 82 percent in the sole university in Korea.[5] Talented and adventurous, many of these Japanese teachers in the colonial outpost displayed strong desires to engage in scientific research, like the Korean natural history teachers that Kamita criticized. By combining the urge to practice science with their profession of teaching, they made quite a successful and typical attempt at scientific professionalization.

At first, the number of Japanese teachers was small, thanks to the colonial education policy, but they moved forcefully from the beginning, organizing themselves and proposing a systematic investigation of Korean nature. In 1911, the natural history teachers in Seoul, four in all, gathered together to create a research group to investigate Korean nature, claiming that Korean animals and plants displayed interesting differences from Japanese ones. They drew up a clear nine-year plan, aiming to "collect and investigate all animals, plants, and minerals in all the provinces of Korea." Mori Tamezo (森爲三, 1884–1962) and Toi Hironobu (土居寛暢, 1884–?), teachers of middle and high schools in Seoul since the protectorate period, were earnest pioneers in natural history research in colonial Korea.[6] The GGK welcomed this initiative of talented teachers, sponsoring their vigorous collection trips to Cheju island, Mount Kŭmkang and T'ongyŏng in 1911, Mount Chiri, Kyŏngsang province and Kanghwa island in 1912, and Mount Paekdu in 1913.[7]

Yet this fortunate situation, unthinkable in Japan did not last, and their trip to Paekdu was last collection trip together. According to Toi, what caused this abrupt end to their serious research plan was the dramatic reduction of the GGK's funding. At first, the GGK provided ample resources to natural history teachers, supporting the teachers' conviction that natural history was the key subject for "scientific imperialism" to civilize unenlightened Korea. It unquestioningly endorsed Toi's ambitious proposal to expand the natural history facilities in his middle school, which eyed "the eventual promotion to university level facilities." It sponsored the expansion of the classroom, a preparation room, and a specimen room on Toi's request and provided enough money to buy microscopes, a photomicroscope, a microtome from Germany and many foreign books including rare copies of Western books, even on subjects such as Indian flora. Toi knew that this was a level of support not given in mainland Japan. Colonial teachers had worked "with a dream-come-true feeling," he remembered.[8]

However, after thus producing a showcase to demonstrate Japan's high standard in science education, the GGK withdrew its support from both natural history education and research. The job was done. Furthermore, for the investigation of natural resources in Korea, the GGK found a more promising alternative that year, Nakai Takenoshin, the best-credentialed researcher at Tokyo Imperial University. The change of situation was so palpable and disappointing that two assiduous investigators in the research group returned to Japan. It seemed that even in Korea, the life of scientists was not an option for school teachers.

Mori and Toi chose to stay in Korea. They continued to search for ways to make a life of research possible in the land where they decided to make their new lives. They soon came to feel less lonely as an increasing number of Japanese educators relocated to Korea, thanks to the pressure of the colonized for more schools. They relaunched their research programs. Their strategies to make their research possible and meaningful were strikingly similar to those of Ishidoya Tsutomu, their fellow settler expert. They criticized Japanese mainland practice while emphasizing the need for the localization of their research and education.

Mori and Toi did not target the whole scientific enterprise in mainland Japan. They focused their critique on natural history education. Mori, who would become the leader of natural history teachers in colonial Korea, asserted: "The conventional natural history education in our country [Japan] is Western-reliant." Especially notable was "the particular influence of the German scholarly tradition." He claimed that they mindlessly translated Western textbooks, even copying the illustrations. Schools in Japan sometimes boasted expensive specimens like those of skunk, anteater, sloth, or penguin imported from abroad. Yet, those were just "a curiosity unrelated to students' real life."[9] Furthermore, elementary and middle school textbooks were just simplified versions of university textbooks. Natural history education in Japan was inadequate to the aim of teaching Japanese children. Colonial educators who would not compromise with such a low standard had to find a proper way to teach natural history based on their own research.

Although a certain exaggeration and simplification seems obvious in their attack, neither Mori nor even Ishidoya invented this critique of Japan's reenactment of Western modernity. Their criticism of Japanese education strongly resonated in Japan, too, creating a movement for "locally rooted" studies and education. Yet, the locally rooted education movement in colonial Korea was distinctive in its earliness and its specific focus on natural history education. In fact, these natural history teachers led this call for locally rooted teaching as soon as they arrived, as early as the protectorate period, while similar calls only became loud in Japan in the 1920s.[10] Natural history teachers in colonial Korea maintained their initiative so strongly that one observer lamented that they distorted locally rooted education in Korea. It was "confined to natural history, the investigation of minerals, animals, and plants" without much concern for history and other social sciences.[11] Indeed, in Japan, what initiated the movement were teachers of geography, and what most animated them was ethnological research, since its foremost leader, Yanagita Kunio (柳田國男, 1875–1962), had inherited Nitobe Inazo's ethnographic interest.[12]

As a result of this early and strong initiative of natural history educators, natural history textbooks in Korea were distinctive throughout the colonial period. Protectorate era texts already contained things not seen in Japanese textbooks, like silkworms, mulberries, cottons, tea, tobacco, rape flower, dandelion, swallow, kelp, carp, octopus, etc. Notably, these protectorate era efforts at localization did not involve much research into colonial nature, since those were widely cultivated and observable things not just in Korea but in Japan. In fact, this was a localization with a specific colonial tinge, aiming at a practical education suitable for the supposed intellectual level of Koreans. Protectorate era Japanese advisors felt that the usual purpose of science education, the scientific understanding of "natural objects and phenomena," did not make sense in the colony. If any education in science was necessary it would be only on practical levels like something "in relation to agriculture, fishery, manufacturing, and home management considering the special situation of Korea." Silkworms, mulberries, cottons, tea, and tobacco entered colonial textbooks less for their regional specificity to Korea but for their strategic value to the Japanese developmental plan in Korea.[13]

Although Japanese natural history teachers in Korea had suffered a serious setback when the ambitious nine-year plan of Mori's team was stalled in 1913, they kept making various efforts for their scientific lives in colonial Korea. Constantly revised colonial textbooks, which often impressed educators in all parts of Japan, show their earnestness. The elementary school natural history textbook published by the GGK in 1931 received especially high praise from educators throughout the Japanese empire. It stood out when it was presented side by side with similar textbooks from all parts of the empire at a conference of all Japanese educators. Mainland teachers envied it for being "progressive" in its educational approach and "locally rooted," although such praise seemed to have its own motive. They used this case of progressive localization in colonial Korea to criticize the tight control of the Ministry of Education, which would not allow such a development in the mainland. Such possible bias and politics aside, this assessment of colonial natural history education became certified and widely circulated by being included in the Japanese historiography of education early on.[14]

Although natural history teachers and educators in Korea showed pride in these recognized achievements, there were constant voices of dissatisfaction and disagreement among them. Ishimura Toshio (岩村俊雄, in Korea 1914–1945), a teacher who guided the writing of a textbook in 1925, expressed his pride: "It is much locally rooted as a Korean textbook in comparison to mainland textbooks; it deals with natural objects commonly seen in Korea."[15] Yet, even for that highly praised 1931 textbook, some educators, like Kamita, raised concerns. He admitted that the colonial textbook was different from Westernized Japanese ones. Nonetheless, he made the curious remark that "students in Korea are in a strange state where they know well about animals and plants in the Japanese mainland while not knowing those in Korea."[16] He claimed that it was because colonial educators had to rely on research from Japan as they could not conduct their own research into Korean animals, plants, and minerals. It made their praised localization mere Japanization, he lamented. What was necessary, he asserted, was to conduct locally rooted research in the colonial field for truly advanced localization that could deal with Korean animals and plants, not Japanese ones. Mori strongly concurred with this need of on-site field research, claiming "Korea's nature is different from Japan's nature."[17]

These natural history teachers raised their voice about this ever urgent need for on-site field research in the late 1930s, sensing a unique chance emerging in the war time emergency. As the GGK considerably increased its investment in science education and research for war, it was a good time to make their demands clear. They compiled an inventory of necessary facilities, devices, and things required for efficient and locally rooted natural history research and education. It seems that they inherited the long wish list that Toi once had made for a university-level research environment. Their list included: a natural history teaching room, a preparation room, a specimen room, a research and demonstration room, specimens and models, machines and equipment, a botanical garden, a nursery for aquatic creatures, a greenhouse, an animal house and a farm, references and teachers' boards, etc., and a budget for supplies, field trips, and specialized analyses.[18] The list demonstrates that they did not just want to be amateurs dabbling with Korean plants and animals while traveling in their leisure time, or just to be the best natural history teachers in the empire. They wanted to become serious scientific professionals who could correct the problems of mainland practice through their research into colonial nature. They wanted to be "scientists before they were educators."

Collaboration on Offer or Collaboration by Need?

Natural history teachers publicized their demands through the journal of the Teachers' Association, launched in 1935. The formation of the Teachers' Association led by Mori and Toi represents the final success of their long organizing efforts begun in 1911. Before creating this forceful organization that was ready to submit such a proposal when the opportunity arose, they had many failures. While this Teachers' Association, with a large Korean membership, indeed displayed the unique openness to Koreans that Kamita had claimed, such openness was not what Japanese teachers showed in the beginning.

At first, it seems to have mostly Koreans who wanted to build relationships with these Japanese teachers-cum-researchers. Yet, Japanese teachers had reasons to become amenable to such Korean initiatives. Toi's and in relation Mori's cases were exemplary. Toi, sending home his fellow researchers except Mori, decided to move to Pyŏngyang, a city central to northern Korea that he thought offered him a new research opportunity. Despite the lack of support from the GGK, he tried to organize a new research group. By gathering local Japanese teachers, he launched a research group, whose ambitious purpose was to "survey and investigate things related to natural history or agriculture in Korea." He named it the "Amoeba Group." He made many efforts to make it a vibrant research group, inviting other Japanese researchers, including temporary visitors from Tokyo for lectures and so on. But it was not successful, and by 1919, it had dwindled away.[19] There were simply not enough Japanese who were serious about such research. At a loss after failing to keep the Amoeba group active in 1919, Toi received a visit from a Korean student, Cho Pok-Sung (趙福成, 1905–1971). Cho brought with him an impressive set of insect specimens that he had himself collected. Cho admired "this Japanese teacher who was also a scientist," and tried to become a teacher like him. Toi showed his appreciation for this serious student, who he made his companion for his weekend collection trips. Cho, the pioneer of Korean entomology, showed great talent as a naturalist, including exquisite drawing skills. A mutually beneficial relationship began.

Mori remained in Seoul. Instead of organizing any group, Mori assiduously published his investigations of Korean plants and animals through official GGK gazettes like the *Korean Repository* (朝鮮彙報), while impressing and seemingly establishing good relationships with GGK officials and other settler naturalist like Ishidoya. In 1920, the GGK requested the compilation of *The Collection of Korean Plant Names* (朝鮮植物名彙) from him, the task unsuitable for Nakai, who was busy displaying the GGK's scientific imperialism to an international audience.[20] This work, compiled by Mori and edited by Ishidoya, made clear their familiarity with both the regional flora and Nakai's work. Mori said the book was "completed on the foundation of a list of plants which I have collected continuously… since my arrival in [Korea]." Yet, he did not neglect to compare his collection with Nakai's, who had already described about 3,000 Korean plants, utilizing 30,000 specimens, "collected and forwarded by colonial collectors" like himself, he noted. Mori in fact sought permission and advice from Nakai to study those specimens in

Tokyo, and finished his work by spending two months at Nakai's lab, "the center of calculation" for Korean flora. Mori added not just Korean names but local information like habitat and use information, in collaboration with Ishidoya, to complete his task for the GGK in 1922.[21]

The GGK rewarded Mori's labor by giving him a two-year study trip to the US and the UK. After his return, he joined the faculty of the newly opened Keijo Imperial University in 1924 as a professor of a general introductory course, making him a hero of natural history teachers in Korea. He decided to specialize in studying animals instead of plants, which Nakai had monopolized. Mori finally received a doctoral degree in 1935 from the Kyoto Imperial University based on his colonial research.[22] Yet he had struggled in his research at Keijo Imperial University, as his Japanese assistants at the university continually deserted him, showing little enthusiasm for his study. So when Mori visited Pyŏngyang, he was deeply impressed on meeting Toi's talented and diligent Korean assistant Cho. Upon learning about his wish to become a teacher to practice science, he eagerly helped Cho get a teaching job in Seoul, also hiring him as his assistant. Although some saw Mori's nurturing of a colonial talent, as an act of generosity, it was also an act of self-help. Cho stayed on and helped Mori's doctoral work immensely.[23] As much as Koreans needed guidance and support in their scientific pursuits, Japanese teachers also discovered the crucial benefit of working with Korean disciples.

In 1923, Mori and Toi must have been really excited to see the impressive organizing of the Natural History Association of Korea (hereafter NHA). The leading organizers of this Association seem to have been Kawasaki Shigetaro (川崎繁太郎, 1878–?), a Tokyo Imperial University graduate at the Geological Survey Institute within the GGK, and Kobayashi Harujiro (小林晴治郎, 1884–1969), also a Tokyo Imperial University graduate working as a physician and parasitologist in the GGK hospital. Kawasaki was the first president and Kobayashi the vice-president. Ishimura Toshio, the proud textbook editorial officer in the GGK, took up a secretarial post while all important ministers in the GGK received honorary membership. The number of founding members was 129, with half of them working as teachers in colonial schools.[24] Toi and Mori joined as founding members. In 1924, it launched a specialized journal to publish the work of their members on Korean animals, minerals, and plants. Ishidoya was also very active both through administrative duties and his contribution to annual meetings and the journal. The NHA was an impressive organization that enlisted all Japanese officials and teachers interested in the enlightening practice of natural history in Korea.

The NHA, despite the august appearance of its honorary membership list, chose a membership policy less typical of a professional organization. The usual policy of limiting membership to a qualified few contradicted with the lessons drawn from the earlier organizing failures of its founding members. Active members alone were still too few to keep the association going. A similar organization in colonial Taiwan, the Natural History Society of Formosa (臺灣博物學會, 1911), like the more prestigious organizations in Europe, such as the Linnaean Society, required recommendation by current members as a means of controlling their membership. The Korean NHA had no such restrictive condition, except for payment of the membership fee and an interest in natural history. In addition to the monthly and annual conferences for members, they organized events such as public lectures and specimen exhibitions to make their presence known to the public and to recruit more people.[25] It aimed not at the elevation of its status by excluding unqualified amateurs, but at visibility and sustainability by building up from a broad foundation to keep the organization going.[26] In response to this open strategy, membership increased gradually, from 174 in 1926 to 308 in 1938. Koreans responded to this effort: the number of Korean members was 21 in 1926 and 43 in 1938, increasing more than in proportion with the overall membership of the association.

Most Koreans who were interested in natural history research gradually joined the NHA. Some Korean founding members, including Cho and some of Ueki's students from the Suwŏn Agricultural and Forestry School, seem to have been recruited by their Japanese teachers. It was a good chance for serious Korean researchers to further their research by joining the association's organized field trips and lectures. They could also present their research and receive feedback at meetings and through journals. Although there were only seven Koreans who decided to publish something through its journal, the productivity of those seven resulted in their work constituting 10 percent of the articles published.[27] Koreans clearly needed this Japanese-centered organization in order to build up their research experience and credentials.

In addition, teachers at the NHA working in Seoul decided to create a separate organization for themselves in 1935, while maintaining their active membership in the NHA. Reportedly, this group was spontaneously organized during a teaching workshop for natural history teachers held in a Korean middle school. Ishimura, now a school inspector, made the proposal for a new organization of natural history teachers. The purpose was to organize cooperative studies and field trips to develop locally rooted teaching materials and methods. According to him, the teachers at the workshop, including Koreans, gave their unanimous support and asked Mori, present as a guest speaker, to be their President. Mori, "the most respected scholar" among these teachers, accepted the offer and became a very active leader. Ishimura, Kamita, and Toi all worked vigorously for this organization of natural history teachers in Seoul.[28]

It seems that their aim was to form a tight-knit group with a similar research agenda to the NHA, but at a smaller and more tightly focused regional level. The Teachers' Association was in a sense exclusive by admitting only natural history teachers in middle schools or above in Seoul, while also admitting Korean natural history teachers in private Korean middle schools. The specimen collections at the Korean school where this new group met greatly impressed Japanese teachers. In fact, without the assurance that they could rely on these active and smoothly Japanese speaking Koreans trained in colonial education, Japanese teachers would not have made this attempt at a specialized organization. Indeed, Korean members made up almost half of its membership, with 32 Korean and 35 Japanese members at its foundation.[29]

Most Korean natural history teachers from Korean middle schools appeared to welcome this initiative and the openness of the Japanese organizers. Koreans fully embraced the purpose that Ishimura proposed: collaborative research to secure locally rooted teaching material and to improve natural history education. They also joined in 1939 when the organization made its proposal for ambitious research funding.[30] The colonizers and the colonized joined together in a strong group with the common need and purpose to further their scientific lives.

Uncommon Ground

Nonetheless, the harmony shown in such organizing initiatives went no further. Although these naturalists in colonial Korea seemed to have freely crossed the notorious dividing line between the colonizer and the colonized, it was only on the surface. In actual locally rooted research, the alleged common goal of their organization, it is hard to find any joint work. Moreover, despite the Japanese teachers' rhetoric about the important difference between Korean and Japanese nature, it is difficult to find specific examples of this discussed in the official publications of the locally rooted studies conducted by Japanese colonial teachers.

Given the outward harmony between Japanese and Koreans in the natural history field, it is striking that it is impossible to list any co-studies of plants and animals between Japanese and Korean teachers. The regions that produced natural history teachers' locally rooted studies were North Hamkyŏng province, Ch'ŏngjin in that province, South Hamkyŏng province, Hwanghae province, Inchŏn, Kwangju, Mokp'o, South Chŏlla province, Pusan, and South Kyŏngsang province, covering six out of eight provinces. Few of the resulting texts are still extant, but to judge from the authorship and prefaces, all these works were exclusively done by Japanese teachers at public Japanese schools.[31] In Seoul in particular, there were no teacher groups or school committees that had undertaken locally rooted research, whereas Japanese and Korean teachers in Seoul were the most active and numerous in both the NHA and the Teachers' Association. Since Mori, their leader, published his textbook on plants in 1937, teachers in Seoul appeared to have cordially accepted Mori's localization as their own, a topic that will be analyzed after we have looked at collective works in other provinces.[32]

The extant works that aimed at locally rooted studies of Korean plants demonstrate why these Japanese teachers did not require collaboration from Koreans on their actual studies. Studies of plants from two regions, South Kyŏngsang and South Chŏlla, both produced in the 1930s, reveal the varying strategies and aims of Japanese teachers in producing their work without Korean collaboration. First, those who produced the work from South Kyŏngsang, commissioned by the regional board of education, were all Japanese teachers in the public Pusan middle school. Pusan is the closest port city to Japan, and had the largest Japanese population. Already in 1910, there were 23,900 Japanese in Pusan, making up one-third of the city's population, and in 1944, there were more than 60,000 Japanese.[33] For this big Japanese community in Pusan, the study that the board aimed was to see whether the current textbooks were localized enough to teach Japanese students in their city.[34] In terms of localization, the textbooks' animal and plant-related content was already very satisfactory, they reported. Most plants discussed in colonial textbooks, 338 species in 145 families, were species that could be observed in their region. What they found lacking in this satisfactory list of local plants was good locally rooted explanations. They decided to correct this by discussing those plants' relationships to local people and the region, adding gardening specifics and their use in food and medicine. The gardening plants highlighted included chrysanthemums, morning glories, and evening primroses, all of which were particularly prized plants in Japan. For edible plants, they provided recipes for common Japanese dishes with 'local' horseradish that came to be widely cultivated.[35] Although Kamita had complained about the strange situation that children in Korea learned only about Japanese plants, the plants discussed turned out to be both Japanese and Korean, owing to the floral similarity between the two lands. For such plants, what seemed lacking for Japanese teachers in Pusan was the cultural understanding adequate for proper Japanese. They seemed to be worried that Japanese students born and raised in Korea would forget Japanese culture, and become kimch'i-eating Koreans.

The Plants of South Chŏlla was a much more ambitious work. Although initiated in 1931 by a small plant study group in Kwangju, this project became a ten-year study involving 300 elementary schools in the region, and secured the support of Ishimura, who was then the Provincial School Inspector, and Mori. They planned to make a new list of regional plants by collecting plants through teachers. They "requested" teachers in 300 provincial schools to send specimens for 100 summer and 100 autumn plants from their vicinity. By 1938, the provincial education office duly accumulated 63,000 dried specimens, which in quantity almost matched the collections at the Tokyo Imperial University. They were making a new "center of calculation" for their regional flora. The study group in Kwangju, led by Mori, completed the examination of all their specimens by the following year to produce their impressive volume.[36]

Both Mori and Ishimura showed satisfaction and pride at this achievement, not just as regional education material but as a scholarly contribution. Ishimura argued that the province, "in a special location regarding the plant distribution of the peninsula," had many precious materials that made it possible to produce new knowledge with scholarly implications. It was a claim that closely echoed Nakai's claim about Korean flora, though they could not find any new species to name. Mori defined its "great scholarly contribution," as that of "discovering new facts about plant geography and confirming the existence of many valuable plants in the province."[37]

Mori and Ishimura seemed to have thought it important to assert their scholarly contributions by differentiating their work on the plants of South Chŏlla from that of Nakai. For one thing, they excluded Nakai's numerous works on Korean flora from their long reference list. Although they strictly followed Nakai's classification and arrangement in cataloging 1,939 kinds of plants in South Chŏlla, they made no attempt to compete with Nakai. Instead, they emphasized other "regional" scholarly contributions. They listed all the elements that the settler naturalists like Mori and Ishidoya had long been cultivating, such as knowledge about plant habitats, seasonal changes, life cycles, and the local uses of plants. They made a separate list dividing plants by use: edible plants, garden plants, poisonous plants, medicinal plants, etc. They even included "folk remedies," comprising about 200 local plants.[38]

The South Chŏlla team seemed to be deepening their expertise on colonial nature and culture. But an examination of the content of their book implies otherwise. For example, the folk remedies that they offered for those 200 "wild and cultivated plants" in South Chŏlla came almost verbatim from one Japanese book on folk remedies, published in Tokyo in 1925.[39] It was just one of the Japanese books that they cited instead of using Nakai's works to obtain local knowledge about South Chŏlla plants. In fact, they did not even include the Korean names of plants in their work. For their localization using Japanese references, Japanese names were enough. It was a localization for which neither Korean collaboration nor understanding about local nature and culture were necessary. These ambitious Japanese scholars seeking to present their work as some locally based scholarly achievement different from Nakai's internationally recognized universal knowledge did so by what was in effect a further Japanizing of their work, as the Pusan teachers had done. It was not a localization for understanding local plants or cultures but a means of introducing Japanese meanings for plants common in Japan and South Chŏlla.

Mori took this Japanese localization in Korea further, both through his textbook published in 1937 and while making the funding proposal with Korean teachers in 1939. The common and familiar emphasis in both was that Korea was a land connected to China, highlighting its strategic importance for Japan's expansion to Manchuria and beyond. For settler teachers, as for Ishidoya and Nakai, the war emergency was thus an opportunity to reveal their expertise cultivated in the colonial land to powerful imperial players leading the expansion, which might lead to

Figure 7.1 The cover of *Newly Revised Plant Textbook*, Tokyo: Sanseido, 1938 (National Museum of Korean Contemporary History).

research funding for their "continental" research. Mori's textbook demonstrated the connection of Korea and Manchuria with its cover design, as he himself explained. It was a map of Korea and Manchuria, decorated with the image of *Pinus koraiensis*, a species of pine trees, which, he pointed out, was "distributed throughout Korea and Southern Manchuria" (see Figure 7.1).

While this nut-bearing pine tree, commonly called Korean pine in both Japanese and English, was also securely established in Japan, Mori mentioned only Korea and Southern Manchuria as its habitat. He added that this tree has five-needle shaped leaves, as shown in his illustration, unlike the Japanese pine, which has two-needle shaped leaves.[40] This presentation appears to be an attempt to differentiate his expertise gained from observing this "unique" pine tree in Korea from that of scholars in mainland Japan, while promising the extension of his expertise on Korean plants to the continent. The textbook also discussed some unique Korean mountain flowers and praised Korean wild cherry trees, which may perhaps be evidence of his attachment to the colonial land that had so far given him all the research and career opportunity unthinkable at home.[41]

Yet in 1939, he proposed a new way of looking at Korea. Instead of emphasizing the regional specificities of Korea and Manchuria compared with Japan, he solemnly declared that Korea was "a peninsula within East Asia." The great task for the educators in the peninsula, which they should carry out with "the resolution equal to that of the Imperial Army," was to recognize Korea's location within this Japanese East Asian empire so as to benefit the empire's development.[42] The view of Korea as a conduit for Japan's expansion to the continent was unchanged, but there was now no need for Korea's special characteristics to be emphasized, as all the regions of Japanese East Asia were to unite in the service of the Japanese empire. Teachers living in the imperial territories of Korea, Manchuria, and China might contribute teaching materials from their own lands, not to show the distinctiveness of those regions but to show "what is Japanese" from them, as teachers in South Chŏlla and Kyŏngsang had done. Such contributions should have "characteristics fitting Japanese climate and culture."[43]

Kamita, who boasted of the natural history field's openness to Korean scientists-cum-teachers, made it clear the next year that in this increasingly important task, they could no longer have to keep Koreans as their colleagues. He clarified the reason through lessons from nature, that he claimed "penetrated the biological world." The first small lesson from nature was that a part exists only because there is a whole and thus the part should contribute something to the whole. A more important lesson from the biological world was that "for the whole, the parts within it should accept extreme inequality among themselves." If every part attempted to become the head, the whole could not sustain itself. So, a certain part should accept the role of hand and follow the head's order. He corrected the mistaken belief that the call for the "Uniting of Japan and Korea (內鮮一體)," which forced people to change their Korean names to Japanese ones, meant an equal pairing of the two. This union would be based on the extreme inequality dictated by natural law, which made Japanese "first-class citizens" with Koreans forming the "second-class." Kamita ingeniously presented these important lessons obtained from the biological world as a reason why the GGK had to prioritize support for biology, instead of physics or chemistry, in increasing its support for science education.[44] Their prize, securing more support for their natural history field, was the one thing unchanged in the constant Japanese reformulations of the role of local research. Kamita and his colleagues simply found a new strategy in this war situation, which did not require uncomfortable collaboration with "second-class citizens." In a way, it was a renewal of their early commitment as proud representatives of the Japanese civilizing mission, but paradoxically they now claimed that this mission could be sustained only by failing, i.e., by keeping the "second-class citizen" permanently "second-class."[45]

In spite of their long sojourns in Korea, very few Japanese teachers learned Korean and few Japanese married Koreans. In this geographically closest colony to the empire, Japanese experts came in large numbers and created a separate society in Japanese quarters instead of mingling with the uncivilized and unwelcoming colonized, except in some public settings. It was not necessary for them to Koreanize. They could easily bring Japanese culture to this colony, in addition to their favorite vegetables and garden plants. Displaying Japanese culture, while strongly opposing any real assimilation of Koreans, was their way of keeping their privileged lives inside their "little Japan."[46] The meaning of Korea to these Japanese teachers was simply a territorial expansion that served to and elevate their own way of life. Thus, their aim in localization was not to learn about the country in which they lived, but to transform and appropriate it for their own by dismissing its peculiar nature and culture.

Conclusion: The Blinding Privilege of Becoming Japanese

These Japanese natural history teachers were not the colonial ideologues who wrote the discriminative or exploitative laws to manufacture the artificial superiority of imperial Japan; they were not the colonial police who harassed and interrogated any Koreans who seemed to doubt these "natural" laws; they were not the jingoistic military men who instigated and expanded wars on fraudulent pretenses. They were law-abiding teachers who enjoyed certain privileges of colonial life and accepted the honor that their imperial nation granted them for their ventures in this foreign land. They chose to accept their civilizing mission in various degrees, for different purposes, and with varying sincerity. Some of them even showed the courage to criticize Japanese education as Western-oriented and sterile. From a Japanese perspective, they could even be seen as progressive, subtly criticizing the Japanese government for not permitting more liberal experimentation in science education.

Nonetheless, the final outcomes of their locally rooted plant studies resemble a mechanistic assertion of imperial ideology, as if crafted by a jingoistic ideologue. In that sense, their work exposes the power of social and intimate boundaries formed by what seemed like 'natural' and 'individual' preferences. They simply chose to surround themselves with people who spoke the same language and had the same tastes in food, cloths, hairdos, habits, and garden plants, as much as many of us do. They did not need to overtly express racial prejudice against Koreans. Instead, they maintained their comfort zones through seemingly innocent personal preferences. But were these choices innocent or natural? John Dewey, the American philosopher who visited Japan and China in the late 1910s, offers a different perspective, writing:

> I still believe in the genuineness of the Japanese liberal movement there, but they lack moral courage. They, the intellectual liberals, are almost as ignorant of the true facts as we are, and enough aware of them to wish to keep themselves in ignorance.[47]

Dewey was referring to the liberal movement in Japan and its silence about Japan's imperial exploitation in China or in Korea conducted in the guise of benevolent assistance to its neighbors. If the fact that the immoral national entity of Japan was using all manner of underhanded tactics for its self-interest was unknown to these liberals, he suggests that they actively guarded their ignorance because they already knew enough to wish to avoid knowing more. They endeavored to keep what might be called an active ignorance. The inhibitions of the settler experts against mingling with the colonized allowed them not to really know any of them personally, and not to recognize the similarity or even equality of these 'different' people with themselves. While colonial regimes everywhere showed awareness of the porous and ambiguous boundaries between the inferior colonized and the superior imperial beings by limiting real contact and fretting about racial mixture or reverse assimilation, or the naturalization of imperial subjects, the danger must have looked greater in this colony that was a closest neighbor of imperial Japan.[48] Their self-imposed confinement was an active effort not to recognize the true significance of their actions and to keep their privileges as colonizers, at the cost of their professed aim as researchers—to conduct truly engaged local exploration and research. These once adventurous Japanese in colonial Korea compromised their own values by confining themselves to the narrow limits of their imperial identity as Natsume Soseki and many other critical minds deplored at the time.

Fortunately for these Japanese natural history teachers working in colonial Korea, they did not seem to find the price of their privileged colonial lives too high, enjoying these advantages well into later years. They took pride in the localization that they achieved and post-war historians often validated their pride as well-deserved. By seemingly erasing any references to biology's role in defining the place of the second-class citizen, they were also able to maintain a posture of political neutrality regarding their localization efforts, presenting their work in retrospect as purely academic, educational, and commendable. However, there was something incomplete about their contentment: an underlying anxiety about ambitious Koreans who wanted to practice science rather than just being teachers. Once again, there may have been an unspoken awareness that these Korean researchers could accomplish what the Japanese could not—true localization—and in doing so, expose the weakness of their own claims.

Notes

1 Chung Yountae 정연태, *Everyday Colonial Discrimination in Japanese Korea* (식민지 차별의 일상사), Seoul: Purŭnyŏksa, 2021, 233–43.
2 Natsume Soseki, *My Individualism and Others* (나의 개인주의 외). Translated by Kim Chŏnghun, Seoul: Ch'aeksesang, 2004, 73–74. For the English translation, see Soseki Natsume, *My Individualism and the Philosophical Foundations of Literature*. Translated by Sammy I. Tsunematsu, North Clarendon: Tuttle Publishing, 2011.
3 By natural history teachers, I mean those who taught natural history (博物) in middle schools. There were also elementary school teachers who taught one unified science course *rika* (理科, literally disciplines about principles), heavily governed by natural history, as mentioned in Chapter 1. Kamita Tsuneichi, "Outline of Science Education in Korea (朝鮮の理科教育史概説)," *Bulletin of Korean Studies* (朝鮮學報) 9 (1956): 337–46, 346.
4 Komagome Takeshi 駒込武, *The Cultural Integration of the Colonial Empire of* Japan (식민지제국 일본의 문화통합). Translated by Oh Sŏngchŏl 오성철 et al., Seoul: Yŏkbi, 2008, 19.
5 Koreans in higher education mostly worked as assistant or temporary teachers. This statistic is taken from Shindo Toyoo 新藤東洋男, *Japanese Teachers in Colonial Korea* (在朝日本人教師: 反植民地教育運動の記録), Tokyo: Shiraishi Shoten, 1981, 29. It discusses the work of Joko Yonetaro (上甲米太郎, 1902–1987), who tried to construct a socialist anti-imperial education criticizing Japanese colonial education and working with Korean activists.
6 Mori Tamezo graduated from the Temporary Teachers' College affiliated with Tokyo Imperial University in 1904 and came to Korea in 1909. He stayed in Korea until the end of the War as Professor at Keijo Imperial University. For him, see Kamita Tsuneichi, "Memories of Prof. Mori Tamezo (森為三先生の追憶)," *Bulletin of Korean Studies* 26 (1963): 114–19. Toi seems to have graduated from Tokyo Higher Normal School. He worked as the vice principal in Pyŏngyang High School (1916–1921) and the principal in Shinŭiju High School (1922–1928) and stayed in Korea working in other private schools and the Science Museum in Korea, according to GGK Personnel Records in the Korean History Database. The number of members was only four due to the slow build-up of colonial education. The other two worked in the Seoul Normal School and the Seoul Foreign Language Institute, although they soon left Korea.
7 Toi Hironobu, "Thinking of Old Days as a Natural History Teacher (三昔前の博物教員としての思出)," *Journal for the Association of Natural History Teachers in Seoul* (京城博物教員會誌, *JNHT* hereafter) 1 (1938): 14–22.
8 Toi Hironobu, "Thinking of Old Days as a Natural History Teacher (三昔前の博物教員としての思出)," *Journal for the Association of Natural History Teachers in Seoul* (京城博物教員會誌, *JNHT* hereafter) 1 (1938): 16, 17.
9 Mori Tamezo, "Thoughts on Natural History Textbooks in Middle School (中等學校博物教材に對する所感)," *JNHT* 3 (1939): 1–2, 1.
10 In Japan, the movement for localized education was launched in the 1920s with the Taisho-era Free Education Movement. Kano Masanao 鹿野正直, *The Popular Movement and its Ideas in Modern Japan* (近代日本の民衆運動と思想), Tokyo: Yuhikaku, 1977, 39–172. This popular movement became official policy in 1930 as the Japanese government began to sponsor it by establishing the Alliance for Localized Education. History of Science Society of Japan 日本科学史学会, *A Compendium of History of Science in Japan* (日本科學技術史大系), vol. 10, Tokyo: Daiichi Hoki Shuppan, 1965, 17–31. Notably, the movement for "localized" teaching or studies, which often aimed at the alienated countryside under rapid urbanization, was quite prevalent in other industrialized countries as well at the time, although each country had its own variations. Its American the equivalent would be the "Country Life Movement," led by the likes of Liberty Hyde Bailey and President Theodore Roosevelt. L. H. Bailey, *The Country-Life*

Movement, New York: MacMillan Co., 1911. In the US also, many natural history teachers, especially women, were active participants in this movement. Sally G. Kohlstedt, *Teaching Children Science: Hands-on Nature Study in North America, 1890-1930*, Chicago, IL: University of Chicago Press, 2010.

11 The criticism was made in 1931 when the GGK made it official policy, following the Japanese government. Fukushi Suenosuke 福士末之助. "Localized Studies as the Foundation of the Educational Approach to Promote Popular Mores (民風作興上教育的方法としての郷土調査)," *Korea of Art and Education* (文教の朝鮮, *KAE* hereafter) 73 (1931): 6–16, 11–12.

12 Nitobe, the author of *Bushido: The Soul of Japan*, defined "locally rooted studies" as the studies of the regions where one lives. To know oneself, one has to know where one lives, the geography, history, and natural environment of the place. It also makes children love their town and then extend that love to the wider world. Yanagita Kunio founded the journal *Locally Rooted Studies* in 1913, inspired by Nitobe's call. He adopted the Western ethnographical method for the purpose and promoted studies of folklores, fables, customs, everyday lives. Nitobe Inazo, "Studies of a Region," in *Essays* (随想錄), Tokyo: Teimi Shuppansha, 1907, 269–82, 275. Ito Junro 伊藤純郎, *Studies on the Localized Education Movement* (郷土教育運動の研究), Kyoto: Shibunkaku Shuppan, 2008.

13 The first revision made in 1913 removed some items from this "regionalized" textbook, such as dandelions and butterflies, not because they were not regional but because they did not give practical information. Those "impractical" things came back when they divided the curriculum for vocational and regular education. Song Minyoung 송민영, "The Implication for 'The Seventh Extracurricular Curriculum' by the Comprehensive Science Education of the GGK Era (조선총독부시대의 종합적인 이과교육이 '제7차 특별활동 교육과정' 에 주는 시사점 고찰)," *Studies on Education* (교육학연구) 41 (2003): 117–46, 125; Kamita also detailed the history of the regionalization of natural history textbooks. Kamita Tsuneichi, "The Guidelines for the Elementary School Science Education for the Promotion of Science (科學振興に伴ふ小學校理科教育の指針)," *KAE* 181 (1940): 38–56, 39–41. Interestingly, the textbook published by Koreans during the protectorate periods, *New Natural History for Elementary School* (新撰小博物學, 1907), also emphasized the regionalization of textbooks. "The animals, plants, and minerals in the book are something familiar to us, easy to study and experiment on, and with strong relevance to our lives." Its relationship to textbooks by Japanese teachers seems worth investigating. Hŏ Chaeyŏng 허재영 ed. *The Textbooks of the Protectorate Period* (통감시대 교과서 자료) vol. 8. Natural History, Kwangmyŏng: Kyŏngjin, 2011, viii.

14 Textbooks in Manchuria were also praised for their progressive nature. These views became common in historiography by entering into the authoritative history written by the History of Science Society of Japan in the 1960s. Kanbe Isaburo 神戸伊三郎. *The Development of Japanese Science Education* (日本理科教育発達史), Tokyo: Keibunsha, 1938; *Compendium*, vol. 10, 24–25. A similar move to use or to promote 'reform' or 'advanced' measures in the colonies to advance the foundation of science in Japan is also seen in population science, and became general in many fields. Aya Homei, *Science of Governing Japan's Population*, Cambridge: Cambridge University Press, 2023, 56.

15 Ishimura Toshio, "The Science Education Method in Elementary School (普通學校に於ける理科教授法)," *KAE* 10 (1925): 55–60, 55, 56.

16 Kamita Tsuneichi, "Comments on New Natural History Textbook: Farm Animals (新理科書の教材の解說 家畜)," *KAE* 75 (1931): 65–75; "On Important Sea Creatures in the New Natural History Textbook Volume 2 (新理科書巻二にある朝鮮の重要なる海生産物に就いて)," *Studies on Korean Education* (朝鮮の教育研究) 5 (1932): 103–12; "The Reality of Localized Education in Natural History in Korea (第五講朝鮮に於ける 理科郷土教育」の實際)," *Studies on Korean Education* 72 (1934): 62–65.

17 Mori Tamezo, "Thoughts on Natural History Textbooks in Middle School (中等學校博物教材に對する所感)," *JNHT* 3 (1939): 1.
18 Toi Hironobu, "Small Words on Natural History (博物小言)," *JNHT* 3 (1939): 3–7; Teachers' Association, "The Survey of Natural History Education Facilities (博物教育設備調査書)," *KAE* 12 (1939): 69–81.
19 Toi Hironobu, "The Now Gone Amoeba Group (今は故きアメバ會)," *JNHT* 2 (1938): 10–16.
20 Mori's English title is "An Enumeration of Plants hitherto Known from Corea." I use a more literal translation. Mori Tamezo, *The Collection of Korean Plant Names*, Seoul: GGK, 1922.
21 Ishidoya took up the final edits of *The Collection of Korean Plant Names* because Mori had to leave for his study trip. Mori Tamezo, *The Collection of Korean Plant Names*, Seoul: GGK, 1922, 1, 5.
22 Kamita Tsuneichi, "Memories of Prof. Mori Tamezo (森為三先生の追憶)," *Bulletin of Korean Studies* 26 (1963): 114–19
23 Doctor Cho Pok-Sung Memorial Committee 관정조복성박사기념사업회, *The Record of Dr. Cho Pok-Sung's Insect Collecting* (조복성곤충채집여행기), Seoul: Korea University Press, 1975, 30–31, 38–39, 44–51; Kim Sungwon 김성원, "The Context of a Korean Naturalist's Career-Building in Colonial Korea: Cho Pok Sung as an Example of Colonial Entomologist (식민지시기 조선인 박물학자 성장의 맥락: 곤충학자 조복성의 사례)," *Journal of the Korean History of Science Society* 30(2) (2008): 353–82.
24 Checked by the institutional affiliation in the member roster.
25 There were two Western members, who also contributed to the journal. One Russian, B. W. Skvortzow, and one missionary elementary school principal, D. J. Cuming. "Founding KANH Amassing All Interested Parties, Including Teachers in Colleges and Middle Schools (斯界의 有志를 網羅하야 博物學會를 創立, 각 전문학교와 중등학교의 교수 교육가 전부 참가하여 박물학회)," *Dong-a Daily*, October 18, 1923; "Popular Lectures by Korean NH Association (博物通俗講演, 朝鮮博物學會總會 박물학회)," *Dong-a Daily*, October 22, 1927; "Natural History Class by Korean NH Association (博物學講習會)," *Dong-a Daily*, October 14, 1931; "Korean NH Association Briefs (회무보고)," *Journal of the Natural History Association of Korea* (朝鮮博物學會雜誌, *JNHK* hereafter).
26 In allegedly better assimilated Taiwan, the membership of the similar Natural History Society of Formosa was by referral and through publication. No active Taiwanese members are observed.
27 For the analysis of the articles in the journal, see Lee Byŏnghun 이병훈, and Kim Chint'ae 김진태. "On the Introduction of Western Modern Biology: Focusing on Animal Systematics (西洋 近代 生物學의 國內 導入에 관한 연구: 동물분류학을 중심으로)," *Animal Systematics, Evolution and Diversity* 10(1) (1994): 85–95.
28 Kamita Tsuneichi, "The History of the ANHT (京城傳教員會活動史)," *JNHT* 1 (1938): 5–13.
29 Including the Keijo Botanical Society (京城植物會) which changed its name into the Research Group for Korean Plants in 1935 (朝鮮植物研究會), founded by Chang Hyŏngdu, Park Mankyu and three Japanese in 1934, there were some more small organizations. It published its own periodical entitled "Korean Regional Plants (朝鮮鄉土植物)." None of its publications remain. "The Foundation of the Keijo Botanical Society (경성식물회창립)," *Donga-a Daily*, April 21, 1934; Lee Deok-Bong 이덕봉, "History of Recent Development of Korean Botany: The Institutes and Status of Botany in Colonial Korea (최근세한국식물학연구사: 일제통치하 한국에 있어서의 식물학 연구에 관한 시설과 그 실태)," *The Journal of Asiatic Studies* (亞細亞研究) 4(2) (1961): 101–49, 116–17.
30 The funding proposal was signed by two Japanese, including Kamita, and two Korean members.

31 The initiators of these localized studies vary. Some works were orchestrated by the Provincial Board of Education, some only by local schools and some by local research groups. Kamita Tsuneichi, "The Guidelines for the Elementary School Science Education for the Promotion of Science (科學振興に伴ふ小學校理科教育の指針)," *KAE* 181 (1940): 47.

32 His textbook was authorized by the GGK but published by a Tokyo publisher. Given the number of remained copies, it must have been widely used in colonial schools. Mori Tamezo, *Newly Revised Plant Textbook* (新修 植物教科書), Tokyo: Sanseido, 1937.

33 Although the Japanese population in Pusan was rising continuously, the proportion gradually came down after a peak of 50 percent in 1914. They were about 20 percent of the city population in the 1940s. Hong Soon-Kwon 홍순권, "Population and Social Stratification of the Japanese Society in Pusan under the Rule of Japanese Imperialism (일제시기 부산지역 일본인 사회의 인구와 사회계층구조)," *History & the Boundaries* (역사와 경계) 51 (2004): 43–73.

34 South Kyŏngsang Province Board of Education 경상남도교육회, *Locally Rooted Studies: Natural History* (鄕土硏究 博物), Pusan: South Kyŏngsang Province Board of Education, 1934.

35 In the list of plants, there are a few foreign plants coveted by the Japanese. Rosewood with "beautiful reddish violet timber" and aerides, an expensive decorative plant, are examples. South Kyŏngsang Province Board of Education 경상남도교육회, *Locally Rooted Studies: Natural History* (鄕土硏究 博物), Pusan: South Kyŏngsang Province Board of Education, 1934, 12–13, 21–22, 141, 145, 149, 192, 196–97.

36 The procedure of the project is described in the foreword written by one of the teachers. It was priced to be sold. South Chŏlla Province Board of Education 全羅南道教育會, *Plants of South Chŏlla* (全羅南道植物), Keijo: South Chŏlla Province Board of Education, 1940. Foreword. Bruno Latour, *Science in Action: How to Follow Scientists and Engineers through Society*, Cambridge, MA Harvard University Press, 1987.

37 The procedure of the project is described in the foreword written by one of the teachers. It was priced to be sold. South Chŏlla Province Board of Education 全羅南道教育會, *Plants of South Chŏlla* (全羅南道植物), Keijo: South Chŏlla Province Board of Education, 1940. Foreword. Bruno Latour, *Science in Action: How to Follow Scientists and Engineers through Society*, Cambridge, MA: Harvard University Press, 1987, prefaces.

38 It cannot be an accident that they excluded all of Nakai's relevant works from their long list. Bruno Latour, *Science in Action: How to Follow Scientists and Engineers through Society*, Cambridge, MA: Harvard University Press, 1987.

39 Maruyama Fusao 丸山房雄, *Folk Remedies and their Usages, with Scientific Explanations* (民間藥及其利用法), Tokyo: Ueda Insatsujo Shuppanbu, 1925. Maruyama seems to have been a most prolific investigator of folk medicines in Japan. *The Plants of South Chŏlla* states that it provides the dosages of those folk remedies in traditional measurement units instead of modern ones "because Korean people still use traditional units for things like herbal medicines," as if to confirm the Korean origin of its information. But Maruyama also used traditional units for his Japanese folk remedies. South Chŏlla Province Board of Education 全羅南道教育會, *Plants of South Chŏlla* (全羅南道植物), Keijo: South Chŏlla Province Board of Education, 1940, 219–27, 219.

40 Franz von Siebold had named it *Pinus koraiensis* probably because Japanese works that he referenced all recorded its origin as Korea. Korean delegates gave these pine nuts as a gift when they visited Japan and China, and Japanese came to name this tree Chosen matsu 朝鮮松, Chosŏn pine, and Chinese Sea Pine (海松) as they often named things from abroad with the epithet 'sea,' according to Ono Ranzan. It had existed in both China and Japan. Ono Ranzan, *Revised Enlightenment of Systematic Materia Medica* (重修本草綱目啓蒙), vol. 22, Kyoto: Hishiya Kichibei, 1844, 91.

41 There were also inner covers with colorfully illustrated Korean mountain and wild flowers. Mori Tamezo, *Newly Revised Plant Textbook* (新修 植物教科書), Tokyo: Sanseido, 1937, 16.
42 Mori Tamezo, "Thoughts on Natural History Textbooks in Middle School (中等學校博物教材に對する所感)," *JNHT* 3 (1939): 2.
43 Tamezo Mori, "Thoughts on Natural History Textbooks in Middle School (中等學校博物教材に對する所感)," *JNHT* 3 (1939): 2.
44 Kamita Tsuneichi, "The Guidelines for the Elementary School Science Education for the Promotion of Science (科學振興に伴ふ小學校理科教育の指針)," *KAE* 181 (1940): 38–56, 38, 44, 44, 45.
45 After this, their survey for science education facilities in 1940 was conducted only by Japanese and the number of Korean members decreased to 17 and, in general, the ANHT lost its vigor. Kamita Tsuneichi, "The Guidelines for the Elementary School Science Education," 42 *JNHT* 3 (1940): 4. Kamita was just one of many who used biology to justify certain political rules. Julia Adeney Thomas, *Reconfiguring Modernity*, Berkeley: University of California Press, 2001; Aya Homei, *Science of Governing Japan's Population,* Cambridge: Cambridge University Press, 2023, 47–48.
46 On Japanese settlers' stances on the "assimilation" of Koreans and the city structure that sustained their privileged lives, see Todd A. Henry, *Assimilating Seoul: Japanese Rule and the Politics of Public Space in Colonial Korea 1910–1945*, Berkeley: University of California Press, 2014; Lee Jeongseon 이정선, *Assimilation and Exclusion: Japanese Assimilation Policies and the Korean-Japanese marriage* (동화와 배제: 일제의 동화정책과 내선결혼), Koyang: Yŏksabipyŏngsa, 2017.
47 John Dewey et al., *Letters from China and Japan*, New York: E.P. Dutton Co., 1920, 169.
48 Frederick Cooper and Ann Laura Stoler, eds., *Tensions of Empire: Colonial Cultures in a Bourgeois World*, Berkeley: University of California Press, 1997.

8 Liberating through Provincial Botany

The emerging knowledge practitioners shaped in various colonial interactions of the early twentieth century displayed many kinds of provincialities in their practices. The colonizers themselves, paradoxically, showed what may be called 'outward provincialities' by asserting that their own standards and tastes were the only ones acceptable, and imposing them on others. The form of provinciality expressed by Korean researchers in Korean, the threatened language of the colonized, and then in Japanese, as it became in time the only language allowed to them, took the form of an inwards turn. The colonized looked deep into themselves, and their society and nature, for the meaning of being "second-class citizens" in their own land while actively engaging with foreign knowledge.

To appreciate these different kinds of provincialities, this last chapter observes the collective botanical practices of Korean researchers that they adopted upon establishing a Korean-only research group in 1933. Notably, this was not a time known for much unity or for the any kind of strong collective movements among Koreans. The chasm between Korean socialist and non-socialist groups became more serious under the GGK's policing, and more and more Koreans lost hope in the possibility of independence and doubted the utility of resistance movements that seemed to offer nothing but sacrifices to those who dared to participate in them. Korean botanical researchers like Toh Bong-Syup and Chung Tyaihyon found themselves increasingly limited to the role of native informants, so that they were unable to define the aim of their own research.

However, the efficiency of the Japanese regime in producing Korean consciousness had the effect of bringing these downtrodden Koreans together and rather than making them just feel shame in their inferior status, led them to see their Korean identity as something to actively embrace. In the 1930s, there arose a number of cultural movements variously searching for and expressing Koreanness, producing the so-called "Korean studies" movement. Korean botanical researchers could be said to have been searching for this Koreanness, leaving aside the creation of 'international' Latin names for plants, and instead concentrating on investigating the natural world of their native land.[1] The solidarity thrust upon them as wretched Koreans brought them together and provided the energy to go on.

Admittedly, the fact that Korean researchers studied Korean plants, whether alone or in groups, did not necessarily result from a conscious choice, since Korean

DOI: 10.4324/9781003511755-9

researchers, whatever their academic credentials and ambitions, were not allowed to carry out any extensive transnational research. But these Korean researchers took the decision to transform this inevitable provinciality into their own conscious choice. They formulated various provincialities that revealed an embracement of their provincial lots. At the height of colonial hopelessness, they found new hope in their inward provincialities.

Finding Each Other at the Lowly End

In 1933, Lee Fui-Jai (李徽載, 1903–1986), Lee Deok-Bong (李德鳳, 1898–1987), Kang Chinhyŏng, Han Ch'ang-u, and Yu Sŏkjun, all alumni of the Suwŏn Agricultural and Forestry School and natural history teachers at private Korean middle schools in the colony, met for discussion. Lee Deok-Bong, a natural history teacher at Paehwa girls' middle school and a member of both the NHA and the Teachers' Association, left a recollection on how one day Korean members of these Japanese-led associations decided that it was time for their own organization. It was too awkward for them to continue to "follow the collecting trips of the NHA at the tail end of the procession," without actually interacting with the Japanese members.[2] They decided to form a Korean-only research group divided into plant and animal sections, to enable them to carry out their own locally rooted research into the Korean natural world. It was named as the Korean Natural History Research Group (朝鮮博物研究會, Korean Group hereafter).

The NHA did not welcome this move of its Korean members. Although most Korean members also kept their memberships in the Japanese-led organizations, the NHA obviously wanted to keep these Koreans at the tail end of their procession. Chung Tyaihyon, whom these teachers from his alma mater sought as a leader, also had reservations about this project. Lee Fui-Jai and Lee Deok-Bong had to make "dozens of visits" to Chung's house to secure his participation in the plant section. Although the timeline cannot be fully established, developments in the Forestry Institute may have helped him make his final decision. Chung, after all, lost his position at the Forestry Institute that year and became a contract researcher researching traditional Korean texts.[3] This Korean Group activity would thus be his only real chance to carry out the modern botanical research he so earnestly wanted.

The plant section of the Korean Group also recruited Toh Bong-Syup as another leader, to their great benefit. This young researcher who had recently returned from Tokyo was well known to Korean botanical researchers and the public, as the Korean media had enthusiastically taken note of Toh's every move, from his study at the imperial university to his appointment to the pharmaceutical college. These Koreans apparently took pride in Toh's achievement, seemingly appreciating Toh's effort as a daring challenge to the imperial restrictions on Koreans' self-enlightenment. Importantly, Korean support for Toh's work in the field of botanical studies did not remain as idle cheerleading. As mentioned in the introduction, Korean pharmaceutical companies, which had not had access to such qualified researchers before, were as eager to sponsor Toh's research as were Japanese companies active in the colony. Before Toh secured support from his Japanese sponsor for his botanical club in 1935, he had already established his own private lab, the Kyenong Laboratory of Herbal Research (桂農生藥研究所), outside his school, due to the largesse of one Korean pharmaceutical company.[4] In that same year of 1933, he was invited to join the plant section of the Korean Group. By securing Toh, the Korean Group could begin their collective research with the Kyenong Laboratory as their lab. Given the struggle of the animal section of the Korean Group, which failed to produce any collective work, this stable research base to work together must have been a boon for the plant section.[5] Thus came into being a Korean-only botanical research team that did away with the guidance of their Japanese counterparts. They created their own comfort zone where they did not have to feel awkward about silently following the Japanese.

Staying Provincially Incomplete

In their comfort zone, Korean researchers found new ways to shape and transform their research. All these ways aimed at creating a certain "Koreanness" in the style and content of their work. Toh, the most fortunate Korean researcher in the field, came up with an idiosyncratic way of presenting his scientific results, whose serious meaning he would reveal by his insistence on it.

He began to apply this unobtrusive strategy to the reports he sent to Japan as a native informant. It first appeared in his three-part report on the plants of central Korea sent to the *Pharmaceutical News of Japan* after his teacher Asahina's visit in 1934. In concluding the report, he specified that the report was limited to "the plants that we [the Botanical Club] could collect ourselves."[6] He reported just 226 kinds of central Korean plants as a result, including 29 purely Korean ones and 126 plants not found in Japan. He admitted that many more plants of central Korea had already been studied and reported but said that he would "report about those plants only after he himself had collected them."[7] Did he mean that he could not trust the Japanese reports already produced by the likes of Nakai? Or was he just trying not to overstate what he really knew by not discussing what he had not examined himself? Whatever the reason, the above statement shows that it was his conscious choice not to build upon previous research done by Japanese scholars, with which he was fully familiar. He repeated this strategy the next year in his report on the plants of South Hamgyŏng for the Pharmaceutical Association of Korea. He again clarified: "I am only reporting what I have personally collected this time and promise to complete the list with future collection trips."[8] What Nakai had already covered was again not accounted for, although this would have been an easy task that could have made his report more comprehensive and authoritative.

On the flora of Ullŭng Island in 1938, Toh appears to have stopped this unique practice by providing the full list of plants reported from the island thus far. However, his tabularized list had separate columns to indicate who had and who had not collected each plant. In a space where he could have provided other information like the habitat, shapes, or uses of plants, which he used to provide in most reports, he made five columns to clarify whether that plant had ever been collected by Ishidoya, by Nakai, by Mori Tamezo, by Ueki Homiki, and by his team. Two

都、沈—欝陵島所産薬用植物 61

學名	日本名	石戸谷勉氏採	中井猛之進氏採	森爲三氏記載	植木秀幹氏採	著者等採
Polypodiaceae ウラボシ科						
※Adianthum pedatum *L.*	クジャクシダ		+	+	+	
Asplenium incisum *Thunb.*	トラノヲシダ		+	+		
△Athyrium acutipinnulum *Kodama*	タケシマメシダ		+	+	+	
Athyrium coreanum *H. Christ*	カウライイヌワラビ			+		
Athyrium giganteum *Kodama*	カウライオホメシダ		+	+	+	
Athyrium brebifrons *Nakai* v. angustifrons *Kodama*	ホソバメシダ		+	+	+	
Athyrium Vidalii *Nakai*	ヤマイヌワラビ		+	+	+	
Cyrtomium falcatum *Presl.*	オニヤブソテツ		+	+	+	
Diplazium Oldhami *Christ.*	カウライミゾシダ		+	+		
※Dryopteris crassirhizoma *Nakai*	ヲシダ		+	+	+	

Figure 8.1 A table from Toh and Sim, “Medicinal Plants produced from Ullŭng Island,” *Journal of the Pharmaceutical Association of Korea* 18(2) (1938): 59–81, 61 (National Library of Korea).

other columns of the table contain scientific names of plants, with a mark to indicate whether it was medicinal, indigenous, or cultivated, and the Japanese names of plants (See Figure 8.1).[9] It was certainly an unprecedented and unrepeated way to present a list of regional plants. Toh did not say why it was important for him to provide the information about individual collecting records in such a way. He just chose to make clear again what he had not collected himself, as in his previous, incomplete reports. The almost empty column of his team, which says that he could collect only a small number of plants in his one-time visit to the remote island, vividly displayed the incompleteness of his work.

The above works were all reports and articles published in journals, not monographs. One might perhaps suppose that Toh's conscious choice of exclusively discussing what he had collected himself was a response to external constraints, such as limitations of space and time. But this hardly explains Toh's efforts to include all the collector information as in the flora of Ullŭng Island. His monograph series on Korean flora, beginning with the one on central Korea as the first volume, further suggests that the incompleteness of his list showed that he did not intend to incorporate the results previously produced by Japanese scholars until he himself had verified them. In apparent violation of what might be seen as the basic premises regarding a regional flora, this monograph only listed plants that he and his team “actually collected,” producing an incomplete flora of central Korea. At the end, he attached the maps of 12 collection sites in central Korea, all marked with the collection routes that he and his team “actually” made (See Figure 8.2).[10]

This practice of leaving one's work incomplete by not incorporating the works of others does not make much sense as science. It was a practice that ignored the most basic rules of modern scientific research, proposed by Francis Bacon:

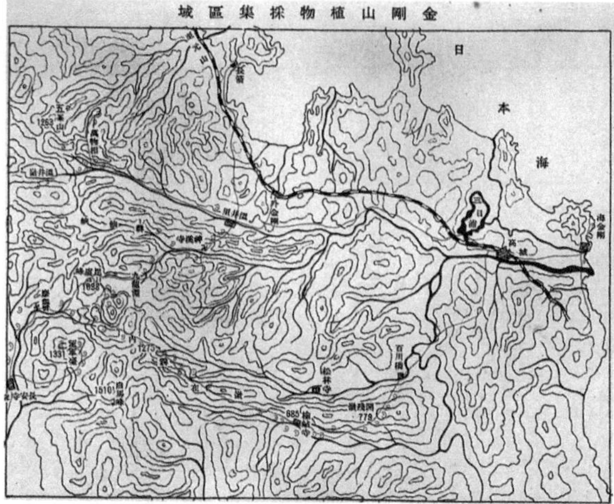

Figure 8.2 Map of the Mount Kŭmkang Collection Route, and on the map, Toh marked their collection routes in red ink. (Courtesy of the Toh family, Toh's second daughter Toh Chung-Ae).

accumulating fact upon fact for a more precise and comprehensive understanding of the natural world. On the other hand, it did exemplify another aspect of the scientific spirit, as expressed in the motto of the Royal Society of London founded on Baconian principles. It was a form of extreme empiricism that doubted any knowledge that he had not verified himself.[11] He chose not to trust Japanese testimonies by actively and persistently excluding knowledge produced by Japanese. By avoiding using Japanese works as a stepping stone for his cataloging project, he managed to produce something certainly different from Nakai's and other Japanese scholars' work, although with incompleteness as its strongest distinction. These works were different from Japanese work, in virtue of being incomplete.

This certainly does not appear a particularly effective way of making one's work distinguished, and it is clearly provincial. In 1936, Toh joined other Korean naturalists in enlightening the Korean people through the public media about the virtue of scientific studies of Korean plants. In a three-part article, he utilized every available scientific study of Korean plants, including Nakai's, to emphasize how rich Korean plant resources were, and how indigenous Korean plants were distinctively beautiful and promising. His problem was that this had only been studied by foreigners in the past and Japanese in the present.[12] "Making a complete investigation of the plants distributed in the Korean land and revealing their relationship with the human life" should be a task for Korean people, he asserted. It would "secure the foundation of the industrial development of one nation (一國)," the foundation for a "scientific Korea."[13] By saying "one nation," he meant that he wanted scientific Korea as a nation, not as part of the Japanese empire.

It was a spirited dream for a colonized Korean. Yet, his proposal of making a complete investigation of Korean plants is not easy to distinguish from what the

GGK had professedly promised for Korea. Nakai's imperial project, sponsored by the GGK to justify and strengthen the imperial regime's control of the colonial mind and colonial resources, was about to be completed. In reclaiming the almost complete imperial project of the GGK as a new Korean project, did he have anything more than a delayed imitation in mind? The clearest distinction that he could present so far for his Korean work was its simple incompleteness, which does not seem a promising sign for the possibility of a different Korean project. However, through his determination to fully reveal the incompleteness of his work with the call for creating a scientific Korea, he showed that it was a task to become "Korean botanists," not just native informants. He invited more Koreans to participate in it.

Making a Simple Task Challenging

The Korean project to which Toh sought to attract more Koreans was the project of the Korean Group that had been carried out in his lab for several years. In fact, the Korean Group had been engaging with a deceptively simple task of Koreanizing botanical studies of Korean plants: making a catalog of vernacular Korean names for plants in Korea.[14] It was a task left undone because Japanese researchers could not and would not do it since Mori's work in 1922. Most Korean researchers had been aware of this negligence and made some kind of preparations on their own, as shown in Chung's collection of vernacular names noted by Nakai and Ishidoya. It was also what the cultural nationalism of the time had been demanding of Korean researchers. Already in 1925, one Korean magazine article harangued:

> No matter how learned you are in philosophy, you are an idiot if you have not probed into the entrenched problems of the Korean people. No matter what you have achieved as a naturalist, you deserve no credit if you do not know Korean plants and cannot recall their names in the Korean language.[15]

In 1926, Lee Deok-Bong dreamed of doing this:

> The In-Wang Mountain becomes a backyard and the Sajikdan Area becomes a front yard. In the backyard, I make an alpine garden by leaving its natural and ecological habitat as it is. In the front yard, I make a herbarium of all Korean plants and a modest greenhouse for teaching and studying tropical plants. I dig a neat pond to raise fish where water plants would flourish by themselves. In a small field, I do genetic and evolutionary experiments on plants. During summer vacations, I explore Korean mountains and fields to collect plants and learn their Korean names. While compiling a Korean plant catalogue and illustrated flora with Korean names, I was suddenly woken by the noise that my students were making. It was all a dream.[16]

He dreamed of a scientific life with a research complex on a grand scale, expressed so idyllically by the addition of his modest dream of cataloging plant names in Korean. The GGK police, however, did not find any idyllic tinge in such a plan from a Korean-only group. They were suspicious of its possibly nationalistic agenda. As the usage of Korean was prohibited and regulated in education and in administration even before its full prohibition in 1938, the police questioned why Korean names were necessary when "Korea and Japan were united as one country." Chung later recalled that they evaded this interrogation by making "an excuse" that it was to help civilize all the people in the colony, even those in the countryside who had not learned Japanese yet.[17] Chung was clearly well aware that their project had political implications that they needed to conceal from the Japanese police. The group of Korean researchers were fully aware that there was nothing idyllic or neutral about their botanical studies. Fortunately, however, the imperial rhetoric of the 'civilizing mission' helped justify their study of Korean plants.

Cataloging the Korean names of Korean plants sounds like an easy task; after all, Nakai had already published descriptions and illustrations of most Korean plants in a language that they used on a daily basis, Japanese, in addition to in English and Latin. However, the Korean Group made their job quite challenging for themselves. Instead of simply translating or giving names for the plants already classified, described, and named by Nakai, they aimed at something more laborious. In the course of more than 100 meetings spread over three years, they tried, in effect, to repeat all the work already done by Nakai before using its results in their work. Toh's Kyenong lab became their center, and their material consisted of the duplicate specimens of Korean plants that Chung had saved in his department and that others had accumulated at their homes and schools through their years of botanizing. Their first task was to examine all those specimens with Nakai's work at hand. They were thus able to grasp the principles underlying Nakai's classification. They matched Korean names only where they were able to confirm the scientifically sanctioned identification by retracing Nakai's process.

For the naming, they again chose a time-consuming approach. They did not want to change names if there were already Korean names in any previous Korean text. If the plant had already been recognized, used, described, mentioned, and named by any Korean through any Korean texts, they would preserve the name. To do that, they compared not only their field notes, where most of them, including Chung, had put down various vernacular names by region, but also traditional texts that contained plant names. The textual search was laborious, as it was not easy to recognize the overall characteristics of plants from texts that usually did not aim at the description of plants, but instead concentrated on their uses or cultural meanings. They had to rummage through texts for various clues. From the outset, they decided to widen their search to the earlier period by exploring texts produced in the Koryŏ period, especially early compilations of local herbs from the twelfth century on. For any given plant, they tried to choose the earliest name given by Koreans. They were cautious in this exhaustive search because they knew that the Chinese character names for a single plant might differ by period and country. In accordance with their agreed guideline, only when they failed to find existing Korean names did they discuss creating new names. Owing to these self-imposed restrictions, they finished only half of the list after over three years' continuous collaboration of about ten researchers. Their *Vernacular Name Catalogue of Korean*

Plants (朝鮮植物鄕名集) published in 1937 contained only about half of the Korean plants reported by Nakai.[18]

The Korean Group seems to have shared Toh's strategy of making his work different from Japanese work by leaving it incomplete. To look at their work, that incompleteness seems one distinction that they could not overlook, as there were so little else that they could accomplish, except for the vernacular names whose scientific values were unrecognized. Having little access to works not in Japanese, their explorations in classificatory issues were limited. Unlike Ishidoya, they did not question the internationally authorized classification of Nakai. If modern plant classification was a science that had produced many different answers, the only answer they had was Nakai's.

Also, although their careful search for Korean names through traditional texts can be seen as an endorsement of Korean traditions, the limitations of their endorsement were also clear. In the preface to the *Vernacular Name Catalogue of Korean Plants*, Chung asserted: "We Koreans have never studied plants scientifically, although we have directly and indirectly utilized rich plant sources, and leisurely appreciated and lyricized them." He thus presented their work as the first scientific work on plants by Koreans.[19] By fully embracing Nakai's imperial studies and denigrating their own traditions, they became the first Koreans who produced modern scientific knowledge of Korean plants. If their work was a challenge to Japanese research, it was only by proving that which did not need to be proved: that Koreans were equally capable of science as Japanese. They could reenact the whole research process on their own without Japanese guidance.

Provincializing Knowledge to Enliven the Land

In spite of these limits, however, within their plan of becoming doers instead of observers of Japanese modernity, this collaboration through Korean solidarity was most effective in the way that it emboldened these researchers. It certainly produced autonomous Korean researchers, who came to realize that in order to practice science it was not in fact necessary to follow their Japanese counterparts. Especially for Chung, who had been given only one chance to be in charge of any knowledge product in the department, this experience of leading a Korean project seemed to have really been transformative. After this collective reenactment, both Chung and Toh assumed much more ambitious projects for themselves, which were not directly related to their work in the forestry department or in the pharmaceutical college. They decided to produce fully illustrated two volume floras of Korean plants, one for woody plants and one for herbaceous plants, with the support of the Kyenong Lab and under the aegis of the Korean Group. Chung was in charge of the volume on woody plants and Toh that on herbaceous plants.

Chung's manuscript on woody plants was published in 1943.[20] As Nakai had not published a one-volume work on Korean plants, except for a small book published in 1914, his refusal to write a preface for this publication, described in Chapter 5, may be understandable. Furthermore, Chung expressed doubts about Nakai's classification in the work, although only minimally. Chung's challenge could not be substantial, as he had absorbed Nakai's taste for detail most fastidiously. Chung seemed to have believed that Nakai's ability to articulate minute details was an essential capacity for a systematist. Chung claimed two new species and one new variety in his work but only one of his new taxa was accepted; a majority of botanists did not think the characteristics pointed out by Chung significant enough for a new taxon.[21] However, unlike Nakai, Chung paid serious attention to the relationship between plants and soil, a concern he obviously acquired from his work with Ishidoya. Chung had shown his awareness of the possibility that some characteristics would not be stable enough to make a new taxon. When he asked Nakai to report his new species *Abelia mosanensis* in 1926, he specifically mentioned that he had verified the durability of its distinctive characteristics by transplanting it to his Seoul nursery, far from its original habitat. In his new independent work, Chung applied this concern consistently. As a result, he found approximately 100 plants about which he could not agree with Nakai's classification, mostly new species and variations claimed by Nakai. For these 100 plants, Chung adopted the scientific names previously proposed by other botanists while treating Nakai's new names as synonyms with these older names. At times, he accepted Nakai's claims but made the tentativeness of his judgment clear by promising more observations in the field that might positively verify Nakai's claims.[22] That Chung differed with Nakai on about 100 out of 1,918 plants that he discussed in this work may not seem very significant. However, it clearly shows that he was not uncritically copying what Nakai had produced. Even on the taxonomic issue, he developed his own principles that he believed to be more reasonable than Nakai's.

Furthermore, for Chung this taxonomic discussion was no more than a side issue in his work. His goal was not to discover more new species to which he could attach his name or to present any universal or regional taxonomic principles. He wanted his study to be useful in addressing one of the "entrenched problems" of Korea: constant deforestation. With this aim, he decided to continue the unfinished project for reforesting Korea, the Right Tree for the Right Land investigation, which he had started with Ishidoya and Asakawa in the 1920s but failed to carry through in the new expanded lab. His illustrated flora provided detailed information to identify trees proper for certain regions, soil types, geography, and uses. The lack of such locally attached useful information was a major weakness in modern classifications that mostly used dried specimens or DNA samples to analyze the traits mostly fixed on a plant. To the information derived by such means, he added all the habitat characteristics for each tree: how high in the mountain or in what types of soil or with how much sunshine it could survive. He tried to indicate whether a tree could survive on the waterside and how much wind it could withstand. He discussed whether a given tree could survive on rocky soil. He used maps every two pages to indicate each tree's geographical distribution within the Korean peninsula, clearly showing his intention to offer a guide for reforestation (See Figure 8.3).[23] He wanted his knowledge to contribute to this immanent problem of Korea that came to be more and more obviously neglected by colonial and imperial governments of Japan as they mobilized for a series of expanding wars.

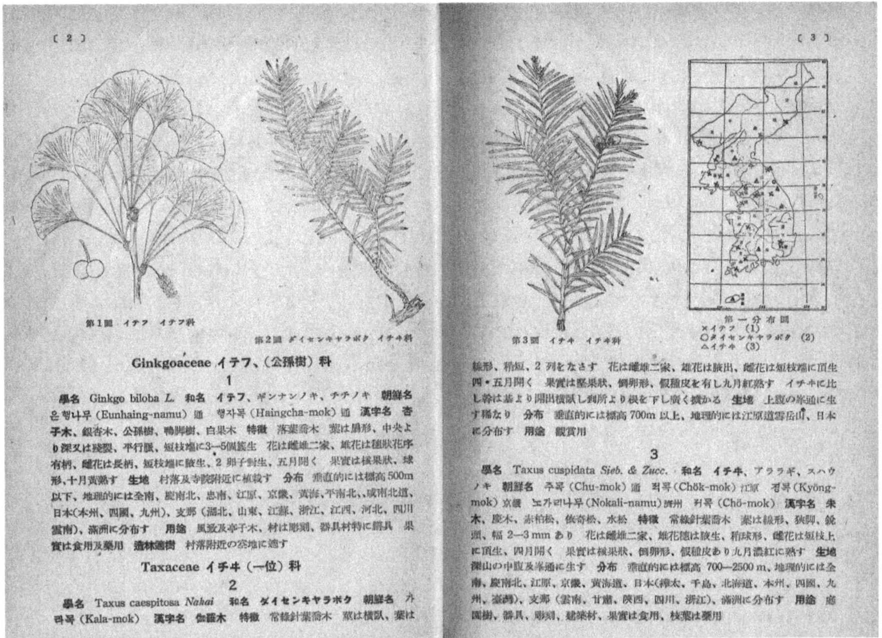

〔2〕

第1圖 イテフ イテフ科

第2圖 ダイセンキヤラボク イチヰ科

Ginkgoaceae イテフ、(公孫樹) 科

1

學名 Ginkgo biloba *L.* **和名** イテフ、ギンナンノキ、チチノキ **朝鮮名** 은행나무 (Eunhaing-namu) 通 행자목 (Haingcha-mok) 通 **漢字名** 杏子木、銀杏木、公孫樹、鴨脚樹、白果木 **特徴** 落葉喬木 葉は扇形、中央より深叉は淺裂、平行脈、短枝端に3—5個簇生 花は雌雄二家、雄花は穗狀花序有柄、雌花は長柄、短枝端に腋生、2 卵子對生、五月開く 果實は核果狀、球形、十月黃熟す **生地** 村落及寺院附近に植栽す **分布** 垂直的には標高 500m 以下、地理的には全南、慶南北、忠南、江原、京畿、黃海、平南北、咸南北道、日本(本州、四國、九州)、支那(湖北、山東、江蘇、浙江、江西、河北、四川雲南)、滿洲に分布す **用途** 風致及亭子木、材は彫刻、器具材特に膳具 果實は食用及藥用 **造林適樹** 村落附近の空地に適す

Taxaceae イチヰ (一位) 科

2

學名 Taxus caespitosa *Nakai* **和名** ダイセンキヤラボク **朝鮮名** 가라목 (Kala-mok) **漢字名** 伽羅木 **特徴** 常綠針葉喬木 幹は橫臥、葉は

〔3〕

第3圖 イチヰ イチヰ科

第一分布圖
×イテフ (1)
○ダイセンキヤラボク (2)
△イチヰ (3)

線形、稍短、2 列をなさず 花は雌雄二家、雄花は腋出、雌花は短枝端に頂生 四・五月開く 果實は堅果狀、倒卵形、假種皮を有し九月紅熟す イチヰに比し幹は基より開出橫臥し到所より根を下し廣く擴かる **生地** 上腹の峯通に生す稀なり **分布** 垂直的には標高 700m 以上、地理的には江原道雪岳山、日本に分布す **用途** 觀賞用

3

學名 Taxus cuspidata *Sieb. & Zucc.* **和名** イチヰ、アララギ、スハウノキ **朝鮮名** 주목 (Chu-mok) 通 적목 (Chŏk-mok) 江原 경목 (Kyŏng-mok) 京畿 노가리나무 (Nokali-namu) 濟州 저목 (Chŏ-mok) **漢字名** 朱木、慶木、赤柏松、依奇松、水松 **特徴** 常綠針葉喬木 葉は線形、狹脚、銳頭、幅 2—3 mm あり 花は雌雄二家、雄花穗は腋生、稍球形、雌花は短枝上に頂生、四月開く 果實は核果狀、倒卵形、假種皮あり九月濃紅に熟す **生地** 深山の中腹及峯通に生す **分布** 垂直的には標高 700—2500 m、地理的には全南、慶南北、江原、京畿、黃海道、日本(樺太、千島、北海道、本州、四國、九州、臺灣)、支那(雲南、甘肅、陝西、四川、浙江)、滿洲に分布す **用途** 庭園樹、器具、彫刻、建築材、果實は食用、枝葉は藥用

Figure 8.3 Botanical illustrations including the rounded leaves of the gingko tree. Chung, *The Illustrated List of Korean Forest Trees*, Seoul: Chosŏn Pangmul Yŏn'guhoe, 1943 (Seoul National University Library).

What is also significant in Chung's practical orientation is that he now could present his effort not just as something stemming from the acceptance of powerful Japanese science, but also as an inheritance of a positive tradition in Korea, which Chung recently "discovered" in his work for the forestry department. While investigating various traditional texts to offer more useful plants for colonial development, he came to think that in some areas there were Korean works that showed a remarkable "modernity" ahead of their time. His main example of this was a work for famine relief produced in the seventeenth century, where he found "concrete examples based on 'experiments'" that allowed people to produce meal supplements with wild and cultivated plants obtainable even during harsh climatic conditions that had led to famine. He seemed to think that this book exemplified the "practical studies" tradition that grew up as a significant minority amongst intellectuals in late Chosŏn.[24] While this selective and modernized embrace of traditions had its limitations, he succeeded in using it as a basis for feeling some pride in his own tradition. In this regard, he was evidently prepared to blur the clear division between modern and traditional knowledge practice that he had made since his acceptance of Japanese tutelage.

In the *Illustrated List of Korean Forest Trees*, Chung chose to list multiple vernacular names for each tree, unlike the *Vernacular Name Catalogue*, which had only one Korean name to pair with the scientific name. If the *Vernacular*

Name Catalogue was making standards for future research, these multiple names seemed to aim at discovering more positive and practical traditions. If one needs to refer to traditional texts or investigate folk knowledge, one standard name cannot be enough. Including different spellings in regional dialects, Chung provided multiple names for a plant while specifying regions for dialects. Unlike the Japanese names provided in localized studies by Japanese researchers, these diverse names may become a practical conduit to the knowledge of the people who lived in that area interacting with the plants, and named and used them. Chung's inclusion of these vernacular names attached to the knowledge circulated among local people or in old texts also suggests his changing idea that those "unscientific" traditional and folk knowledge could be as meaningful as any knowledge from modern labs. These vernacular names and local knowledge deserved to have a place in his botany as much as some uninspiring names like *Prunus yedoensis* and *Hibiscus syriacus*.

Conclusion: From Outward to Inward Provinciality

Colonized Koreans, like most people in the world in that era, were enamored by the power of science and technology and mostly agreed that, as a nation, they needed to learn science and technology. However, not many Koreans actually took up the task. The GGK's careful restrictions on the intellectual pursuits allowed to its second-class citizens, including the remoteness of the possibility of earning a living as a scientist or engineer, made any such pursuit an uphill struggle requiring serious determination. Only a few hundred out of about 20 million Koreans dared to choose scientific careers. Given the lack of social support, most of them failed to leave any significant research results, and ended up working as educators or promoters of science for future generations.

However, in botanical research, many Korean researchers, including others not discussed in this book, such as Park Man-kyu (1907–1977) and Chang Hyŏngdu (1906–1949), were active in producing significant works on Korean plants, despite all the barriers they faced.[25] This unusual success in botany owed much to the modest institutes for botanical education and research built by active Korean. Plants were a resource that could not be ignored in both governmental and private colonial development. Also, the ambition of Japanese settlers to become full-fledged researchers through their colonial research aided these Korean efforts. The collaboration between these Japanese and Koreans eager for new scientific lives was thus a case of mutual reliance and benefit. It was shaped in spite of both parties' clear, if rarely spoken, awareness of their different political stances, and their discomfort about close interactions with the other.

Notably, the ones who first abandoned such politically uncomfortable reliance on the other group were not the Japanese researchers, who needed their Korean disciples at the tail end of their processions until they secured the strong war-time sponsorship from the GGK. But Korean researchers decided to go their own way prior to that. By ignoring Japanese suspicions and making a Korean-only organization, they became Korean botanists themselves on their own terms.

Their tactic of leaving their work incomplete by refusing to use Japanese results without their own verifying process might seem laughable, or at least stubbornly provincial. However, it reflected their clear understanding that the impressive monographs on Korean plants produced by the colonial regime were not intended for the benefit of the colonized. It might easily be taken away from them, if one day the colonizer could be expelled from their land, and their nation finally became independent. They rightly understood that knowledge, however universal or mobile it might appear, does not have its power through codification in written texts but only as carried out, understood, and used by people. So, although they could not conceive a project far removed from the imperial one, and although they might produce exactly the same information already given by the Japanese researchers, their laborious reenactment to make the knowledge their own allowed them to carry out what might be called a 'self-civilizing mission.'

The desire to create a Korean name catalog of plants might be seen as stemming from the obvious violence of having a culture and language other than their own imposed by a colonial power. But their modest provincializing by collecting multiple vernacular names served as a conduit to years and years of collective efforts to know and utilize the plants native to their land. The colonizer's assumptions that Korean cultures were unscientific and useless could at least thus be tested. Chung, who had accepted every unfair conditions of collaboration imposed on him in order to have access to Japanese science, finally made Nakai's project subsidiary to his own concerns, empowered by collaboration with his Korean colleagues. Instead of the superfluous task of proving the intellectual equality of the colonized to the colonizer, he tried to inform the Korean people of the best way to reforest the progressively denuded Korean land, by amassing all the neglected knowledge accumulated at his institute, for which he had labored so hard without proper recognition. By working on a practical task left behind by imperial scientists gloriously marching on into expanding Japanese territory, these colonized people rooted their knowledge in the field and in their collective concerns. For them, this knowledge could make their land livable again for its native plants and for its people, offering an undeniable and perennial hope for the wretched as well as any fortunate of the earth.

Notes

1 Yi Chiwŏn 이지원. *Modern Cultural Thoughts in Korea* (한국 근대 문화사상사 연구), Seoul: Hyean, 2007; Ku Chaejin 구재진et al., *The Formation of 'Things Korean' and Modern Cultural Discourse* ('조선적인 것' 의 형성과 근대문화 담론), Seoul: Somyŏng, 2007.
2 Lee Deok-Bong, "Hearing from Elder Scientists and Engineers (원로과학기술자의 증언)," *The Science & Technology* (과학과 기술) 12(10) (1979): 32–37, 34.
3 Yi Uch'ŏl 이우철, "A Biography of Haŭn Chung Tyaihyon (하은 정태현 박사 전기, A Biography of Chung hereafter)," in *Haŭn Biology Award: The 25th Anniversary* 霞隱生物學賞: 二十五周年, Seoul: Committee on Haŭn Biology Award, 1994, 53–100, 81.
4 Hong Hyŏn-o 홍현오, *A History of Korean Pharmaceutics* 韓國藥業史, Seoul: Handokyakpum Co., 1972, 95–103.

5 The animal section failed to produce a collective result. One participant later published an individual work.

6 Although Japanese students certainly constituted the majority of the botanical club in his pharmaceutical school, Toh came to forge a strong scholarly collaboration with one of his Korean students, Sim Hakchin. Sim worked as Toh's assistant at his lab after graduation and then studied with Asahina at Tokyo Imperial University through the transfer system in the 1940s. Toh and Sim produced many papers together. Sim became professor in pharmaceutics with Toh at Seoul National University after liberation. They were abducted to North Korea. Hong Hyŏn-o 홍현오, *A History of Korean Pharmaceutics* 韓國藥業史. Seoul: Handokyakpum Co., 1972, 95–103.

7 Toh Bong-syup, "Wild Medicinal and Indigenous Plants in Central Korea (中部朝鮮の野生植物及び朝鮮特産植物に就て3)," *Pharmaceutical News of Japan* (日本藥報) 9(18) (1934): 3–7, 7.

8 Toh Bong-syup, "Alpine and Medicinal Plants in the Alpine Area of the South Hamkyŏng Province (咸鏡南道山岳地帯に於ける 高山植物及び藥用植物)," *Journal of the Pharmaceutical Association of Korea* (朝鮮藥學會雜誌, *JPAK* hereafter) 15 (1935): 212–25, 212.

9 Ueki Homiki was introduced in Chapter 5, and Mori Tamezo Chapter 7. Toh Bong-syup and Sim Hakchin, "Medicinal Plants produced from Ullŭng Island (鬱陵島所産藥用植物-附 島勢一班)," *JPAK* 18(2) (1938): 59–81.

10 Botanical Club 京城藥專植物同好會, *Catalogue of Korean Plants: Central-Korea* (朝鮮植物目錄 1: 中部朝鮮編), Seoul: The Botanical Club, 1936.

11 Christopher Cullen informed me that the Royal Society has the Latin motto 'Nullius in verba,' literally 'On the the words of nobody,' or in more natural English 'Rely on nobody's word for anything.'

12 Toh Bong-syup, "Classification of Korean Plants: Field Collections made by Foreigners in the Past (朝鮮産植物의 分類 (上) 過去諸外國人의 實地採集)," *Dong-a Daily*, April 19, 1936; "Classification of Korean Plants: Indigenous Plants 5 Genera, about 500 Species (朝鮮産植物의 分類 (中) 朝鮮固有植物五屬五百餘種)," *Dong-a Daily*, April 21, 1936; "Classification of Korean Plants: Two New Genera Discovered in Korea Last Year (朝鮮産植物의 分類 (下) 昨年度에 發見된 二新屬)," *Dong-a Daily*, April 22, 1936.

13 Toh, "Two New Genera Discovered in Korea Last Year." Notably, Toh tied his concern with the widely supported cultural movement for "Scientific Korea," launched in 1934 and soon terminated by the GGK, alarmed by its popularity. To never officially joined the movement or the related organizations like the Society of Invention and the Association for the Distribution of Scientific Knowledge. For the complex makeup of the Scientific Korea movement, see Jung Lee, "Invention without Science: 'Korean Edisons' and the Changing Understanding of Technology in Colonial Korea," *Technology and Culture* 54(4) (2013): 782–814.

14 This concern about the Korean language is what shows the link of the Korean Group with the "Korean Studies" movement, which included substantial research related to the Korean language as an important section. Seok Joo myung, an entomologist working in the animal section of the Korean Group showed a clear link to active leaders of the Korean studies movement like Chŏng In-bo (鄭寅普), as shown by their correspondence from Chŏng's collections.

15 Kwŏn Tŏkgyu 권덕규, "Finally, Korean People Should Become their Own Pride (마침내 조선 사람이 자랑이여야 한다)," *Kaebyŏk,* July 1925, 18–21, 19.

16 Lee Deok-Bong 이덕봉, "Plants in the School Ground (학교 구내식물 이야기)," *Paehwa* (배화) 2 (1926): 95–110, 110.

17 Yi Uch'ŏl이우철, "A Biography of Haŭn Chung Tyaihyon (하은 정태현 박사 전기, A Biography of Chung hereafter)," in *Haŭn Biology Award: The 25th Anniversary* 霞隱生物學賞: 二十五周年, Seoul: Committee on Haŭn Biology Award, 1994, 81.

18 The process is detailed in the introduction and guidelines. Chung Tyaihyon, Toh Bong-syup, Lee Deok-Bong and Lee Fui-Jai, *Vernacular Name Catalogue of Korean Plants* (朝鮮植物鄉名集), Seoul: Chosŏn Pangmul Yŏn'guhoe, 1937. For more on their references, Lee Deok-Bong, "Considering Korean Names for Korean Native Plants (조선산 식물의 조선명고)," *Hangŭl* (한글) 5(1) (1937): 312–15. That Lee's work was published in *Hangŭl* suggests his strong connection to the Korean studies movement. For naming, they considered morphological characteristics, the meaning of scientific names, related legends or fables, native regions, the name of discoverer, or the color, smell, and taste of the plants.

19 Chung Tyaihyon, Toh Bong-syup, Lee Deok-Bong and Lee Fui-Jai, *Vernacular Name Catalogue of Korean Plants* (朝鮮植物鄉名集), Seoul: Chosŏn Pangmul Yŏn'guhoe, 1937, preface.

20 Toh almost finished his manuscript but, after liberation, was abducted to North Korea before publishing it. Toh's wife Jung Chanyoung (鄭燦英, 1906–1988), a renowned painter who contributed to Toh's study by beautiful illustrations, kept the manuscript. It was published in 1956. Moon Manyong, "Toh Bong-Syup," in *A Biographical Dictionary of Who's Who in Science and Engineering* (과학기술인명사전), Seoul: National Research Foundation of Korea, 2012, 121–27, 125.

21 Yi Uch'ŏl 이우철, "A Biography of Haŭn Chung Tyaihyon (하은 정태현 박사 전기, A Biography of Chung hereafter)," in *Haŭn Biology Award: The 25th Anniversary* 霞隱生物學賞: 二十五周年, Seoul: Committee on Haŭn Biology Award, 1994, 84–85.

22 For example, Chung dismissed Nakai's *Carpinus coreana* for *Carpinus paxii* and *Betula collina* for *Betula chinensis*. For the plants for which he accepted Nakai's identification, he showed the tentativeness of his agreement by promising further observations. For example, for Nakai's claim of *Ulmus macrophylla*, he noted that he had not yet confirmed the characteristics in its flowers that Nakai claimed and without that, it was not distinguishable from *Ulmus macrocarpa*. Chung Tyaihyon, *The Illustrated List of Korean Forest Trees* (朝鮮森林植物圖說), Seoul: Chosŏn Pangmul Yŏn'guhoe, 1943.

23 Chung Tyaihyon, *The Illustrated List of Korean Forest Trees* (朝鮮森林植物圖說), Seoul: Chosŏn Pangmul Yŏn'guhoe, 1943.

24 Hayashi Yasuharu 林泰治 and Kawamoto Tyaihyon 河本台鉉, "Comments on the Concise Reference for Famine Relief 救荒撮要の解說," *The Journal of the Forestry Association of Korea* (山林會報) 209 (1942): 21–28, 21. Looking at the authorship, this work is not obviously a work written by Chung alone. His new boss, Hayashi, did not acknowledge Chung's contribution in any other work, suggesting this was mostly written by Chung, who here had changed his name into the Japanese style, Kawamoto.

25 Park Man-kyu became a middle school natural history teacher in 1933. He was the first Korean who passed the natural history teacher qualifying exam administered by the Ministry of Education in Japan. He accompanied Nakai's collection trips in the 1930s and published works widely, including one in the *Journal of Japanese Botany*. In 1934, Park and Chang, with some Japanese, formed a group to carry out localized studies of Korean plants. The journals that they published are not extant. After Chang's death in 1949, Park inherited Chang's manuscript and produced an illustrated Korean Flora. Kim Choon-Min 김준민, "Prof. Man Kyu Park and His Works (박만규 선생 회갑 기념호: 박만규 선생의 반면,)," *Journal of Plant Biology* (식물학회지) 10(1–2) (1967): 1–2.

Conclusion

Moving beyond Mistaken Names for Connected Provincial Tasks

Nakai Takenoshin, the international authority on Korean flora, who had built up his career in his comfortable Tokyo lab largely owing to many colonial collectors including his Korean assistant, Chung Tyaihyon, showed a defensive response to Chung's intellectual development. When Chung asked him for a preface, he said that it would be impossible for him to do so, as he would need at least three years to review Chung's work to give it his blessing.[1] Although that might not seem like a long time compared to the time that Chung and other Koreans had taken just to add Korean names to what Nakai had already published, Nakai must have had the same kind of thorough re-enactment of Chung's knowledge practice in mind. Of course, the thrust seems entirely different. Korean researchers carried out this re-enactment because they fully trusted the knowledge that Nakai had produced. They just wanted to prove that they were capable of doing the same. Nakai asked for so much time probably because he could not trust anything done by this Korean 'interpreter' who did not know his place.

After proudly serving the International Botanical Congress in 1950, with the permission of the Allied Powers led by the US that occupied Japan after World War II, Nakai provided his own summary of his works on Korean Flora, *A Synoptical Sketch of Korean Flora*, in 1952, the final year of his life. It seems that Nakai wanted to solidify his legacy and guard against challenges from the likes of Chung by providing this final summary of his achievements. It was a simple list of all the scientific names of Korean plants according to his classification system that he had claimed throughout his life, under the subtitle *Plantarum Koreanarum Vascularium Systema Naturale*, a natural system for vascular Korean plants, with introductory remarks and a conclusion in English. In spite of the fact that a majority of Japanese and Korean scholars had chosen other existing scientific names for many of his new taxa, Nakai chose to keep all of his classifications, including those hundred or so names carefully questioned by Chung's wider field observations and transplantation experiments. Nakai also insisted on calling the renamed Tokyo University by its former name, the Tokyo Imperial University, and used Chung's Japanized name, Kawamoto, which was only briefly adopted during the last couple of years of Japan's reign.[2] As articulated in his concluding remarks, he knowingly used obsolete names to express his support for the imperial politics that had produced and sustained his botanical

DOI: 10.4324/9781003511755-10

achievement. The target audience was "foreigners," as it often had been for the imperial center for which he had done his work.

> Occupation of Manchuria and Jahol by [the] Japanese army during 1932–1942 was exceedingly unpopular among the **foreigners**, however, even under occupation, Japanese reformed the countries of tyrannical warlords and bandits to a peaceful Manchoukuo or **the paradise under the rule of right**, as her ruler Pu-I declared. Japanese savants have finished numbers of illustrious investigations on the natural resources of said territories, which were generally published and made great contributions to **the progress of sciences and industries** of East Asia.
>
> [emphasis mine][3]

Nakai was making a final attempt to glorify his nation by listing and supporting all the names that it allowed him to create and apply. Those were exemplary contributions to the "progress of sciences and industries" that produced "a paradise under the rule of right." He apparently believed that the high moral status and scientific achievement he claimed for Japanese imperial rule in East Asia entitled him to continue to use the names it had authorized him to define.

Although many would not go so far as to say that the unwelcome colonial rule created a paradise, many readers of his work appear to have appreciated Nakai's portrayal of his meticulous and tireless naming of thousands of new taxa as a progress of science and Japan's exemplary success in Western modernization. They found it a notable example of botanical achievement made under colonial rule in Korea. The strong dominance of English in scientific communities, a seeming indication of internationalism in science, strongly shaped this notion; it granted an international status to his works, while provincializing other works simultaneously published in Korean and Japanese.[4] The considerable number of works produced in vernacular languages, despite their importance, originality, rigor, and actual importance in shaping our understanding of and relationships with plants, had lower visibility in the realm of 'real science.' They were rather seen as belonging to the realm of routine intellectual labor, irrelevant to the general understanding of nature that proper science should aim for. But it is, on the contrary, precisely those vernacular achievements of Ishidoyas and Chungs that enlighten us about plants' complex relations to their local soil, climate, and society in all-encompassing socio ecology. They may further help us formulate overdue and too-long evaded questions about those names created by people like Nakai that failed to address these complexly intertwined relations in evolving ecology, despite many sound questions from plants and people in every local field.

This book from Japanese colonial Korea has tried to demonstrate those rooted practices in colonial Korea as originally creating, shaping, and deepening our scientific understanding of plants at least as much as Nakai's internationally authorized works that produced and sustained rather too many provincial names. These interactively produced Japanese and Korean botany has made two things clearer. First, it shows who indeed actualized and globalized "Western" and "modern" botany. It could not be Carl Linné, the father of botany in his provincial center, whose supposedly universal system naming plants produced endless confusions in every local field. It was these local Linnés, whose connected practices in local fields were just obscured in the story spotlighting certain metropoles and their thin theories, under wrongly divided names. Our understanding of modern botany and its globalization is quite limited without incorporating these active local Linné's complex initiatives, struggles, doubts, and reconciliations into stories of European theories and theorizers. Second, these concurrent and connected makings of Western, Japanese, and Korean botanies highlight once again how untenable those hierarchical dichotomies are between those ill-divided binaries: the West and the Rest, the center and the periphery, the modern and the traditional, the imperial and the colonized, and theory and practice.

Above all, Japan's "Western" and "modern" botany, initiated by the Japanese center in Tokyo, was built upon Japan's interactions with many cultures, including Linné's Sweden as well as Japan's old neighbors. This Japanese making of its modern botany was hardly a delayed transfer from an already enlightened pure 'West' to a benighted non-West, given the numerous contributions of Tokugawa Japanese scholars to Linnaean botany. The endless discussions at International Botanical Congresses further revealed how the 'West,' which was busy shaping its modernity upon many previous and current inputs from different cultures, presented no complete rules to resolve various concerns and questions about the Linnaean system from these concurrently emerging centers. Korean botanists, too, were not passive recipients of these Western or Japanese botanies, actively mired in this messy cacophony. They eagerly participated in these emerging botanical practices, refusing to stay mere informants and producing their own rules, priorities, and knowledge, emboldened even by their benighted "traditions," which were transculturally shaped like Japan's and Sweden's, though narrowed and fragmented by these competitive emergences of the newly enlightened. What made Engler, the authority at the time of Western plant systematics, admit the imperfections of his criteria was dynamic inputs from these thinking hands in the field that revealed the limitations of the impressive dried plant specimens at his huge center. The Hayatas, Ishidoyas, Moris, Tohs, Chungs, and even Nakais exerted sufficient pressure to reshape and unsettle Engler's system, while forging their own variegated botanical standards. From their provincial yet connected centers, all these local Linnés from Engler to Chung simultaneously shaped their own botanies, mutually and constantly transforming each other's knowledge practices.

Nevertheless, those dichotomies, also repeatedly challenged by the many apt critiques noted in the introduction, still lurk in our works and the public mind. Much credit should go to these powerful local Linnés. These botanists emerging in Japanese colonial Korea constantly evoked those dichotomies that denied their initiatives and obfuscated their complex trajectories and various reformulations. It was their choice to evoke them in building their more or less modest botanical centers in the names of enlightenment and science. By new Latin names like *Prunus yedoensis*, they responded to the political needs of their own emerging nations, which came to exercise an unprecedented force over other polities as well as their own people's life upon joining this world of nations, from the flowers that they praised to professional opportunities intricately related to these emerging knowledge practices. Notably, quite a few botanists had qualms about these often divisive political impositions on their practices, trying to neutralize or work beyond them, though a clear articulation of such qualms was rare and those differently carved out paths were often smoothed out by enlarged opportunities rolled out from powerful parties allying with new nation-states, or ignored by the Nakais. Still, enough botanists in colonial Korea found the imperially charged practice of Nakai uninspiring. They envisioned and pursued a locally rooted botany, while not fully hiding or overcoming their envy about his recognized 'universal' achievements within those hierarchies.

Notwithstanding their hesitation, and the obscurity and limitations of their products, it seems worthwhile to reflect on the meanings of their deeper explorations into their local fields. Those colonized Koreans, by working together at the "tail end of the procession," came to embrace canvassing their lands and cultivating their plants as valuable intellectual exercises responding to the social and ecological needs of their lands; they could set aside the endless debates about mistaken names as irrelevant to the pressing intellectual challenges; they took more and more pride in their effort to grasp the complex interlinkages of plants, soils, climates, cultures, while embracing mixed vernacular traditions.

Consideration of one of those mistaken names, *Ginkgo biloba*, may further help our reevaluation of these connected affairs. *Ginkgo biloba*, despite temporary name changes to *Salisburia biloba* or *Salisburia adianthifolia*, became one of Linné's most enduring names. Yet this seems scarcely due to Linné's scientific criteria.[5] This so-called "living fossil" that survived the extinction of the dinosaurs and whose oldest fossil record is about 200 million years old is just too singular. This tree that produces its seeds by having separate male and female trees has no living relatives. It had its struggles, disappearing from most continents during the Great Ice Ages when our species came into being, except for eastern and south-central China. While it can live over a thousand years, and is resistant to bacteria, insects, pollution, and quite extreme heat and coldness, its heavy seeds unpalatable to most animals made its spread around the globe challenging. It beat the odds about a thousand years ago when some Chinese cultivated it, using its fruits as medicine and food. It came to 'Korea' and 'Japan' probably in the fourteenth and the fifteenth centuries and spread to Europe in the eighteenth century after Kaempfer's introduction. It now lives in most continents as a street tree that can survive harsh urban conditions, although it is listed as Endangered (EN) according to the International Union for Conservation of Nature (IUCN) Red List criteria.[6]

What does *Ginkgo biloba* tell us about modern Linnaean systematics? Linné made an exception by adopting this possibly strange sounding vernacular name, "ginkgo," in his Latin binomial for this tree, unknown in Europe. This naming certainly made his system wider in scope. However, in addition to the spelling, he made an error in the species epithet *biloba*, by noting only its bifurcated leaves, not the well-rounded ones (See Figure 9.1, also Figures 1.2 and 8.3).[7] How Western, modern, and scientific is this naming of the ginkgo tree? In fact, given the Linnaean premise of finding the universal order in nature, as a pristine world separate from human culture, it is perhaps illogical of him to include this tree, which survived in

Figure 9.1 A botanical illustration of the Ginkgo tree. Here, both the rounded and bifurcated leaves of the ginkgo tree are shown. Philipp Franz von Siebold and Joseph Gerhard Zuccarini, *Flora Japonica.*

Source: www.biolib.de.

cultivation through its healing and nurturing relation with the human species, in his natural system.

Yet this bifurcation between nature and culture that Linné often ignored in practice, as further shown in his scientific naming of many garden plants from Clifford's garden like *Hibiscus syriacus*, seems to function strangely, or rather to malfunction, in the conservation project for this endangered species. The *Plants of the World Online* at Kew lists its current habitats as "China North-Central, China South-Central, Illinois, Japan, Korea, Romania," despite many of us having seen it in African cities like Cape Town, American cities like New York, Oceanian cities like Sydney, and European cities like Paris, where this tree displays in brilliant yellow every autumn.[8] *Ginkgo biloba* succeeds only as a mistaken name, failing to combine scientific efforts in plant geography with those made by urban horticulturalists or landscapers, and many other concerned citizens caring for these carbon-absorbing robust plants. Even this little challenged name did not meet our expectation that such standardized names in universal Latin should facilitate such combined intellectual efforts across regions and disciplines. Rather, these "scientific" names, too numerous and changing to be followed with our limited capacity, often overwhelm and isolate us in our small corner without giving us time to engage with each other's efforts, and to pay actual attention to all plants precious and awesome regardless of names. As Ishidoya and Chung had shown, transplanting and rooting plants into a new land is an amazingly creative and intellectually fulfilling task that had to be carried out with utmost attention to changing local ecologies. Perhaps better appreciating these rooted intellectual labor may finally free us from those ill-defined binaries and names and help keep our fragile relations with these life forms that had brought us to this planet by their transformative power.

Names like *Ginkgo biloba*, *Prunus yedoensis*, and *Hibiscus syriacus* simply highlight the confluence of so many cultures, and the deep histories of these plants on earth, histories much longer than our own, in turn reminding us of our inseparable relations with and deep reliance on these living forms. The view of those Latin names as some exceptional example of rationality and intellectual achievement that originated from one corner of the earth should not stop us from seeing the rich reasonings that took place everywhere for a long time before those namings, and the serious contradictions and questions, which any universal rules could not dissipate without locally engaged yet connected endeavors. Those hands-on field practices, shown in this book as influencing Linnés' as Linnés' practice influenced theirs, have already played a vital role within our perennial effort to understand and reshape our relations with nature. By appreciating these rooted practices in various vernacular languages more deeply, we may move beyond unnecessary divisions that do not help our shared predicaments in relations with nature. May we perhaps learn from cherry trees by developing flexibly hybrid approaches across disciplinary boundaries and cultural differences—which may not be so big once we move beyond these mistaken names?

Notes

1 Yi Ch'angbok 이창복, "New Natural History: Plants (신박물기 : 식물)," in *The Life of Yi Ch'angbok* (樹友 李昌福 教授의 발자취), The Commemoration Committee for Prof. Yi Ch'angbok ed., Seoul: Chŏngminsa, 1984, 1–10, 7.
2 Nakai Takenoshin, *A Synoptical Sketch of Korean Flora,* Tokyo: The National Science Museum, 1952; According to Lee Deok-Bong, for whatever reason, even Japanese scholars said that "he was senile" or "it would have been nice if Nakai did not publish it." Lee Deok-Bong, "Hearing from Elder Scientists and Engineers (원로과학기술자의 증언)," *Science & Technology* (과학과 기술) 12(10) (1979): 32–37, 36.
3 Nakai Takenoshin, *A Synoptical Sketch of Korean Flora,* Tokyo: The National Science Museum, 1952, 152.
4 Michael D. Gordin, *Scientific Babel*, Chicago, IL: University of Chicago Press, 2015.
5 Armed with DNA analyses, modern botanists readily condemn the Linnaean system's reliance on morphological characteristics like stamens as "artificial," discarding many taxa that Linné and earlier botanists had named. Yet when the DNA sequences of all individuals are unique, and the differences in billions or millions of sequences are minimal, deciding on the significant parts to compare and the degrees of difference to note requires judgment, just as with shapes or the number of stamens. If we choose the consistency of applying this one "objective" criterion, only visible to experts connected to expensive sequencing facilities, we would have to dissolve much more reasonable, universal, and perennial categories than even "fish," owing to the thankful and creative fecundity of plants which, in the long history of evolution, have crossed all boundaries and produced allopolyploid, i.e., hybrids of heterogeneous species in nature, as Hayata and many other botanists have pointed out. On cladists, who claim that fish can't exist as a group, see Carol Kaesuk Yoon, *Naming Nature: The Clash between Instinct and Science*, New York: W. W. Norton & Company, 2009.
6 Peter Crane, *Ginkgo*, New Haven, CT: Yale University Press, 2013, 3–7.
7 Peter Crane, *Ginkgo*, New Haven, CT: Yale University Press, 2013, 38–39.
8 Plants of the World Online (https://powo.science.kew.org/) accessed 2021.12.15.

Bibliography

For a more exhaustive bibliography of Korean and Japanese sources, see Lee Jung 이정. "Contested Botanizing in Colonial Korea (1910–1945): Conflicting Visions of Modernity Emerging through Colonial Interactions between Korean and Japanese Researchers 식민지 조선의 식물 연구(1910–1945) 조일 연구자의 상호 작용을 통한 상이한 근대 식물학의 형성." Ph.D. Thesis, Seoul National University, 2013.

Websites and DatabasesAcademy of Korean Studies, Biographic Information System [https://people.aks.ac.kr]

Biodiversity Heritage Library [https://www.biodiversitylibrary.org/]Korean Plant Names Index [www.nature.go.kr/kpni/]

Korean Plant Blog by Chŏn Ŭisik [https://blog.daum.net/kplant1/7903276]

National Institute of Korean History, Korean History Database [https://db.history.go.kr]

Plants of Taiwan Database [https://tai2.ntu.edu.tw/PlantInfo.php]

Plants of the World Online at Kew [https://powo.science.kew.org/]

The Plant List [https://www.theplantlist.org/]

The University Museum, University of Tokyo, Type Collection Database [https://umdb.um.u-tokyo.ac.jp/DShokubu/TShokubu.htm]

Commonly Used Korean and Japanese Periodicals

Botanical Magazine, Tokyo, abbreviated *BMT* 植物学雑誌

Bulletin of Korean Studies 朝鮮學報

Dong-a Daily 동아일보, literally East Asian

Government-General of Korea Monthly 朝鮮總督府月報, official organ of the Government-General of Korea (GGK) in 1911–1915, followed by *Korean Repository* 朝鮮彙報 in 1915–1920, and *Korea* 朝鮮 in 1920–1944.*Journal for the Association of Natural History Teachers in Seoul*, abbreviated *JNHT* 京城博物教員會誌*Journal of the Korean History of Science Society* 한국과학사학회지

Journal of the Natural History Association of Korea 朝鮮博物學會雜誌

Journal of the Pharmaceutical Association of Korea 朝鮮藥學會雜誌

Korea & Manchuria 朝鮮及滿洲

Korea of Art and Education KAE 文教の朝鮮

Korean Journal of Medical History 의사학

Korean Journal of Plant Taxonomy 식물분류학회지

The Medical World of Manchuria and Korea 滿鮮之醫界

Pharmaceutical News of Japan 日本藥報

Research on Continental Culture 大陸文化研究**Primary Sources**

Alumni Association of the Keijo Imperial University 京城帝國大學同窓會. *The Deep Blue Yonder: Fiftieth Anniversary of Founding the Keijo Imperial University* 紺碧遙かに: 京城帝國大學創立五十周年記念誌. Tokyo: Alumni Association of the Keijo Imperial University, 1974.

Anesaki, Masaharu. *History of Japanese Religion: With Special Reference to the Social and Moral Life of the Nation*. 1963 reprint ed. London: Kegan Paul, Trench, Trubner, 1930.

"An Outline of Regional Forestation Projects 地方造林事業概要." *The Monthly of the Government-General of Korea* 朝鮮總督府月報 2(11) (1912): 1–11.

Association of Natural History Teachers in Seoul 京城傳物教員會. "The Survey of Natural History Education Facilities 博物教育設備調査書." *Korea of Art and Education* (12) (1939): 69–81.

Berg, Leo S. *Nomogenesis or Evolution Determined by Law*. London: Constable & Company Ltd., 1926.

Botanical Club of the Keijo Higher School of Pharmaceutics 京城藥專植物同好會. *Catalogue of Korean Plants: Central-Korea* 朝鮮植物目錄 1:中部朝鮮編. Seoul: Botanical Club, 1936.

Cassino, Samuel Edson. *The Scientists' International Directory: Containing the Names, Addresses, Special Departments of Study, Etc., of Professional and Amateur Naturalists, Chemists, Physicists, Astronomers, Etc., Etc. In All Parts of the World*. Boston, MA: S.E. Cassino, 1883.

Chang, Ŭngchin 장응진. "On Science 과학론." *T'aekŭk Journal* 태극학보 5 (1906): 9–12.

Chung, Tyaihyon 정태현. "Distribution and the Right Land for Important Korean Trees 朝鮮産主要樹種ノ分布及適地," *Current News of the Forestry Research Institute* 林業試驗場時報5, (1925): 1–48.

Chung, Tyaihyon 정태현. "On the Medicinal Plants Produced in Korean Mountains and Fields 朝鮮の山野より生産する藥科植物に就て." *The Journal of the Forestry Association of Korea* 山林會報 97 (1933): 21–28.

Chung, Tyaihyon 정태현. "On Plants of Kanghwado 江華島所産森林植物に就て." *The Journal of the Forestry Association of Korea* 99 (1933): 40–45.

Chung, Tyaihyon 정태현. "On Plants of Kŭmkang Mountain 金剛山産森林植物に就て." *The Journal of the Forestry Association of Korea* 102 (1933): 19–27.

Chung, Tyaihyon 정태현. "On Plants of Ŏch'ŏng Island 於青島所産植物に就て." *Current News of the Forestry Research Institute* 林業試驗場時報 9 (1933/10/25): 31–43.

Chung, Tyaihyon 정태현. *The Illustrated List of Korean Forest Trees* 朝鮮森林植物圖說. Seoul: Chosŏn Pangmul Yŏn'guhoe, 1943.

Chung, Tyaihyon 정태현. "50 Years of Carrying Field Press 야책을 메고 50년." *Forest and Culture* 숲과 문화 11(3) (2002): 52–62.

Chung, Tyaihyon, Toh Bong-Syup 도봉섭, Deok-Bong Lee 이덕봉, and Lee Fui-Jai 이휘재. *Vernacular Name Catalogue of Korean Plants* 朝鮮植物鄉名集. Seoul: Chosŏn Pangmul Yŏn'guhoe, 1937.

Ch'unp'a 春坡. "What You Have Forgotten in Touring Seoul 서울구경 왓다가 니저버리고 가는 것." *Pyŏlkŏnkon* 별건곤 (1929/09/27): 129–31.

Committee for Commemoration of Dr. Nakai's Works 中井博士功績記念事業会. *The List of Monographs and Articles by Prof. Nakai and the Index of New Groups, Species and Scientific Names by Him* 中井教授著作論文目録並に教授の研究発表による植物新群名,新植物名及新学名總索引. Tokyo: Hokuryukan, 1943.

Crow, William B. "Phylogeny and the Natural System." *Journal of Genetics* 17(2) (1926): 85–155.

Duggar, Benjamin. *Proceedings of the International Congress of Plant Sciences, Ithaca, New York, August 16–23, 1926.* Menasha, WI: George Banta Pub. Co, 1929.

Du Rietz, Gustaf Einar. "The Fundamental Units of Biological Taxonomy." *Svensk Botanisk Tidskrift* 24(3) (1930): 333–428.

Fisher, Mary Jones. "The Morphology and Anatomy of the Flowers of the Salicaceae I." *American Journal of Botany* 15(5) (1928): 307–26.

Forestry Association of Kyŏnggi Province 京畿道林業會. *The Beautification of the Countryside and the Cultivation of Medicinal Plants* 農村美化と藥草栽培. Seoul: The Forestry Association of Kyŏnggi Province, 1936.

Forestry Research Institute 林業試験場. *Wild Medicinal Plants of Korea* 朝鮮産 野生 藥用植物. Seoul: Government-General of Korea, 1936.

Fukushi, Suenosuke 福士末之助. "Localized Studies as the Foundation of the Educational Approach to Promote Popular Mores 民風作興上教育的方法としての鄉土調查." *Korea of Art and Education* 73 (1931): 6–16.

G. K. "On Ishidoya Tsutomu's Chinese Bencao Medicine Research 石戸谷勉氏:支那本草漢薬植物考." *Acta phytotaxonomica et geobotanica* 植物分類 地理 10(2) (1941): 143–50.

GGK. *The Catalogue of Old Books by the Government General of Korea* 朝鮮總督府古圖書目錄. Seoul: GGK, 1913.GGK. *Big, Old, and Named Trees in Korea* 朝鮮巨樹老樹名木誌. Seoul: GGK, 1919.

GGK. "Report on Korean Traditional Medicinal Herbs 朝鮮漢方藥科植物調查書." In *Report on Plants of Nobong Mountain in Korea* 朝鮮鷲峯植物調査書. Seoul: GGK, 1918.

GGK. *The Catalogue of Old Books by the Government General of Korea*. Seoul: GGK, 1921.

Government-General of Korea. Annual Report on Reforms and Progress in Chosen.

Green, J. Reynolds. *A History of Botany 1860–1900*. Oxford: Clarendon Press, 1909.

Hara, Hiroshi. "Takenoshin Nakai 1882–1952." *BMT* 66 (1953): 1–4.

Hayashi, Yasuharu 林泰治, and Kawamoto Tyaihyon 河本台鉉. "Comments on the Concise Reference for Famine Relief 救荒撮要の解說." *The Journal of the Forestry Association of Korea* 209 (1942): 21–28.

Hayata, Bunzo 早田文藏. "On the Distribution of the Formosan Conifers." *BMT* 19 (1905): 43–60.

Hayata, Bunzo 早田文藏. "On Taiwania, a New Genus of Coniferae from the Island of Formosa." *The Journal of theLinnean Society. Botany* 37(260) (1906): 330–331.

Hayata, Bunzo 早田文藏. "On Taiwania and its Affinity to Other Genera." *BMT* 21 (1907): 21–28.

Hayata, Bunzo 早田文藏. *Icones plantarum Formosanarum nec non et contributiones ad floram Formosanam: or, Icones of the Plants of Formosa, and Materials for a Flora of the Island, based on a Study of the Collections of the Botanical Survey of the Government of Formosa*. Taihoku: The Bureau of Productive Industries, Government of Formosa, 1911–1921, vols. I–X.

Hayata, Bunzo 早田文藏. "One New Genus in the Sandalwood Family やどりき科ノ一新屬." *BMT* 29(340) (1915): 166–69.

Hayata, Bunzo 早田文藏. "Type Specimen of Arhangiopteris Henryi Christ et Giesenhagen む k しりうひんたい屬ノ基本種ニ就キテ." *BMT* 43(514) (1929): 560–74.

Hayata, Bunzo 早田文藏. "Succession in the Vegetation of Mt. Fuji and the Formulation of a New Theory, the Succession Theory, in Opposition to the Natural Selection Theory;" "The Succession and Participation Theories and Their Bearings upon the Objects of the

Third Pan-Pacific Science Congress;" "The Relation between Succession and Participation Theories and their Bearings upon the Natural System." In *Proceedings of the Third Pan-Pacific Science Congress*, The National Research Council of Japan ed. 1867–1868, 1869–1875, 1876–1886. Tokyo: Maruzen, 1928, vol. 2.

Hayata, Bunzo 早田文藏—. "On the Dynamic System of Plants 植物の動的分類系に就きて." In *Iwanami Biology* 岩波講座 生物學. Tokyo: Iwanami Shoten, 1931, 95–111.

Hayata, Bunzo 早田文藏. "A. von Humboldt's Letter A. von Humboldt ノ書簡." *BMT* 45(529) (1931): 29–31.

Hayata, Bunzo 早田文藏. "On Some Other Systems Bearing a More or Less Resemblance to the Dynamic System 動的分類系ニ類似スル 他ノ分類系ニ 就キテ." *BMT* 45(535) (1931): 364–68.

Hayata, Bunzo 早田文藏. "On the Dynamic System of Plants Founded on the Theory of Participation因子分配說ニ 基ツイテ 組織セラレタル 植物ノ 動的分類系ニ 就キテ." *BMT* 45(538) (1931): 490–93.

Hayata, Bunzo 早田文藏. "A Piece of the Historical Accounts of Our Botanical Society, So Far Remaining in My Memory私ノ記憶ニ殘ツテヰル本會ノ歷史ノ一片." *BMT* 46(544) (1932): 278–80.

Hayata, Bunzo 早田文藏. "Rejection of the Phylogenetic Classification of the Leguminosae Based on Embryological Observations發生學より觀察して豆科植物の系統分類を否定す." *Botany and Zoology: Scientific and Applied* 植物及動物1(9) (1933): 1265–70.

Hayata, Bunzo 早田文藏. "What Is the So-Called Eternal Life永遠の生命とは如何なるものか." *Botany and Zoology* 1(12) (1933): 1743–50.

Hayata, Bunzo 早田文藏. "Discussing the Concept of Systematics分類學の體系を論ず." *Botany and Zoology* 2 (1934): 79–88.

Hayata, Bunzo 早田文藏. "Principal Aims of Systematic Botany 分類學トハ如何ナルモノカ." *BMT* 47(558) (1933): 461–65.

Hayata, Bunzo, and Satake Yoshisuke佐竹 義輔. "Contributions to the Knowledge of the Systematic Anatomy on Some Japanese Plants日本植物ニ關スル解剖分類學的貢献." *BMT* 43(506) (1929): 73–106.

Hŏ, Chaeyŏng 허재영, ed. *The Textbooks of the Protectorate Period 8. Natural History* 통감시대 교과서 자료-8 박물. Kwangmyŏng: Kyŏngjin, 2011.

International Botanical Congress. *Official Program of the International Congress of Plant Sciences (Fourth International Botanical Congress),* Cornell University, Ithaca, NY. August 16–23, 1926. Ithaca, NY: The Congress, 1926.International Botanical Congress. *Nomenclature: Proposals by British Botanists*. London: His Majesty's Stationary Office, 1929.

International Botanical Congress. *International Rules of Botanical Nomenclature Adopted by the Fifth International Botanical Congress, Cambridge, 1930.* London: Taylor and Francis, 1934.

Ishimura, Toshio 岩村俊雄. "The Science Education Method in Elementary School 普通學校に於ける理科教授法." *Korea of Art and Education* 10 (1925): 55–60.

Ishidoya, Tsutomu 石戸谷勉. "Herbal Plants Produced from Korean Mountains and Fields 朝鮮の山野より生産する藥科植物." *Korean Repository* 2(13) (1916): 75–148.

Ishidoya, Tsutomu 石戸谷勉. "Chinese Characters Used for Tree Names in Korea 朝鮮に於て用いらるる樹木漢字名." *Korean Repository* 3(3) (1917): 91–94.

Ishidoya, Tsutomu 石戸谷勉. "The Growth and Development of the Japanese Cedar and Cypress Transplanted into Korea 朝鮮に移植せられたる「スギ」「ヒノキ」の生育." *Korean Repository* 4(12) (1918): 20–32.

Ishidoya, Tsutomu 石戸谷勉. "Classification of Salix and Populus in Korea-1 朝鮮に於ける柳屬及上天柳屬の分類(上)." *Korea* (1921/04): 37–44.
Ishidoya, Tsutomu 石戸谷勉. "Classification of Salix and Populus in Korea-2." *Korea* (1921/07): 91–104.
Ishidoya, Tsutomu 石戸谷勉. "About Forests, Forest Canopy, Interior, and Floor Viewed from the Ecological Perspective 生態上ヨリ觀察シタル森林ト林冠 、 林幹及林床ニ就テ." *Journal of the Japanese Forestry Society* 日本林學會誌 (10) (1921): 26–29.
Ishidoya, Tsutomu 石戸谷勉. "Foresters' Attitude towards the Forest and the Chŏnnam Province's Gift of Nature 삼림에 대한 조림자의 태도와 전남삼림의 천혜." *The Journal of the Forestry Association of Chŏnnam* 全南山林會報 1 (1922): 18–26.
Ishidoya, Tsutomu 石戸谷勉. "On the Original Plants of the Korean Medicinal Plants 朝鮮ノ 漢方藥ト其ノ原植物ニ就テ." *Journal of the Natural History Association of Korea* 3 (1925): 1–10.
Ishidoya, Tsutomu 石戸谷勉. "Plants of Cheju Island and Expected Problems 濟州島の植物と將來の問題." *Korea of Art and Education* 10 (1928): 71–92.
Ishidoya, Tsutomu 石戸谷勉. "On the Floristic Region of the Ullŭng Island 欝陵島ノ植物區系=閥スル考察 (1)." *Journal of the Natural History Association of Korea* 7 (1928): 21–25.
Ishidoya, Tsutomu 石戸谷勉. "Origins of Chinese Ancient Medicines 支那古代藥物の原産地に關る考察." *Journal of the Pharmaceutical Association of Korea* 8(3) (1928): 97–144.
Ishidoya, Tsutomu 石戸谷勉. "The Firsts in Korean Plant Collections 朝鮮植物採集一番鎗の記." *Korea & Manchuria* 242 (1928): 38–40.
Ishidoya, Tsutomu 石戸谷勉. "A New Species in Korean Corydalis 朝鮮産 延胡索 一新種." *Journal of the Natural History Association of Korea* 6 (1928): 87–91.
Ishidoya, Tsutomu 石戸谷勉. "The Origin of Somei Yoshino Cherry 染井吉野櫻の原産地." *Korea & Manchuria* 258 (1929): 67–68.
Ishidoya, Tsutomu 石戸谷勉. "An Examination of Original Plants for Ginseng, Recorded in *Materia Medica* Works 歴代本草ニ所載スル人蔘ノ原植物ニ關スル考察." *Journal of the Natural History Association of Korea* 9 (1929): 7–16.
Ishidoya, Tsutomu 石戸谷勉. "A Story about Herbal Medicines 藥草ノ話." *Journal of the Natural History Association of Korea* 10 (1930): 1–7.
Ishidoya, Tsutomu 石戸谷勉. "The Exploration of Medicinal Plants and Philosophy 藥草の探險と哲學思想." *Korea & Manchuria* 269 (1930): 50–55.
Ishidoya, Tsutomu 石戸谷勉. "The Past and the Future of Korean Medicinal Plants 朝鮮藥用植物の 過去と將來." *Korea* 195 (1931): 62–74.
Ishidoya, Tsutomu 石戸谷勉. "Miscellany about Manchurian and Korean Landscapes 滿鮮植物風景雜觀." *Korea & Manchuria* 325 (1934): 65–68.
Ishidoya, Tsutomu 石戸谷勉. "The Flora and its Elements in Jilin Region 吉林地方に於ける植物景と其の要素に就いて." *Bulletin of a Cultural Survey of Manchu and Mongol by Keijo Imperial University* 京城帝國大學滿蒙文化研究會會報, (1934): 3–9.
Ishidoya, Tsutomu 石戸谷勉. *Chinesische Drogen*. Vol. I–IV. Keijo: Pharmakologischen Institut der Kaiserliohen, 1934–1941.
Ishidoya, Tsutomu 石戸谷勉. "Mongolian Medicines and their Original Plants Used in Manchukuo Area 蒙疆地方に行る蒙古藥とその原植物." *Journal of the Pharmaceutical Association of Korea* 19(3) (1939): 97–110.
Ishidoya, Tsutomu 石戸谷勉. "Plant Resources of North Asia 北亞細亞の植物資源." In *Research on Continental Culture*. Tokyo: Iwanami Shoten, 1940, 457–90.

Ishidoya, Tsutomu, and Toh Bong-Syup 都逢涉. "Florula Seoulensis 京城附近植物小誌." *Journal of the Natural History Association of Korea* 14 (1932): 1–48.

Ishidoya, Tsutomu, and Sugihara Noriyuki 杉原德行. "Korean Traditional Medicine (朝鮮の漢藥)." *The Medical World of Manchuria and Korea* 68 (1926/11): 17–24.

Ishidoya, Tsutomu, and Sugihara Noriyuki 杉原德行. "Korean Traditional Medicine 朝鮮の漢藥 (十六)." *The Medical World of Manchuria and Korea* 89 (1928/08): 45–48.

Ishidoya, Tsutomu, and Asakawa Takumi 浅川巧. "Reporting the Cultivation and Forestation Result of the American Catalpa in Korea 朝鮮に於けるカタルペ-スペシオサの養苗及造林成績を報す." *The Journal of the Forestry Association of Great Japan* 414 (1917/05): 32–34.

Ishidoya, Tsutomu, and Asakawa Takumi 浅川巧. "Reporting the Successful Cultivation of the Korean Larch テウセンカラマツの養苗成功を報す." *The Journal of the Forestry Association of Great Japan* 419 (1917/06): 36–40.

Ishidoya, Tsutomu, and Chung Tyaihyon. *The Identification Keys of Korean Forest Trees* 朝鮮森林樹木鑑要. Seoul: The Government-General of Korea, 1923.

Ishidoya, Tsutomu, and Tojinkai Research Section. *Medicinal Plants in North China* 北支那の藥草. Tokyo: Tojinkai, 1931.

Kaempfer, Engelbert. *Amoenitatum Exoticarum.* Lemgoviae: Typis & impensis Henrici Wilhelmi Meyeri, aulae Lippiacae typographi, 1712.

Kamita, Tsuneichi 上田常一. "Comments on New Natural History Textbook: Farm Animals 新理科書の教材の解說 家畜." *Korea of Art and Education* 75 (1931): 65–75.

Kamita, Tsuneichi 上田常一. "On Important Sea Creatures in the New Natural History Textbook Volume 2 (新理科書卷二にある朝鮮の重要なる海生産物に就いて)." *Studies on Korean Education* (朝鮮の教育研究) 5 (1932): 103–12.

Kamita, Tsuneichi 上田常一. "The Reality of Localized Education in Natural History in Korea (第五講 朝鮮に於ける「 理科鄉土教育 」の實際)." *Studies on Korean Education* 72 (1934): 62–65.

Kamita, Tsuneichi 上田常一. "The History of the ANHT 京城傳教員會活動史." *JNHT* 1 (1938): 5–13.

Kamita, Tsuneichi 上田常一. "The Guidelines for the Elementary School Science Education for the Promotion of Science 科學振興に伴ふ小學校理科教育の指針." *Korea of Art and Education* 181 (1940): 38–56.

Kamita, Tsuneichi 上田常一. "Outline of Science Education in Korea 朝鮮の理科教育史概說." *Bulletin of Korean Studies* 9 (1956): 337–46.

Kamita, Tsuneichi 上田常一. "Memories of Prof. Mori Tamezo 森為三先生の追憶." *Bulletin of Korean Studies* 26 (1963): 114–19.

Kanbe, Isaburo 神戸伊三郎. *The Development of Japanese Science Education* 日本理科教育発達史. Tokyo: Keibunsha, 1938.

Koidzumi, Genichi 小泉源一. "The Habitat of the Somei Yoshino Cherry そめゐよしのざくらノ自生地." *BMT* 27(320) (1913): 395.

Koidzumi Genichi 小泉源一. "Prunus Yedoensis Matsum Is a Native of Quelpaert! 染井吉野櫻の天生地 分明す." *Acta Phytotaxonomica et Geobotanica* 植物分類地理 1(2) (1932): 177–79.

Krause, Kurt. "Eine Neue Form Des Natürlichen Systems." *Naturwissenschaften* 11(4) (1923): 60–63.

Kure, Shuzo 呉秀三. *Siebold: His Life and Achievements* シーボルト先生: その生涯及び功業. Tokyo: Tohodo Shoten, 1926.

Kwŏn, Tŏkgyu 권덕규. "Finally, Korean People Should Become their Own Pride 마침내 조선 사람이 자랑이여야 한다." *Kaebyŏk* (1925/07/01): 18–21.

Lanjouw, Joseph. *Synopsis of Proposals Concerning the International Rules of Botanical Nomenclature, Submitted to the Seventh International Botanical Congress*. Stockholm, 1950. Utrecht: A. Oosthoek, 1950.

Lanjouw, Joseph. *Seventh International Botanical Congress, Section Nomenclature: Report*. Uppsala: International Bureau for Plant Taxonomy and Nomenclature of the International Association for Plant Taxonomy, 1953.

Lee, Deok-Bong 이덕봉. "Plants in the School Ground 학교 구내식물 이야기." *Paehwa* 배화 2 (1926): 95–110.

Lee, Deok-Bong 이덕봉. "Considering Korean Names for Korean Native Plants 조선산 식물의 조선명고." *Hangŭl* 한글 5(1) (1937): 312–15.

Lee, Deok-Bong 이덕봉. "Hearing from Elder Scientists and Engineers 원로과학기술자의 증언." *Science & Technology* 과학과 기술 12(10) (1979): 32–37.

Linné, Carl. *Species Plantarum*. Holmiæ [Stockholm]: Impensis Laurentii Salvii, 1753, vol. 2.

Linné, Carl. *Mantissa Plantarum Altera Generum* Editionis VI & Specierum Editionis II. Stockholm: Impensis Laurentii Salvii. 1771.

Lotsy, Johannes Paulus. *Evolution by Means of Hybridization*. Hague: M. Nijhoff, 1916.

Makino, Tomitaro 牧野富太郎. *Makino's New Illustrated Flora of Japan* (牧野) 新日本植物圖鑑. Tokyo: Hokuryukan, 1965.

Masamune, Genkei 正宗嚴敬. "Bibliography of Dr. B. Hayata's Work早田博士の著書及論文." *Transactions of the Natural History Society of Formosa* 臺灣博物學會會報24 (135) (1934): 478–85.

Matsumura, J. "Cerasi Japonicæ Duæ Species Novæ." *BMT* 15(174) (1901): 99–101.

Meikle, R. D. "The History of the Index Kewensis." *Biological Journal of the Linnean Society* 3(3) (1971): 295–99.

Merill, E. D. "To the Memory of Dr. Bunzo Hayata." *Transactions of the Natural History Society of Formosa* 24(135) (1934): 389–90.

Miki, Sakae 三木榮, and Togashi Naojiro 富樫直次郞. "On Farm Management 山林經濟考." *Korea* 262 (1937): 1–19.

Mori, Tamezo 森爲三. *The Collection of Korean Plant Names* 朝鮮植物名彙. Seoul: GGK, 1922.

Mori, Tamezo 森爲三. "Acknowledgement upon Publication 發刊の辭." *JNHT* 2 (1938): 1–3.

Mori, Tamezo 森爲三. "Thoughts on Natural History Textbooks in Middle School 中等學校博物教材に對する所感." *JNHT* 3 (1939): 1–2.

Nakai, Takenoshin. "An Observation on Japanese Aconitum-I." *BMT* 22(259) (1908): 127–33.

Nakai, Takenoshin. "An Observation on Japanese Aconitum-II." *BMT* 22(260) (1908): 133–40.

Nakai, Takenoshin. *Flora Koreana*. Tokyo: Imperial University of Tokyo, 1909.

Nakai, Takenoshin. "Cornaceæ in Japan." *BMT* 23(266) (1909): 35–45.

Nakai, Takenoshin. "On Naming a New Genus of Korean Native Lilies, Derauchia (Preliminary Report) 朝鮮産百合科植物ノ一新屬てらうちそう(新稱)ニ就テ (豫報)." *BMT* 27(322) (1913): 441–43.

Nakai, Takenoshin. *Korean Plants* 朝鮮植物. Tokyo: Seibido, 1914.

Nakai, Takenoshin. "Japanese Celtis." *BMT* 28(330) (1914): 261–69.

Nakai, Takenoshin. *Report on the Plants of Cheju and Wan Islands* 濟州島竝莞島植物調査書. Seoul: GGK, 1914.

Nakai, Takenoshin. *Flora Sylvatica Koreana* 朝鮮森林植物編. Seoul: GGK, 1915.

Nakai, Takenoshin. "An Outline of Dr. Matsumura Jinzo's Achievement 理學博士松村任三氏植物學上ノ事績ノ概略." *BMT* 29(346) (1915): 342–48.
Nakai, Takenoshin. "Aconitum of Yeso, Saghaline and the Kuriles." *BMT* 31(368) (1917): 219–31.
Nakai, Takenoshin. "Notulæ Ad Plantas Japoniæ Et Koreæ Xviii." *BMT* 32(382) (1918): 215–32.
Nakai, Takenoshin. *Florula of M't Paik-Tu-San* 白頭山植物調査書. Seoul: GGK, 1918.
Nakai, Takenoshin. *Report on the Vegetation of the Island Ooryongto or Dagelet Island, Corea* 鬱陵島植物調査書. Seoul: GGK, 1919.
Nakai, Takenoshin. "Chosenia, a New Genus of Salicaceæ." *BMT* 34(401) (1920): 66–69.
Nakai, Takenoshin. "Notulæ Ad Plantas Japoniæ Et Koreæ Xxvii." *BMT* 36(426) (1922): 61–73.
Nakai, Takenoshin. "Classification of Japanese Cynoglossum日本産おほるりさうハ四種アリテ皆新種デアル." *BMT* 37(434) (1923): 68–70.
Nakai, Takenoshin. "About the Last Names of Western Scholars with 'De'[De]ノ附ク植物學者ノ姓ニ就テ." *BMT* (411) (1925): 154–55.
Nakai, Takenoshin. "Notulæ Ad Plantas Japoniæ & Koreæ Xxxi." *BMT* 40(472) (1926): 161–71.
Nakai, Takenoshin. *Suggestions for an Amendment to Be Made to the Rules of Botanical Nomenclature*. Tokyo, 1926.
Nakai, Takenoshin. "Research on Korean Flora 朝鮮植物の研究." *Oriental Art and Science Magazine* 東洋學藝雜誌 43(534) (1927): 561–71.
Nakai, Takenoshin. "The Comparison of Floras of the Isolated Islands of in the East and West Part of the Korean Peninsula 朝鮮半嶋東西ニ孤立スル鬱陵島ト大黑山島トノ植物帶ノ比較." *Oriental Art and Science Magazine* 43(4) (1927): 214–27.
Nakai, Takenoshin. "The Floras of Tsusima and Quelpaert as Related to those of Japan and Korea." In Proceedings of the Third Pan-Pacific Science Congress, Tokyo, October 30th–November 11th, 1926. Tokyo, 1928, 893–911.
Nakai, Takenoshin. "The Vegetation of Dagelet Island: Its Formation and Floral Relationship with Korea and Japan." In *Proceedings of the Third Pan-Pacific Science Congress,* Tokyo, October 30th–November 11th, 1926. Tokyo, 1928, 911–14.
Nakai, Takenoshin. "Notulæ Ad Plantas Japoniæ & Koreæ Xxxvii." *BMT* 43(513) (1929): 439–59.
Nakai, Takenoshin. *Regarding the Botanical Nomenclature* 植物命名規則に就いて. Iwanami Biology Series. 東京: Iwanamishoten, 1930.
Nakai, Takenoshin. "To Asakawa Takumi 浅川巧君へ." *Craft Arts* 工藝 (1934/04): 81–84.
Nakai, Takenoshin. *East Asian Plants* 東亞植物. Vol. 52. Tokyo: Iwanamishoten, 1935.
Nakai, Takenoshin. *Newly Revised Plant Textbook* (新修 植物教科書). Tokyo: Sanseido, 1938.
Nakai, Takenoshin. "A New Classification of the Genus Lonicera in the Japanese Empire, Together with the Diagnoses of New Species and New Varieties." *JJB* 14(6) (1938/06): 359–76.
Nakai, Takenoshin. "A New Classification of the Sino-Japanese Genera and Species which Belong to the Tribe Camellieae I 日華兩國產 つばき族 植物ノ 新分類法 (其一)." *JJB* 16(11) (1940/11): 659–67.
Nakai, Takenoshin. "A New Classification of the Sino-Japanese Genera and Species which Belong to the Tribe Camellieae II 日華兩國產 つばき族 植物ノ 新分類法 (其二)." *JJB* 16(12) (1940/12): 691–708.
Nakai, Takenoshin. *Essential Results Obtained from My Observation on Tropical Plants in Java, Galang Island of Rio Archipelago, and on Japanese Plants in the Surroundings of Beppu Hotspring, Province of Bungo, Kyusyu*. Tokyo: Tokyo Science Museum, 1948.

Nakai, Takenoshin. "Reexamination of Questioned Scientific Names of Plants 問題にされた学名の再検討." *JJB* 26(11) (1951/11): 321–28.

Nakai, Takenoshin. "Essential Results Obtained from My Observations on Tropical Plants in Java, Galang Island of Rio Archipelago, and on Japanese Plants in the Surroundings of Beppu Hot Springs, Province of Bungo, Kiusie." *The Bulletin of the Tokyo Science Museum* 22 (1952): 1–43.

Nakai, Takenoshin. *A Synoptical Sketch of Korean Flora*. Tokyo: The National Science Museum, 1952.

Nakai, Takenoshin, and Honda Masaji 本田正次. *The List of Vascular Plants in Jehol Province* 熱河省ニ自生スル高等植物目錄. 第一次滿蒙學術調査研究團報告. Vol. 4.4, Tokyo: The First Scientific Expedition Team to Manchukuo 第一次滿蒙學術調査研究團, 1936.

National Research Council of Japan. *Scientific Japan: Past and Present*. Kyoto: Maruzen Co., 1926.

Natsume, Soseki夏目漱, *My Individualism and Othes* 나의 개인주의 외. Translated by Kim Chŏnghun. Seoul: Ch'aeksesang, 2004.Nitobe, Inazo 新渡戸稲造. *Bushido: The Soul of Japan*. New York: G.P. Putnams Sons, 1905.

Nitobe, Inazo 新渡戸稲造. *Essays* 隨想錄. Tokyo: Teimi Shuppansha, 1907.

Palibin, J. *Conspectus Florae Koreae*. Petropoli: Acta Horti Petropolitani, 1901.

Park, Mankyu 박만규. "A Historological Survey on the *Prunus yedoensis* in Korea 韓國왕벚나무의 調査研究史." *Journal of Botany* 식물학회지 8(3) (1965): 12–15.

Ridgway, Robert. *Color Standards and Color Nomenclature*. The Author, 1912.

Sato Sinichi 佐藤信一. "On Two or Three Things Regarding Japanese and Korean Plant Exchanges 日鮮植物交換史實の二三に就て." *The Journal of the Society of Forestry* 15(10) (1933): 884–93.

Seward, A. *Abstracts of Communications: 5th International Botanical Congress*. Cambridge, 16–23 August, 1930. Cambridge: University Press, 1930.

Siebold, Philipp Franz von. *On the Status of Japanese Botany* 日本植物學ノ現狀ニツキテ. Translated by Nakai Takenoshin. Tokyo: Shokubutsu Bunken, 1938.

Siebold, Philipp Franz von and Joseph Gerhard Zuccarini. *Flora Japonica*. Lugduni Batavorum: Apud auctorem, 1835.

Skottsberg, Carl. "Remarks on the Relative Independency of Pacific Floras." In *Proceedings of the Third Pan-Pacific Science Congress,* Tokyo, October 30th–November 11th, 1926, 914–20. Tokyo, 1928.

South Chŏlla Province Board of Education 全羅南道教育會. *Plants of South Chŏlla* 全羅南道植物. Seoul: South Chŏlla Province Board of Education, 1940.

South Kyŏngsang Province Board of Education 경상남도교육회. *Locally Rooted Studies: Natural History* 鄕土研究 博物. Pusan: South Kyŏngsang Province Board of Education, 1934.

Sugihara, Noriyuki. "Scientific Examination of Chinese Traditional Medicine 漢方醫學の科學的檢討." *Korea and Manchuria* (1939/01/05): 41–43.

Sugihara, Noriyuki. "Scientific Examination of Chinese Traditional Medicine (2/2)." *Korea & Manchuria* (1939/02/05): 34–37.

Sugihara, Noriyuki. "On the Pharmacological Section 京城帝國大學 藥理學教室." In Alumni Association of the Keiji Imperial University 京城帝國大學同窓會, *The Deep Blue Yonder: Fifty Years of the Foundation of the Keijo Imperial University* 紺碧遙かに: 京城帝國大學創立五十周年記念誌. Tokyo: Alumni Association of the Keiji Imperial University 京城帝國大學同窓會, 1974, 212–14.

Tanaka, Shigeo 田中茂穂. "Critique on Professor Hayata's Dynamic Classification of Plants 早田教授の「植物の動的分類系」を評す." *Zoological Magazine*動物學雜誌45(539) (1933/09): 378–88.

Terada, Torahiko. "On the Bathymetric Features of Japan Sea." *The Bulletin of the Earthquake Research Institute* 12(4) (1934): 650–55.

Theis, Nina, Michael J. Donoghue and Jianhua Li. "Phylogenetics of the Caprifolieae and Lonicera (Dipsacales) based on Nuclear and Chloroplast DNA Sequences." *Systematic Botany* 33 (2008): 776–83.

Thunberg, Carl Peter. *Flora Japonica*. Lipsiae: Bibliopolio I.G. Mülleriano, 1784.

Toh, Bong-syup 都逢涉. "Reporting a New Korean Site Producing Salicornia Herbacea L.あつけしさうノ朝鮮新産地 ヲ報ズ." *JJB* 8(9–10) (1933/04): 477–79.

Toh, Bong-syup 都逢涉. "Wild Medicinal and Indigenous Plants in Central Korea (Pt. 1) 中部朝鮮の野生植物及び朝鮮特産植物に就て(一)." *Pharmaceutical News of Japan* 9(16) (1934/08/20): 4–6.

Toh, Bong-syup 都逢涉. "Wild Medicinal and Indigenous Plants in Central Korea (Pt. 2)." *Pharmaceutical News of Japan* 9(17) (1934/09/05): 5–7.

Toh, Bong-syup 都逢涉. "Wild Medicinal and Indigenous Plants in Central Korea (Pt. 3)." *Pharmaceutical News of Japan* 9(18) (1934/09/20): 3–5.

Toh, Bong-syup 都逢涉. "Investigation of Traditional Medicines in Korea 朝鮮産漢藥調査." *Pharmaceutical News of Japan* 9(21) (1934/11/05): 15–17.

Toh, Bong-syup 都逢涉. "Alpine and Medicinal Plants in the Alpine Area of the South Hamkyŏng Province 咸鏡南道山岳地帯に於ける 高山植物及び藥用植物." *Journal of the Pharmaceutical Association of Korea* 15(3) (1935): 212–25.

Toh, Bong-syup 都逢涉. "Classification of Korean Plants: Field Collections Made by All Foreigners in the Past 朝鮮産植物의 分類 (上) 過去諸外國人의 實地採集." *Dong-a Daily* 1936/04/19.

Toh, Bong-syup 都逢涉. "Classification of Korean Plants: Indigenous Plants 5 Genera, About 500 Species 朝鮮産植物의 分類 (中) 朝鮮固有植物五屬五百餘種." *Dong-a Daily* 1936/04/21.

Toh, Bong-syup 都逢涉. "Classification of Korean Plants: Two New Genera Discovered in Korea Last Year 朝鮮産植物의 分類 (下) 昨年度에 發見된 二新屬." *Dong-a Daily* 1936/04/22.

Toh, Bong-syup 都逢涉. "Plants in Korea 朝鮮産植物." *Pharmaceutical News of Japan* 12(1) (1937/01/05): 24–26.

Toh, Bong-syup 都逢涉. "A Researcher's Note: The Future of Traditional Medicine Seen through Medicinal Plants 學究一家言: 藥草를 通해서 본 漢醫學의 將來[下]." *Dong-a Daily*, 1938/06/17.

Toh, Bong-syup, and Sim Hakchin 沈鶴鎭. "A Comparative Study of the Korean Medicinal Plants from Granite and Limestone Regions 朝鮮に於ける花崗岩地帯と石灰岩地帯とに分布せる藥用植物の比較研究 ". *Journal of the Pharmaceutical Association of Korea* 17(3) (1937/06): 91–98.

Toh, Bong-syup, and Sim Hakchin 沈鶴鎭. "Medicinal Plants Produced from Ullŭng Island 鬱陵島所産藥用植物-附 島勢一班." *Journal of the Pharmaceutical Association of Korea* 18(2) (1938): 59–81.

Toi, Hironobu 土居寛暢. "Thinking Old Days as a Natural History Teacher 三昔前の博物教員としての思出." *JNHT* 1 (1938): 14–22.

Toi, Hironobu 土居寛暢. "The Now Gone Amoeba Group 今は故きアメバ會." *JNHT* 2 (1938): 10–16.

Toi, Hironobu 土居寛暢. "Small Words on Natural History 博物小言." *JNHT* 3 (1939): 3–7.

Turrill, W. B. "The Expansion of Taxonomy with Special Reference to Spermatophyta." *BRV Biological Reviews* 13(4) (1938): 342–73.

U Hoik 우호익, "About Hibiscus 無窮花考(上)." *Eastern Light* 동광 13 (May 1927): 2–15.
Willis, John C. *Age and Area: A Study in Geographical Distribution and Origin of Species*. Cambridge: The University Press, 1922.
Willis, John C. "The Origin of Species by Large, Rather than by Gradual, Change, and by Guppy's Method of Differentiation." *Annals of Botany* 37(4) (1923): 605–28.
Wilson, Ernest Henry. *The Cherries of Japan*. Cambridge, MA: Printed at the University Press, 1916.
Yatabe, Ryokichi. "A Few Words of Explanation to European Botanists 泰西植物学者諸氏に告ぐ." *BMT* 4(44) (1890): 355–56.
Yi, Kilyong 이길용. "Mountain Climbing and Swimming: Self-Cultivation in Clear Sky 炎天의 修養 登山과 水泳, 夏期休暇特輯." *Tongkwang* 동광 (1927/07/05): 20–22.
Yi, Sukwang 이수광. *Chibong's Classified Commentaries* 지봉유설精選. Translated by Chŏng Hyeryŏm 정해렴. Seoul: Hyŏndaesilhaksa, 2000.
Yu, Fujikawa. "A Brief Outline of the History of Medicine in Japan." In *Scientific Japan: Past and Present*. Kyoto: Maruzen Co., 1926, 229–42.
Yu, Chin-o 유진오. "A Genealogy of Brilliant Wishes (華想譜)." In Yi Hyosŏk 이효석 and Yu Chin-o eds., *The Complete Collection of Korean Literature* (한국문학전집) vol. 8. Seoul: Minchungsŏkwan, 1959, 219–500.
Z, Y. "T.Ishidoya: Chinesische Drogen, Teil I, 1933." *Acta phytotaxonomica et geobotanica* 2(4) (1933/11/11): 315.
Yasuda, Ken 安田健. *Collection of the Nation-Wide Local Product Surveys* 諸国産物帳集成. Tokyo: Kagaku Shoin, 1996.

Secondary Sources

Korean and Japanese WorksBae, Jae-Soo 배재수. *The Changes in the Forest Ownership in Korea* 한국의 근・현대 산림소유권 변천사. Seoul: Imŏp Yŏn'guwŏn, 2001.
Ch'a, Munsŏng 차문성. *The Formation and the Transformation of Modern Museums* 근대 박물관 그 형성과 변천 과정. Seoul: Hankukhaksulchŏngbo, 2008.
Chang, Chin Sung 장진성. "A Reconsideration of Nomenclatural Problems on Korean Plants and the Korean Woody Plant List 韓國樹木의 目錄과 學名에 대한 再考." *Korean Journal of Plant Taxonomy* 24(2) (1994): 95–124.
Chang, Hyechŏng 장혜정. "Life of Asakawa Takumi and 'Korea' 아사카와 타쿠미의 삶의 궤적과 '朝鮮'." *Comparative Studies of World Literature* 世界文學比較研究 14 (2006): 103–22.
Ch'oe, Chaech'ŏl 최재철. "Abe Yoshishige, Keijo Imperial University, and Intelletuals in the Colonized Seoul 경성제국대학과 아베 요시시게, 그리고 식민지 도시 경성의 지식인." *Japan Research* 日本研究 42 (2009): 261–79.
Chŏi, Hwan 최환. *Overview of Culture of Classified Writings in China and Korea* 한・중 유서문화 개관. Kyŏngsan: Youngnam University Press, 2008.
Chŏng Chaech'ŏl 정재철. *The History of Education Policy of the Japanese Colonial Regime* 日帝의 對韓國 植民地 教育政策史. Seoul: Ilchisa, 1985.
Chŏng, Min 정민. *Discovery of the 18th Century Korean Intellectuals* 18세기 조선 지식인의 발견. Seoul: Humanist, 2007.
Chŏng, Minsŏng 정민성. *History of our Medicine and Pharmaceutics* 우리 醫藥의 역사. Seoul: Hakminsa, 1990.
Chŏng, Mungi 정문기. "My Life: Dr. Chŏng Mungi Who Devoted His Life to the Study of Fishes in Korea 나의 인생회고: 한반도 물고기 연구에 생을 바친 정문기박사." *United Korea* 통일한국 22 (1985): 80–87.

Cho, Hyŏngrae 조형래. "The Science of *Youth* 소년의 과학." *SAI* 사이 6 (2009): 273–302.
Chŏng, Yŏng-ho 정영호. *An Outline of History of Plant Classification in Korea* 한국 식물 분류학사 개설. Seoul: Academy Books, 1986.
Chung, Kyu-Young 정규영. "Colonialism and the Politics of Knowledge: The Continent Research by Keijo Imperial University 콜로니얼리즘과 학문의 정치학 -15년전쟁하 경성제국대학의 대륙연구." *Educational History Research* 교육사학연구 9 (1999): 21–36.
Chung, Kyu-Young 정규영. "A History of the Normal School of Korea during the Japanese Colonial Ruling 日帝時代 師範學校의 歷史." *Theses Collection* 論文集 41 (2003): 29–56.
Chung, Yountae 정연태. *Everyday Colonial Discrimination in Japanese Korea* 식민지 차별의 일상사. Seoul: Purŭnyŏksa, 2021.
Chung, Tyaihyon. "50 Years of Carrying Field Press 야책을 메고 50년." *Forest and Culture* 숲과 문화 11(3) (2002): 52–62.
College of Agriculture and Life Sciences 농업생명과학대학. *100 Years of Agricultural Education: 1906–2006* 농학교육 100년: 1906–2006. Seoul: College of Agriculture and Life Sciences, Seoul National University, 2006.
Compendium of History of Modern Culture in Korea 한국현대문화사대계, Vol. III History of Science and Technology. Seoul: Research Institute of Korean Studies, Korea University, 1977.
Department of Forest Sciences 산림과학부. *One Hundred Years of Forestry Science: 1906–2006* 산림과학 백년사: 1906–2006. Seoul: Department of Forest Sciences, Seoul National University, 2006.
Doctor Cho Pok-Sung Memorial Committee 관정조복성박사기념사업회. *The Record of Dr. Cho Pok-Sung's Insect Collecting* 조복성곤충채집여행기. Seoul: Korea University Press, 1975.
Forestry Association of Great Japan 大日本山林會. *80 Years of Japanese Forestry* 日本林業八十年史. Tokyo: Dainihon Sanrinkai, 1962.
Ha, U-Bong 하우봉. *Koreans' Understanding of Japan in Joseon Dynasty* 조선시대 한국인의 일본인식. Seoul: Hyean, 2006.
Hagino, Toshio 萩野敏雄. *On the Development of Forestry in Japanese Korea, Manchuria and Taiwan* 朝鮮・満州・台湾林業発達史論. Tokyo: Rin'ya Kosaikai, 2005.
Hiroshige, Tetsu 廣重徹. *Social History of Science* 科學の社會史. Tokyo: Iwanamishoten, [1973] 2002.
History of Korean Agricultural Education 한국농업교육사. Seoul: Taehankyokwasŏ, 1994.
History of Science Society of Japan 日本科学史学会. *A Compendium of History of Science in Japan* 日本科學技術史大系. Vols. 1, 8, 9, 10, 15, Tokyo: Daiichi Hoki Shuppan, 1965.
Hŏ, Chaeyŏng 허재영 ed. *The Textbooks of the Protectorate Period* 통감시대 교과서 자료. vol. 8. Natural History, Kwangmyŏng: Kyŏngjin, 2011.
Hŏ, Suyŏl 허수열. *Development Without Development* 개발 없는 개발 : 일제하, 조선경제 개발의 현상과 본질. Seoul: ŭnhangnamu, 2005.
Hŏ, Suyŏl 허수열. *Korean Agriculture in the Early Japanese Era* 일제초기 조선의 농업 : 식민지근대화론의 농업개발론을 비판한다. Paju: Hangilsa, 2011.
Hong, Hyŏn-o 홍현오. *A History of Korean Pharmaceutics* 韓國藥業史. Seoul: Handokyakpum Co., 1972.
Hong, Munhwa 홍문화. "The Influence of Korean Transliterated Names of Local Medicines on Japanese Bencao Studies 우리의 이두향약명이 일본의 본초학에 미친 영향." *Korean Journal of Pharmacognosy* 생약학회지 3(1) (1972): 1–10.

Hong, Munhwa 홍문화. *Scattered Thoughts on History of Pharmaceutics* 藥史散攷. Seoul: Dongmyŏngsa, 1980.

Hong, Soon-Kwon 홍순권. "The Status of Ginseng Cultivation in the Kaesŏng Area before the Colonial Period 한말시기 開城地方 蔘圃農業의 전개 양상 (上)." *Journal of Korean Studies* 韓國學報 13(4) (1987): 33–67.

Hong, Soon-Kwon 홍순권. "Population and Social Stratification of the Japanese Society in Pusan under the Rule of Japanese Imperialism 일제시기 부산지역 일본인사회의 인구와 사회계층구조." *History & the Boundaries* 역사와 경계 51 (2004): 43–73.

Hur, Kyung Jin 허경진. "Japanese Travels of Joseon Medicine and the Aspects of Publication of Collections of Medical Written Conversations 조선 의원의 일본 사행과 의학 필담집의 출판 양상." *Korean Journal of Medical History* 19(1) (2010): 137–56.

Im, Jong Hyok 임정혁. "On Works of Physicist Toh Sangrok during the Colonial Era 연구노트: 식민지시기 물리학자 도상록의 연구활동에 대하여." *Journal of the Korean History of Science Society* 27(1) (2005): 109–26.

Ito, Junro 伊藤純郎. *Studies on the Localized Education Movement* 郷土教育運動の研究. Kyoto: Shibunkaku Shuppan, 2008.

Jeon, Chan-Mi 전찬미. "Plans for Science Education in Colonial Korea: The Establishment of the Science Department of Chosen Christian College 식민지시기 연희전문학교 수물과의 설립과 과학 교육." *JKHSS* 32(1) (2010): 43–68.

Jeon, Sang-woon 전상운. "Yi Kyukyŏng and his Broad Studies 李圭景과 그의 博物學." *Research Journal* 4 (1972): 89–96.

Ji, Su-Gol 지수걸. "The Militaristic Fascism of the Japanese Empire and the Rural Development Movement in Colonial Korea 일제의 군국주의 파시즘과 '조선농촌진흥운동'." *History Review* 47 (1999): 16–36.

Kamimura, Minoru 上村登. *Loving Flowers* 花と恋して: 牧野富太郎伝. Kochi: Kochi Shinbunsha, 1999.

Kang, Hyesun 강혜선. "The Naturalistic Taste in the Late Chosŏn Period and Kim Ryŏ's Chinese Poetry 조선후기 박물학적 취향과 김려의 한시." *Theses on Korean Literature* 韓國文學論叢 43 (2006): 37–68.

Kang, Myŏng-gwan 강명관. *History of Books and Knowledge in Chosŏn Korea* 조선시대 책과 지식 의 역사: 조선의 책과 지식은 조선 사회와 어떻게 만나고 헤어졌을까? Seoul: Ch'ŏnnyŏn ŭi Sangsang, 2014.

Kang, Segu 강세구. "On Kang Hŭian's Small Notes on Cultivating Flowers 姜希顔의 <養花小錄>에 관한 一考察." *The Journal of Korean History* 한국사연구 60 (1988): 37–56.

Kano, Masanao 鹿野正直. "Cultural Movements under the 'Crossroads of Reform' '改造の十字路' 下の文化運動." In *The Popular Movement and its Ideas in Modern Japan* 近代日本の民衆運動と思想. Tokyo: Yuhikaku, 1977, 139–72.

Kato, Shigeo 加藤茂生. "Science and Imperialism: Boundaries of Science Seen from History of Colonial Science 科学と帝国主義 科学の外延–植民地科学史の視点から." *Contemporary Philosophy* 現代思想 29(10) (2001/08): 176–85.

Kawamura, Yutaka 河村豊 et al. "The Status and Task for War-time Science in Japan: The Report from the 2003 Annual Conference 日本戦時科学史の現状と課題 (2003年度年会報告)." *Journal of History of Science, Japan. Series II* 科学史研究. 第II期 43(229) (2004): 45–56.

Kim, Boumsoung 金凡性. *Meiji and Daisho Seismology: Beyond Local Science* 明治・大正の日本の地震学 ：「ローカル・サイエンス」を超えて. Tokyo: Tokyo Daigaku Shuppankai, 2007.

Kim, Choon-Min 김준민. "Prof. Man Kyu Park and His Works 박만규 선생 회갑 기념호: 박만규 선생의 반면 (半面)." *Journal of Plant Biology* 식물학회지 10(1–2) (1967): 1–2.

Kim, Doo-Jong 김두종. *History of Korean Medicine* 한국의학사 全. Seoul: T'amgudang, 1966.

Kim, Ho 김호. "The Life of Medical Historian Miki Sakae, and the History of Korean Medicine and of Diseases in Korea 醫史學者 三木榮의 생애와 朝鮮醫學史及疾病史." *Korean Journal of Medical History* 14(2) (2005): 101–22.

Kim, Hui 김휘, Chang Gae-Sun 장계선, Chang Chin-Sung 장진성, and Choi Byoung Hee 최병희. "Re-Examination on Foreign Collectors' Sites and Exploration Routes in Korea: Nakai Takenoshin 외국인의 한반도 식물 채집행적과 지명 재고(2)-나카이 다케노신." *Korean Journal of Plant Taxonomy* 식물분류학회지 36(3) (2006): 227–55.

Kim, Geun-Bae 김근배. *The Emergence of Modern Techno-Scientific Manpower in Korea* 한국 근대 과학기술 인력의 출현. Seoul: Munji, 2005.

Kim, Kŭnsu 김근수. "Chosŏn Korea's Practical Studies and Studies about Everything 한국 실학과 명물도수학." *Studies of Intellectual Cultures* 정신문화연구 5(1) (1982): 86–97.

Kim, Nak-nyŏn 김낙년. *Korean Economy under the Japanese Rule* 일제하 한국경제. Seoul: Haenam, 2003.

Kim, Nak-nyŏn 김낙년. *Korean Economic Growth, 1910–1945* 한국의 경제성장 1910–1945. Seoul: Seoul National University Press, 2006.

Kim, Ock-Joo 김옥주. "Physical Anthroplogy Studies at Keijo Imperial University Medical School 경성제대 의학부의 체질인류학 연구." *Korean Journal of Medical History* 의사학 17(2) (2008): 191–203.

Kim, Sŏkkwŏn 김석권. "Reflection of Asakawa Takumi from Forest Historical Point of View 아사카와 타쿠미의 임업사적 재평가." *Forest and Culture* 숲과 문화 20(5) (2011): 45–56.

Kim, Sungwon 김성원. "The Context of a Korean Naturalist's Career-Building in Colonial Korea: Cho Pok Sung as an Example of Colonial Entomologist 식민지시기 조선인 박물학자 성장의 맥락: 곤충학자 조복성의 사례." *JKHSS* 30(2) (2008): 353–82.

Kim, Yŏngjin 김영진. "The Influence of Chinese Agriculture in Korea through Farm Management References 산림경제의 인용문헌으로 본 분석을 통한 중국 농학의 영향." *Rural Economy* 농촌경제 15(3) (1992): 59–70.

Kim, Yongsup 김용섭. *Agricultural History of the Late Chosŏn Era* 朝鮮後期農學史研究: 農書와 農業 관련 文書를 통해 본 農學思潮. Paju: Jisiksanŏpsa, 2009.

Kimura, Yojiro 木村陽二郎. *Works on History of Biology* 生物学史論集. Tokyo: Yasaka Shobo, 1987.

Kimura, Yojiro 木村陽二郎. *Naturalists of the Edo Era* 江戸期のナチュラリスト. Tokyo: Asahi Shinbunsha, 2005.

Komagme, Takeshi 駒込武 고마고메 다케시. *The Cultural Integration of the Colonial Empire of Japan* 식민지제국 일본의 문화통합. Translated by Oh Sŏngchŏl 오성철, Lee Myŏngsil 이명실, and Kwon Kyŏnghi 권경희. Seoul: Yŏkbi, 2008.

Komori, Yoichi 小森陽一 고모리 요이치. *Postcolonial: Colonial Sub-Consiousness and Colonial Consciousness* 포스트콜로니얼: 식민지적 무의식과 식민주의적 의식. Translated by Song Tae-uk 송태욱. Seoul: Samin, 2002.

Koyasu, Nobukuni 子安宣邦. *To-a, Great to-a, East Asia: Orientalism in Modern Japan* 동아・대동아・동아시아: 근대 일본의 오리엔탈리즘. Translated by Yi Sŭngyŏn 이승연. Seoul: Yŏkbi, 2005.

Ku, Chaejin 구재진 et al. *The Formation of 'Things Korean' and Modern Cultural Discourse* '조선적인 것' 의 형성과 근대문화 담론. Seoul: Somyŏng, 2007.

Kwon, Tae-Eok 권태억. "Japan's Assimilation Policy 동화정책론." *The Journal of the Korean Historical Association* 역사학보 172 (2001): 335–65.

Kwon, Tae-Eok 권태억. "Japan's Assimilation Policy of the 1920s and 30s 1920, 30년대 일제의 동화정책론." *Han'guk saron*韓國史論 53 (2007): 403–42.

Kwon, Tae-Eok 권태억. *Japanese 'Civilizing Mission' in Korea (1904–1919)* 일제의 한국 식민지화와 문명화 (1904–1919). Seoul: Seoul National University Press, 2014.

Lee, Byŏnghun 이병훈, and Kim Chint'ae 김진태. "On the Introduction of Western Modern Biology: Focusing on Animal Systematics 西洋 近代 生物學의 國內 導入에 관한 연구: 동물분류학을 중심으로." *Animal Systematics, Evolution and Diversity* 10(1) (1994): 85–95.

Lee, Deok-Bong 이덕봉. "History of Recent Development of Korean Botany: The Institutes and Status of Botany in Colonial Korea 최근세한국식물학연구사: 일제통치하 한국에 있어서의 식물학 연구에 관한 시설과 그 실태." *The Journal of Asiatic Studies* 亞細亞研究 4(2) (1961): 101–49.

Lee, Jeongseon 이정선. *Assimilation and Exclusion: Japanese Assimilation Policies and the Korean-Japanese marriage* 동화와 배제: 일제의 동화정책과 내선결혼. Koyang: Yŏksabipyŏngsa, 2017.

Lee, Jung 이정, "Dreamland for Japanese Bureaucrats: Harmonious Conflicts in Japanese Colonial Korea (관료들의 천국: 일제강점기 약초재배운동의 조화로운 동상이몽)." *The Journal of the Korean Historical Association* 238 (2018): 299–342.

Lee, Won-pil 이원필. "A Study on Teacher Training System in Korea under Japanese Imperialism 日帝統治期의 中等教員養成法制에 관한 研究." *The Journal of Educational Idea* 教育思想研究 5 (1996): 7–41.

Lim, Jongtae 임종태. "'도리'의 형이상학과 '형기'의 기술: 19세기 중반 한 주자학자의 눈에 비친 서양 과학 기술과 세계: 이항로(1792–1868) Yi Hang - No's Neo - Confucian Critique of Western Science and Technology and His Natural Philosophy of Civilizations in the Mid -19th Century Korea." *Journal of the Korean History of Science Society* 한국과학사학회지 21(1) (1999): 58–91.

Lim, Jongtae 임종태. *Understanding of Western Geography in Seventeenth- and Eighteenth-Century China and Korea: Fables about the Globe and the Five Continents* 17, 18 세기 중국과 조선의 서구 지리학 이해: 지구와 다섯 대륙의 우화. Paju: Changbi, 2012.

Lim, Jongtae 임종태. *Travel, Statecraft Reform, and Science and Technology in Eighteenth-Century Korea* 여행과 개혁, 그리고 18 세기 조선의 과학기술. Paju: Tŭllyŏk, 2022.

Marius, Jansen 마리우스 잔센. *Making of Modern Japan* 현대일본을 찾아서. Translated by Kim U-yŏng et al. Seoul: Yeesan, 2006.

Miki, Sakae 三木榮. *History of Medicine and Diseases in Korea* 朝鮮醫學史及疾病史. Osaka: Miki Sakae, 1963.

Mitani, Hiroshi 三谷博. *Meiji Restoration and Nationalism* 明治維新とナショナリズム: 幕末の外交と政治変動. Tokyo: Yamakawa Shuppansha, 1997.

Mitani, Hiroshi 三谷博. *The Visit of Commodore Perry* ペリー来航. Tokyo: Yoshikawa Kobunkan, 2003.

Moon, Manyong 문만용. "'조선적 생물학자' 석주명의 나비분류학 Butterfly-Taxonomy of 'the Korean Biologist', Seok Joo-Myung." *JKHSS* 21 (1999): 157–193.

Moon, Manyong 문만용. "Toh Bong-syup 도봉섭." In *A Biographical Dictionary of Who's Who in Science and Engineering* 과학기술인명사전. Seoul: National Research Foundation of Korea, 2012, 121–127.

Moon, Myung-ki 문명기. "The Gap Between the 'Colonial Modernity' of Taiwan and Korea: Comparing the Police Systems 대만・조선의 '식민지근대' 의 격차." *Studies on Chinese Modern History* 중국근현대사연구 59 (2013): 69–99.

Moon, Myung-ki 문명기. "A Comparative Study on Annual Expenditures of Taiwan and Chosun Government General during Japanese Colonial Period 일제하 대만·조선총독부의 세출구조 비교분석." *Theses of Korean Studies* 한국학논총 44 (2015): 409–454.

Mun, Jung-Yang 문중양. "Science and Technology in King Sejang's Era: Questioning their Independence from the Chinese Ones 세종대 과학기술의 '자주성' 다시 보기." *The Journal of the Korean Historical Association* 189 (2006): 39–72.

Nagakubo, Hen'un 長久保片雲. *The Life of the World-Class Botanist Matsumura Jinzo* 世界的植物学者松村任三の生涯. Tokyo: Akatsuki Inshokan, 1997.

Nakashima, Koji 中島弘二 ed. *The Japanese Empire and Forests: Conservation and Development of Forest Resource in Modern East Asia* 帝国日本と森林: 近代東アジアにおける環境保護と資源開発. Tokyo: Keiso Shobo, 2023.

Noh, Dae-Hwan 노대환. *Studies on the Making of the Eastern Way and Western Means Idea* 동도서기론 형성 과정 연구. Seoul: Ilchisa, 2005.

Oba, Hideaki 大場秀章. *History of Botanical Research in Japan: 300 Years of Koishikawa Botanical Gardens* 日本植物研究の歴史 ：小石川植物園三〇〇年の歩み. 東京大学コレクション. Vol. 4, Tokyo: Tokyo Daigaku Sogo Kenkyu Hakubutsukan, 1996.

Oba, Hideaki 大場秀章. *History of Botany/Cultural History of Botany* 植物学史・植物文化史. Vol. 1. Tokyo: Yasakashobo, 2006.

Oh, Chae-kun 오재근. "Behind the Naming of Herbal Section as the Decoction Section in Treasured Mirror of Eastern Medicine 조선 의서 『동의보감』은 왜 본초 부문을 「탕액편」이라고 하였을까." *Korean Journal of Medical History* 20(2) (2011): 263–90.

Oh, Chae-kun, and Kim Yong-jin 김용진. "A Study on the Classification of Materia Medica in Medicinal Part of Treasured Mirror of Eastern Medicine 동의보감「탕액편 」의 본초 분류에 대한 연구." *Journal of Oriental Medical Classics*大韓韓醫學原典學會誌 23(5) (2010): 55–66.

Oh, Chae-kun, and Yoon Chang-yeol 윤창열. "A Study on the Nature of Medicinals in Rhymes of Medical Books in Chosun Dynasty 조선 의서 중의 약성가(藥性歌)에 대한 연구 -『제중신편』,『의종손익』을 중심으로." *Journal of Oriental Medical Classics* 24(3) (2011): 49–64.

Oh, Sŏngchŏl 오성철. *The Formation of Colonial Elementary Education* 식민지 초등교육의 형성. Seoul: Kyoyukkwahaksa, 2000.

Oka, Eiji 岡衛治. *History of Korean Forestry* 조선임업사. Translated by Lim Kyeong-Bin 임경빈. Kwachŏn: Korea Forest Service, 2000.

Omori, Minoru 大森實. *Ph. Fr. Von. Siebold and the Modernization of Japan* Ph.Fr.Vonシーボルトと日本の近代化. Tokyo: Hosei Daigaku, 1992.

Osato, Hiroaki 大里浩秋. "Dojinkai and Dojin (同仁会と『同仁』)." *Institute for Research in Humanities Brief* 人文学研究所報 40 (2007): 47–105.

Otsuka, Yasuo 大塚恭男. *Oriental Medicine in Japan* 일본의 동양 의학. Translated by Yi Kwangchun. Seoul: Sohwa, 2000.

Park, Changsung 박찬승. *The Era of Nationalism* 민족주의의 시대. Seoul: Kyŏngin Ch'ulp'ansa, 2007.

Pak, Kihwan 朴己煥. "History of Cultural Exchange between Modern Japan and Korea: Koreans' Study Abroad in Japan 近代日韓文化交流史研究: 韓國人の日本留學." Ph.D. Thesis, Osaka University, 1998.

Park, Keong-Suk 박경숙. "Population Dynamics of Korea during the Colonization Period (1910–1945) 식민지 시기(1910년-1945년) 조선의 인구 동태와 구조." *Korea Population Studies* 한국인구학 32(2) (2009): 29–58.

Park, Man-kyu 박만규. "Overview of the Flora of South Chŏlla Province 전라남도 식물상의 개관." *Korean Journal of Plant Taxonomy* 식물분류학회지 1(2) (1969): 9–12.

Park, Seong-Rae 박성래. "Western Science Introduced in the Late Chosun Period 개화기의 서양과학." *JKHSS* 10(1) (1988): 169–72.

Park, Seong-Rae 박성래. "Ueki Homiki, Who Introduced Pinus Rigida to Korea 국내에 리기다소나무 보급한 임학자 植木秀幹." *Science & Technology* 과학과 기술 452 (2007): 106–07.

Park, Yun-Jae 박윤재. *The Origin of Korean Modern Medicine* 한국 근대의학의 기원. Seoul: Hyean, 2005.

Park, Yun-Jae 박윤재. "Joseon Government-General's Policy of Local Medicine and Local Consumptions of Medical Service 조선총독부의 지방 의료정책과 의료 소비." *Critical Studies on Modern Korean History* 역사문제연구 21 (2009): 161–83.

Research Group on Japanese Colonialism 日本植民地研究会 ed. *The Status and Task of Japanese Colonialism Studies* 日本植民地研究の現状と課題. Tokyo: Atenesha, 2008.

Sakano, Toru 坂野徹. *Investigating the Empire: A History of Science of Colonial Fieldwork* 帝国を調べる: 植民地フィールドワークの科学史. Tokyo: Keiso Shobo, 2016.

Sakano, Toru, and Tsukahara Togo 塚原東吾. *Intellectual History of Science in Imperial Japan* 帝国日本の科学思想史. Tokyo: Keiso Shobo, 2018.

Setoguchi, Akihisa 瀬戸口明久. *The Birth of Insect Pests: Japanese History through Insects* 害虫の誕生: 虫からみた日本史. Tokyo: Chikuma Shobo, 2009.

Shin, Chang-Geon 愼蒼健, "Institutionalization of Research on Traditional Medicine in Keijo Imperial University 경성제국대학에 있어서 한약연구의 성립." *Society and Culture* 사회와 역사 76 (2007): 105–139.

Shin, Dong-won 신동원. *A History of Modern Public Health and Medicine in Korea* 한국 근대 보건 의료사. Seoul: Hanul, 1997.

Shin, Dong-won 신동원. "The Medicine and Medical Practice in Japanese Colonial Period in Korea 일제 강점기의 의학과 의료." In Park, Seong-Rae 박성래, Shin Dong-won 신동원, O Tonghun 오동훈 eds., *One Hundred Years of Our Science* 우리과학 100년. Seoul: Hyŏnamsa, 2001, 120–41.

Shin, Dong-won 신동원. *Hŏ Chun, A Man of Chosŏn* 조선사람 허준. Seoul: Hangyereh Sinmunsa, 2001.

Shin, Dong-won 신동원. "Changes in Late Chosŏn Medical Life: From Gift Economy to Market Economy 조선 후기 의약생활의 변화:선물경제에서 시장경제로 : 『미암일기』,『쇄미록』,『이재난고』,『흠영』의 분석." *History Review* 역사비평 75 (2006): 344–91.Shindo, Toyoo 新藤東洋男. *Japanese Teachers in Colonial Korea* 在朝日本人教師: 反植民地教育運動の記錄. Tokyo: Shiraishi Shoten, 1981.

Sim, Ch'anggu 심창구, Nam Yŏnghŭi 남영희, Chŏng Sŏnguk 정성욱, and Hwang Sŏngmi 황성미. "History of Korean Pharmaceutics 한국약학사." *Journal of the Pharmaceutical Society of Korea* 약학회지 51(6) (2007): 361–82.

Sŏ, Yugu 서유구. *Treatises on Forestry and Economy: The Largest Practical Encyclopedia in Chosŏn Korea* 임원경제지: 조선 최대의 실용백과사전. Translated by Chŏng Myŏnghyŏn 정명현. Seoul: Ssiasŭlp'urinŭnsaram, 2012.

Song, Minyoung 송민영. "The Implication for 'The Seventh Extracurricular Curriculum' by the Comprehensive Science Education of the GGK Era 조선총독부시대의 종합적인 이과교육이 '제7차 특별활동 교육과정' 에 주는 시사점 고찰." *Studies on Education* 교육학연구 41 (2003): 117–46.

Sun, You-Jeong 선유정. "The Trajectory of Hyun Sin-Kyu's Forestry Research 현신규의 임학연구 궤적: 과학연구의 사회적 진화." Ph.D. Thesis, Chŏnbuk University, 2012.

Suzuki, Zenji 鈴木善次. *Biology-The Beginning* バイオロジ-事始. Tokyo: Yoshikawa Kobunkan, 2005.

Takatsu, Takashi 高津孝. *East Asia of Natural History and Texts: Maritime Exchange between Satsuma and Ryukyu*博物学と書物の東アジア: 薩摩·琉球と海域交流. Okinawa: Yoju Shorin, 2010.

Takasaki, Soji 高崎宗司 다카사키 소지. *Becoming the Part of Korean Soil: Asakawa Takumi* 조선의 흙이 되다: 아사카와 다쿠미 평전. Translated by Kim Sunhi 김순희. Paju: Hyohyŏng, 2005.

Takeuchi, Yoshimi 竹内好, *Japan and Asia* (일본과 아시아), Translated by Sŏ Kwangtŏk 서광덕 and Paek Chiun 백지운. Seoul: Somyŏng, 2004.

Tanaka, Akira 田中彰. *The Idea of Small Japan: Re-Reading Japanese Modernity* 소일본주의 : 일본의 근대를 다시 읽는다, Translated by Kang Chin-a 강진아. Seoul: Sohwa, 2002.

Tashiro, Kazui 田代和生. *An Investigation of Korean Medicines in the Edo Period* 江戸時代朝鮮藥材調査の研究. Tokyo: Keio Gijuku Daigaku Shuppankai, 1999.

Tateiwa, Iwao 立岩巌. *The Early History of Geological Research in Korea* 韓半島 地質學의 初期研究史: 朝鮮·日本列島地帶 地質構造論考. Translated by Yang Sŭngyŏng 양승영. Taegu: Kyŏngbuk University Press, 1996.

Toh, Chung-Ae 도정애. *Toh Bong-Syup: Celebrating the 100th Anniversary of His Birth* 都逢涉誕生百週年記念資料集. Seoul: Chayŏnmunhwasa, 2003.

Tsukiashi, Tatsuhiko 月脚達彦 쓰키아키 다쓰히코. "Korean Intellectuals for Civilization an Enlightenment and Fukuzawa Yukichi 朝鮮開化派와 후쿠자와 유키치." *Hankukhakyŏnku* 한국학연구 26 (2012): 307–35.

Tsurumi, Shunsuke 鶴見俊輔 쓰루미 슌스케. *Conversions: Lectures on Japanese Ideas during the War Period, 1931–1945* 전향: 쓰루미 슌스케의 전시기 일본정신사 강의 1931–1945. Translated by Ch'oe Yŏngho 최영호. Seoul: Nonhyŏng, 2005.

Ueno, Masuzo 上野益三. *A History of Natural History in Japan*日本博物学史. Tokyo: Heibonsha, 1973.

Wŏn, Chiyŏn 원지연. "Nitobe Inazo and Orientalism 니토베 이나조와 오리엔탈리즘." *Tongamunhwa* 동아문화 40 (2002): 47–72.

Yamamoto, Hikaru 山本光. *History of Forestry/Forest Geography* 林業史·林業地理. Tokyo: Meibundo, 1961.

Yang, Jeongpil 양정필. "Modern Medicine Environment and Adaptation of Korean Trader for Medicinal Herbs 한말-일제 초 근대적 약업 환경과 한약업자의 대응." *Korean Journal of Medical History* 15(2) (2006): 189–209.

Yang, Jeongpil 양정필. "The Commercialization and Promotion of White Ginseng in the 1910s and 1920s 1910–20년대 개성상인의 백삼(白蔘) 상품화와 판매 확대 활동." *Korean Journal of Medical History* 20(1) (2011): 83–118.

Yeo, In-sok여인석. "Kim Doo-jong: A Life for the History of Korean Medicine 一山 金斗鍾 선생의 생애와 학문." *Korean Journal of Medical History* 7(1) (1998): 1–12.

Yeo, In-sok, Yi Hyŏnsuk 이현숙, Kim Sŏngsu 김성수, Sin Kyuhwan 신규환, Park Yunhyŏng 박윤형, and Park Yun-jae 박윤재. *History of Korean Medicine* 한국의학사. Seoul: Korean Medical Association Research Institute for Healthcare Policy, 2012.

Yi, Ch'angbok 이창복. "New Natural History: Plants 신박물기: 식물." In The Commemoration Committee for Prof. Yi Ch'angbok ed., *The Life of Yi Ch'angbok* 樹友 李昌福 教授의 발자취. Seoul: Chŏngminsa, 1984, 1–10.

Yi, Ch'angbok, and Yi Munho 이문호. "Korean Plant Research of a French Priest 프랑스 선교사의 한국식물 연구." *Studies in Church History* 教會史研究 (5) (1987): 149–207.

Yi, Chiwŏn 이지원. *Modern Cultural Thoughts in Korea* 한국 근대 문화 사상사 연구. Seoul: Hyean, 2007.

Yi, Chongmuk 이종묵, and Kang Hŭian 강회안. *Yanghwasorok: A Scholar Befriending Flowers and Trees* 양화소록: 선비, 꽃과 나무를 벗하다. Seoul: Acanet, 2012.
Yi, Ch'ung-u 이충우. *Keijo Imperial University* 경성제국대학. Seoul: Tarakwŏn, 1980.
Yi, Ch'unyŏng 이춘영. *Pioneers' Footprints in Korean Agriculture: Cho Paek-hyŏn* 韓國農學 巨聖의 발자취-조백현. Suwŏn: Hwanong Yŏnhak Chaedan, 2002.
Yi, Kyŏngjun이경준. *Pioneers' Footprints in Korean Agriculture: Hyun Shin-Gyu* 韓國農學 巨聖의 발자취-현신규. Suwŏn: Hwanong Yŏnhak Chaedan, 2006.
Yi, Uch'ŏl 이우철. "A Biography of Haŭn Chung Tyaihyon 하은 정태현 박사 전기." In *Haŭn Biology Award: The 25th Anniversary* 霞隱生物學賞: 二十五周年. Seoul: Committee on Haŭn Biology Award, 1994, 53–100.
Yi, Ŭnung 이은웅. *Pioneers' Footprints in Korean Agriculture: Chi Yŏnglin* 韓國農學 巨聖의 발자취-지영린. Suwŏn: Hwanong Yŏnhak Chaedan, 2002.
Yoshida Mitsukuni 吉田光邦. *History of Science in Japan* 日本科學史. Tokyo: Kodansha, 1987.

English Works

Adas, Michael. *Machines as the Measure of Men: Science, Technology, and Ideologies of Western Dominance*. Ithaca, NY: Cornell University Press, 1990.
Allen, David Elliston. *The Naturalist in Britain*. Princeton, NJ: Princeton University Press, 1994.
Anderson, Benedict. *Imagined Communities: Reflections on the Origin and Spread of Nationalism*. Revised ed. London: Verso, 1991.
Anderson, Warwick. "Introduction: Postcolonial Technoscience." *Social Studies of Science* 32(5/6) (2002): 643–58.
Anker, Peder. *Imperial Ecology*. Cambridge, MA: Harvard University Press, 2001.
Arnold, David. "Europe, Technology, and Colonialism in the 20th Century." *History and Technology* 21(2005): 85–106.
Arnold, David. *The Tropics and the Traveling Gaze: India, Landscape, and Science, 1800–1856*. Seattle: University of Washington Press, 2006.
Ash, Eric H. "Introduction: Expertise and the Early Modern State." *Osiris* 25(1) (2010): 1–24.
Bartholomew, James R. *The Formation of Science in Japan*. New Haven, CT: Yale University Press, 1993.
Basalla, George. "The Spread of Western Science." *Science* 156(3775) (1967): 611–22.
Bayly, C. A. *The Birth of the Modern World, 1780–1914: Global Connections and Comparisons*. Malden, MA: Blackwell Pub., 2004.
Bleichmar, Daniela. *Visible Empire: Botanical Expeditions & Visual Culture in the Hispanic Enlightenment*. Chicago, IL; London: University of Chicago Press, 2012.
Bonneuil, Christophe. "The Manufacture of Species: Kew Gardens, the Empire and the Standardisation of Taxonomic Practices in late 19th Century Botany." In M.- N. Bourguet, C. Licoppe, and O. Sibum, eds., *Instruments, Travel and Science: Itineraries of Precision from the 17th to the 20th Century*. London: Routledge, 2002, 189–215.
Bowler, Peter J. *The Eclipse of Darwinism: Anti-Darwinian Evolution Theories in the Decades around 1900*. Baltimore, MD: Johns Hopkins University Press, 1983.
Bowler, Peter J. *The Non-Darwinian Revolution: Reinterpreting a Historical Myth*. Baltimore, MD: Johns Hopkins University Press, 1988.
Bowler, Peter J. *Life's Splendid Drama: Evolutionary Biology and the Reconstruction of Life's Ancestry, 1860–1940*. Chicago, IL: University of Chicago Press, 1996.

Bowler, Peter J. *Darwin Deleted: Imagining a World without Darwin*. Chicago: University of Chicago Press, 2013.

Bray, Francesca. "Only Connect: Comparative, National, and Global History as Frameworks for the History of Science and Technology in Asia." *East Asian Science, Technology and Society* 6 (2012): 233–241.

Brockway, Lucile. *Science and Colonial Expansion: The Role of the British Royal Botanic Gardens. Studies in Social Discontinuity*. New York: Academic Press, 1979.

Brockway, Lucile H. *Science and Colonial Expansion: The Role of the British Royal Botanic Gardens*. New Haven, CT: Yale University Press, 2002.

Browne, E. J. *The Secular Ark: Studies in the History of Biogeography*. New Haven, CT: Yale University Press, 1983.

Buchwald, Jed Z., and Mordechai Feingold. *Newton and the Origin of Civilization*. Princeton, NJ: Princeton University Press, 2012.

Bullock, Mary Brown. *The Oil Prince's Legacy: Rockefeller Philanthropy in China*. Washington, DC: Stanford University Press, 2011.

Chan, Alan Kam-leung, Gregory K. Clancey, and Hui-Chieh Loy. *Historical Perspectives on East Asian Science, Technology,and Medicine*. Singapore: Singapore University Press, 2001.

Ching, Leo T. S. *Becoming "Japanese": Colonial Taiwan and the Politics of Identity Formation*. Berkeley: University of California Press, 2001.

Cittadino, Eugene. *Nature as the Laboratory: Darwinian Plant Ecology in Thegerman Empire, 1880–1900*. Cambridge; New York: Cambridge University Press, 1990.

Clancey, Gregory. *Earthquake Nation: The Cultural Politics of Japanese Seismicity, 1868–1930*. Berkeley: University of California Press, 2006.

Clancey, Gregory. "Japanese Colonialism and Its Sciences: A Commentary." *East Asian Science, Technology and Society: An International Journal* 1(2) (2007): 205–11.

Clancey, Gregory. "The History of Technology in Japan and East Asia." *East Asian Science, Technology and Society: An International Journal* 3(4) (2009): 525–30.

Cohen, Paul A. *Discovering History in China: American Historical Writing on the Recent Chinese Past*. New York: Columbia University Press, 1984.

Collins, Harry. *Changing Order: Replication and Induction in Scientific Practice*. Chicago, IL: University of Chicago Press, 1985.

Conrad, Sebastian. "Enlightenment in Global History: A Historiographical Critique." *The American Historical Review* 117(4) (2012): 999–1027.

Cook, Harold. *Matters of Exchange: Commerce, Medicine, and Science in the Dutch Golden Age*. New Haven, CT: Yale University Press, 2007.

Cooper, Alix. *Inventing the Indigenous: Local Knowledge and Natural History in Early Modern Europe*. Cambridge: Cambridge University Press, 2007.

Cooper, Frederick and Ann Laura Stoler, eds. *Tensions of Empire: Colonial Cultures in a Bourgeois World*. Berkeley: University of California Press, 1997.

Crane, Peter. *Ginkgo*. New Haven, CT: Yale University Press, 2013.

Cumings, Bruce. *The Origins of the Korean War*. V 1: Studies of the East Asian Institute, Columbia University. 2 vols. Princeton, NJ: Princeton University Press, 1981.

Cunningham, Andrew and Perry Williams. "De-Centring the 'Big Picture': "The Origins of Modern Science" and the Modern Origins of Science." *The British Journal for the History of Science* 26(4) (1993): 407–32.

Daston, Lorraine. "Afterword: The Ethos of Enlightenment." In William Clark, Jan Golinski, and Simon Schaffer eds., *The Sciences in Enlightened Europe*. Chicago, IL: University of Chicago Press, 1999, 495–504.

Daston, Lorraine and Peter Galison. *Objectivity*. New York: Zone Books, 2007.
Daston, Lorraine and Katharine Park. *Wonders and the Order of Nature, 1150–1750*. New York: Zone Books, 2001.
Dear, Peter. "The History of Science and the History of the Sciences: George Sarton, Isis, and the Two Cultures." *Isis* 100(1) (2009): 89–93.
Dewey, John, Harriet A. C. Dewey and Evelyn Dewey. *Letters from China and Japan*. New York: E.P. Dutton Co., 1920.
Drayton, Richard Harry. *Nature's Government*. New Haven, CT: Yale University Press, 2000.
Duara, Prasenjit. *Rescuing History from the Nation Questioning Narratives of Modern China*. Chicago, IL: University of Chicago Press, 1996.
Du Bois, W. E. B. *The Souls of Black Folk: Essays and Sketches*. Chicago, IL: A.C. McClurg & Co., 1903.
Duus, Peter. *The Abacus and the Sword: The Japanese Penetration of Korea, 1895–1910*. Berkeley: University of California Press, 1998.
Eckert, Carter J. *Offspring of Empire: The Koch'ang Kims and the Colonial Origins of Korean Capitalism, 1876–1945*. Seattle: University of Washington Press, 1996.
Elman, Benjamin. ""Universal Science" Versus "Chinese Science": The Changing Identity of Natural Studies in China, 1850–1930." *Historiography East and West* 1(1) (2003): 68–116.
Elman, Benjamin. *On their Own Terms: Science in China, 1550–1900*. Cambridge, MA: Harvard University Press, 2005.
Elman, Benjamin. "Sinophiles and Sinophobes in Tokugawa Japan: Politics, Classicism, and Medicine during the Eighteenth Century." *East Asian Science, Technology and Society: An International Journal* 2(1) (2008): 93–121.
Elshakry, Marwa. "When Science Became Western: Historiographical Reflections." *Isis* 101(1) (March 2010): 98–109.
Elshakry, Marwa. *Reading Darwin in Arabic, 1860–1950*. Chicago, IL : The University of Chicago Press, 2013.
Endersby, Jim. "'From Having No Herbarium' Local Knowledge versus Metropolitan Expertise: Joseph Hooker's Australasian Correspondence with William Colenso and Ronald Gunn." *Pacific Science* 55(4) (2001): 343–358.
Endersby, Jim. *Imperial Nature: Joseph Hooker and the Practices of Victorian Science*. Chicago, IL: University of Chicago Press, 2008.
Endersby, Jim. "Lumpers and Splitters: Darwin, Hooker, and the Search for Order." *Science* 326(5959) (2009): 1496–99.
Fan, Fa-ti. *British Naturalists in Qing China*. Cambridge, MA: Harvard University Press, 2004.
Fan, Fa-ti. "The Global Turn in the History of Science." *East Asian Science, Technology and Society* 6(2) (2012): 249–58.
Fan, Fa-ti. "Modernity, Region, and Technoscience: One Small Cheer for Asia as Method." *Cultural Sociology* 10(3) (2016): 352–368.
Fan, Fa-ti. "'Mr. Science', May Fourth, and the Global History of Science," *East Asian Science, Technology and Society: An International Journal* 16(3) (2022): 279–304.
Fanon, Frantz. *Black Skin, White Masks*. New York: Grove Press, 1967.
Fanon, Frantz. *Towards the African Revolution: Political Essays*. Translated by Haakon Chevalier. New York: Grove Press, 1988.
Findlen, Paula. *Possessing Nature*. Berkeley: University of California Press, 1996.
Finlay, Robert. "China, the West, and World History in Joseph Needham's Science and Civilisation in China." *Journal of World History* 11(2) (2000): 265–303.

Foucault, Michel. *The Order of Things*. New York: Pantheon Books, 1971.
Franz, Edgar. *Philipp Franz Von Siebold and Russian Policy and Action on Opening Japan to the West in the Middle of the Nineteenth Century*. Munich: IUDICIUM Verlag, 2005.
Fedman, David. *Seeds of Control: Japan's Empire of Forestry in Colonial Korea*. Seattle: University of Washington Press, 2020.
Fujitani, Takashi. *Race for Empire: Koreans as Japanese and Japanese as Americans during World War II*. Berkeley: University of California Press, 2011.
Fukuoka, Maki. *The Premise of Fidelity Science, Visuality, and Representing the Real in Nineteenth-Century Japan*. Stanford, CA: Stanford University Press, 2012.
Garon, Sheldon. "Transnational History and Japan's 'Comparative Advantage'." *Journal of Japanese Studies* 43(1) (2017): 65–92.
Godart, Gerard Rainier Clinton. "Darwin in Japan: Evolutionary Theory and Japan's Modernity (1820–1970)." Ph.D. Dissertation, The University of Chicago, 2009.
Goethe, Johann Wolfgang von. *The Metamorphosis of Plants*. Translated by Gordon L. Miller. Cambridge, MA: MIT Press, 2009.
Golinski, Jan. *Making Natural Knowledge: Constructivism and the History of Science*. Cambridge: Cambridge University Press, 1998.
Goodman, Grant Kohn. *Japan and the Dutch, 1600–1853*. Richmond, Surrey: Curzon, 2000.
Gordin, Michael D. *Scientific Babel*. Chicago, IL: University Of Chicago Press, 2015.
Graham, A. C. "China, Europe, and the Origins of Modern Science." In Joseph Needham, Shigeru Nakayama and Nathan Sivin eds., *Chinese Science: Explorations of an Ancient Tradition*. Cambridge, MA: MIT Press, 1973, 45–69.
Grove, Richard. *Green Imperialism: Colonial Expansion, Tropical Island Edens, and the Origins of Environmentalism, 1600–1860*. Cambridge: Cambridge University Press, 1995.
Hagen, Joel B. "Experimentalists and Naturalists in Twentieth-Century Botany: Experimental Taxonomy, 1920–1950." *Journal of the History of Biology* 17(2) (1984): 249–70.
Hagen, Joel B. "Naturalists, Molecular Biologists, and the Challenges of Molecular Evolution." *Journal of the History of Biology* 32(2) (1999): 321–41.
Hagen, Joel B. "The Statistical Frame of Mind in Systematic Biology from Quantitative Zoology to Biometry." *Journal of the History of Biology* 36(2) (2003): 353–84.
Hare, Brian and Vanessa Woods. *Survival of the Friendliest: Understanding Our Origins and Rediscovering Our Common Humanity*. New York: Random House, 2020.
Harootunian, Harry D. *Overcome by Modernity: History, Culture, and Community in Interwar Japan*. Princeton: Princeton University Press, 2000.
Hart, Roger. "Beyond Science and Civilization: A Post-Needham Critique." *East Asian Science, Technology, and Medicine* (16) (1999): 88–114.
Hawksworth, D., ed. *Improving the Stability of Names: Needs and Options: Proceedings of an International Symposium, Kew,* 20–23 February 1991. Königstein/Taunus Germany: Published for the International Association for Plant Taxonomy in conjunction with the International Union of Biological Sciences and the Systematics Association by Koeltz Scientific, 1991.
Headrick, Daniel R. *The Tools of Empire: Technology and European Imperialism in the Nineteenth Century*. New York: Oxford University Press, 1981.
Henry, Todd A. *Assimilating Seoul: Japanese Rule and the Politics of Public Space in Colonial Korea, 1910–1945*. Berkeley: University of California Press, 2014.
Hinz, Petra-Andrea. "The Japanese plant collection of Engelbert Kaempfer (1651–1716) in the Sir Hans Sloane Herbarium at the Natural History Museum, London." *Bulletin of the Natural History Museum. Botany Series* 31(1) (2001): 27–34.

Hobsbawm, Eric J. *The Age of Empire, 1875–1914*. New York: Pantheon Books, 1987.
Hobsbawm, Eric, and Terence Ranger, eds. *The Invention of Tradition*. Cambridge: Cambridge University Press, 2012.
Holton, Gerald. "George Sarton, His Isis, and the Aftermath." *Isis* 100(1) (2009): 79–88.
Homei, Aya. *Science of Governing Japan's Population*. Cambridge: Cambridge University Press, 2023.
Hong, Sungook. "Korean Scientists Look at Japanese Colonization: The Emergence of an Idea of the Value-Neutrality and Objectivity of Science." *Historia Scientiarum* 15 (2006): 268–75.
Hull, David L. *Darwin and His Critics: The Reception of Darwin's Theory of Evolution by the Scientific Community*. Chicago, IL: University of Chicago Press, 1983.
Hull, David L. *Science as a Process: An Evolutionary Account of the Social and Conceptual Development of Science*. Chicago, IL: University of Chicago Press, 1990.
Hung, Kuang-chi. "When the Green Archipelago Encountered Formosa: The Making of Modern Forestry in Taiwan under Japan's Colonial Rule," in Bruce Batten and Philip Brown eds., *Environment and Society in the Japanese Islands: From Prehistory to the Present*. Corvallis: Oregon State University Press, 2015, 174–93.
Hyun, Jaehwan. "Racializing Chōsenjin: Science and Biological Speculations in Colonial Korea", *East Asian Science, Technology and Society: An International Journal* 13 (2019): 489–510.
Hyun, Jaehwan. "Blood Purity and Scientific Independence: Blood Science and Postcolonial Struggles in Korea, 1926–1975", *Science in Context* 32 (2019): 1–22.
Jardine, Nicholas, James A. Secord and Emma C. Spary. *Cultures of Natural History*. Cambridge: Cambridge University Press, 1996.
Johnson, Chalmers. *MITI and the Japanese Miracle: The Growth of Industrial Policy, 1925–1975*. Stanford, CA: Stanford University Press, 1982.
Kay, Lily E. *The Molecular Vision of Life: Caltech, the Rockefeller Foundation, and the Rise of the New Biology*, New York: Oxford University Press, 1993.
Keller, Evelyn Fox. *The Century of the Gene*. Cambridge, MA: Harvard University Press, 2000.
Kim, Boumsoung. "Seismicity within and Beyond the Empire: Japanese Seismological Investigation in Taiwan and Its Global Deployment, 1895–1909." *East Asian Science, Technology and Society: An International Journal* 1(2) (2007): 153–65.
Kim, Hoi-Eun. *Doctors of Empire: Medical and Cultural Encounters between Imperial Germany and Meiji Japan*. Toronto: University of Toronto Press, 2014.
Kim, Jina E. *Urban Modernities in Colonial Korea and Taiwan*. Leiden: Brill, 2019.
Kim, Ock-Joo. "Physical Anthroplogy Studies at Keijo Imperial University Medical School." *KJMH* 17(2) (2008): 191–203.
Kim, Yung-Sik. "Problems and Possibilities in the Study of the History of Korean Science." *Osiris* 13, Beyond Joseph Needham: Science, Technology, and Medicine in East and Southeast Asia (1998): 48–79.
Kim, Yung-Sik. *The Natural Philosophy of Chu Hsi (1130–1200)*. Philadelphia, PA: American Philosophical Society, 2000.
Kim, Yung-Sik. "Problem of Early Modern Japan in the History of Science in East Asia." *Historia Scientiarum* 18 (2008): 49–57.
Kim, Yung-Sik. *Questioning Science in East Asian Contexts: Essays on Science, Confucianism, and the Comparative History of Science*. Leiden; Boston, MA: Brill, 2014.
Kingsland, Sharon E. *The Evolution of American Ecology, 1890–2000*. Baltimore, MD: Johns Hopkins University Press, 2005.

Koerner, Lisbet. *Linnaeus*. Cambridge, MA: Harvard University Press, 1999.
Kohler, Robert E. *Landscapes & Labscapes*. Chicago, IL: University of Chicago Press, 2002.
Kohler, Robert E. *All Creatures: Naturalists, Collectors, and Biodiversity,1850–1950*. Princeton, NJ; Oxford: Princeton University Press, 2006.
Kohlstedt, Sally Gregory. *Teaching Children Science: Hands-on Nature Study in North America, 1890–1930*. Chicago, IL; London: University of Chicago Press, 2010.
Kornicki, Peter F. *The Book in Japan: A Cultural History from the Beginnings to the Nineteenth Century*. Leiden; Boston, MA: Brill, 1998.
Kouwenhoven, Arlette and Matthi Forrer. *Siebold and Japan. His Life and Work*. Leiden: KIT Publishers, 2000.
Kuhn, Thomas. *The Structure of Scientific Revolutions*, 2nd ed. Chicago, IL: University of Chicago Press, 1970.
Kumar, Deepak. "The 'Culture' of Science and Colonial Culture, India 1820–1920." *The British Journal for the History of Science* 29(2) (1996): 195–209.
Kwon, Nayoung Aimee. *Intimate Empire: Collaboration and Colonial Modernity in Korea and Japan*. Durham, NC: Duke University Press, 2015.
Lang, W. H. "Fredrick Orpen Bower. 1855–1948." *Obituary Notices of Fellows of the Royal Society* 6(18) (1949): 347–74.
Latour, Bruno. *Science in Action: How to Follow Scientists and Engineers through Society*. Cambridge, MA: Harvard University Press, 1987.
Latour, Bruno. *We Have Never Been Modern*. Cambridge, MA: Harvard University Press, 1993.
Latour, Bruno. *Pandora's Hope*. Cambridge, MA: Harvard University Press, 1999.
Latour, Bruno, and Steve Woolgar, *Laboratory Life: The Construction of Scientific Facts*. Princeton: Princeton University Press, 1986.
Lean, Eugenia. *Vernacular Industrialism in China: Local Innovation and Translated Technologies in the Making of a Cosmetics Empire, 1900–1940*. New York: Columbia University Press, 2020.
Lee, John S. "Postwar Pines: The Military and the Expansion of State Forests in Post-Imjin Korea, 1598–1684." *The Journal of Asian Studies* 77(2) (2005): 319–332.
Lee, Jung. "Invention without Science: 'Korean Edisons' and the Changing Understanding of Technology in Colonial Korea." *Technology and Culture* 54(4) (2013): 782–814.
Lee, Jung. "Cranes, Cultivating a New Knowledge Practice in Late-Chosŏn Korea: Knowledge Transformations Connected by Things." *East Asian Science, Technology and Society: An International Journal* 18(1) (2024): 47–69. DOI: 10.1080/18752160.2023.2249739.
Lee, Victoria. *The Arts of the Microbial World: Fermentation Science in Twentieth-Century Japan*. Chicago, IL: University of Chicago, 2021.
Lei, Sean Hsiang-lin. *Neither Donkey nor Horse: medicine in the struggle over China's modernity*. Chicago, IL: The University of Chicago Press, 2014.
Lewis, James Bryant. *Frontier Contact between Chosŏn Korea and Tokugawa Japan*. London: Routledge, 2003.
Lewis, Pyenson and Christophe Verbruggen. "Ego and the International: The Modernist Circle of George Sarton." *Isis* 100(1) (2009): 60–78.
Liao, Binghui and Dewei Wang, eds. *Taiwan under Japanese Colonial Rule, 1895–1945: History, Culture, Memory*. New York: Columbia University Press, 2006.
Liu, Lydia He. *Translingual Practice: Literature, National Culture, and Translated Modernity--China, 1900–1937*. Stanford: Stanford University Press, 1995.

Liu, Shiyung. "The Ripples of Rivalry: The Spread of Modern Medicine from Japan to Its Colonies." *East Asian Science, Technology and Society: An International Journal* 2(1) (2008): 47–71.

Lloyd, G. E. R. *Cognitive Variations: Reflections on the Unity and Diversity of the Human Mind.* Oxford: Clarendon Press, 2007.

Lo, Ming-cheng Miriam. *Doctors within Borders: Profession, Ethnicity, and Modernity in Colonial Taiwan.* Berkeley: University of California Press, 2002.

Long, Pamela O. *Artisan/Practitioners and the Rise of the New Sciences, 1400–1600.* Corvallis: Oregon State University Press, 2011.

Low, Morris. *Building a Modern Japan: Science, Technology, and Medicine in the Meiji Era and Beyond.* New York: Palgrave Macmillan, 2005.

Low, Morris. *Japan on Display: Photography and the Emperor*. London: Taylor & Francis, 2006.

MacLeod, Roy. "On Visiting the 'Moving Metropolis': Reflections on the Architecture of Imperial Science." In Nathan Reingold and Marc Rothenberg eds., *Scientific Colonialism: A Cross-Cultural Comparison.* Washington, DC: Smithsonian Institution Press, 1987, 217–49.

MacLeod, Roy M., ed. *Nature and Empire: Science and the Colonial Enterprise*. *Osiris* 15. Chicago, IL: University of Chicago Press, 2000.

Marcon, Federico. *The Knowledge of Nature and the Nature of Knowledge in Early Modern Japan.* Chicago, IL: University of Chicago Press, 2015.

Mayr, Ernst. *One Long Argument: Charles Darwin and the Genesis of Modern Evolutionary Thought.* Questions of Science. Cambridge, MA: Harvard University Press, 1991.

McOuat, Gordon R. "Species, Rules and Meaning: The Politics of Language and the Ends of Definitions in 19th Century Natural History." *Studies in History and Philosophy of Science Part A* 27(4) (1996): 473–519.

McOuat, Gordon R. "The Origins of 'Natural Kinds': Keeping "Essentialism" at Bay in the Age of Reform." *Intellectual History Review* 19(2) (2009): 211–30.

Métailié, Georges. "Bencao Gangmu of Li Shizhen: Innovations in Natural History." In Elisabeth Hsu ed., *Innovation in Chinese Medicine*. Cambridge: Cambridge University Press, 2001.

Métailié, Georges. "Comparative Study of the Introduction of Modern Botany in Japan and China." *Historia Scientiarum* 11(3) (2002): 205–17.

Métailié, Georges. "The Representation of Plants: Engravings and Paintings." In Francesca Bray, Vera Dorofeeva-Lichtmann, and Georges Métailié eds. *Graphics and Text in the Production of Technical Knowledge in China: The Warp and the Weft.* Leiden: Brill, 2007, 487–520.

Miller, Ian Jared. *The Nature of the Beasts: Empire and Exhibition at the Tokyo Imperial Zoo.* Berkeley: University of California Press, 2013.

Milo, Daniel S. *Good Enough: The Tolerance for Mediocrity in Nature and Society.* Cambridge, MA: Harvard University Press, 2019.

Miyagawa, Takuya. "The Meteorological Observation System and Colonial Meteorology in Early 20th-Century Korea." *Historia Scientiarum* 18 (2008): 140–50.

Mizuno, Hiromi. *Science for the Empire*. Stanford, CA: Stanford University Press, 2009.

Moon, Manyong. "Becoming a Biologist in Colonial Korea: Cultural Nationalism in a Teacher-Cum-Biologist." *East Asian Science, Technology and Society: An International Journal* 6(1) (2012): 65–82.

Moore, Aaron S. *Constructing East Asia: Technology, Ideology, and Empire in Japan's Wartime Era, 1931–1945*. Stanford, CA: Stanford University Press, 2013.

Morris-Suzuki, Tessa. *The Technological Transformation of Japan: From the Seventeenth to the Twenty-First Century*. Cambridge: Cambridge University Press, 1994.

Morris-Suzuki, Tessa. "Debating Racial Science in Wartime Japan." *Osiris* 13 (1998): 354–75.

Mueggler, Erik. *The Paper Road: Archive and Experience in the Botanical Exploration of West China and Tibet*. Berkeley: University of California Press, 2011.

Müller-Wille, Staffan. "Joining Lapland and the Topinambes in Flourishing Holland: Center and Periphery in Linnaean Botany." *Science in Context* 16(04) (2003): 461–88.

Myers, Ramon Hawley and Mark R. Peattie. *The Japanese Colonial Empire, 1895–1945*. Princeton, NJ: Princeton University Press, 1984.

Nagata, Toshiyuki et al. "Engelbert Kaempfer, Genemon Imamura and the Origin of the Name Ginkgo," *Taxon* 64 (2015): 131–36.

Nakayama, Shigeru. *The Orientation of Science and Technology: A Japanese View*. Folkestone: Global Oriental, 2009.

Nappi, Carla Suzan. *The Monkey and the Inkpot*. Cambridge, MA: Harvard University Press, 2009.

Needham, Joseph. *Science and Civilisation in China*. Cambridge: Cambridge University Press, 1954.

Ogilvie, Brian W. *The Science of Describing*. Chicago, IL: University of Chicago Press, 2006.

Ohashi, Hiroyoshi. "Bunzo Hayata and His Contributions to the Flora of Taiwan." *Taiwania* 54(1) (2009): 1–27.

Ohnuki-Tierney, Emiko. *Kamikaze, Cherry Blossoms, and Nationalisms: The Militarization of Aesthetics in Japanese History*. Chicago, IL: The University of Chicago Press, 2002.

Orwell, George. *Shooting an Elephant and Other Essays*. San Diego, CA: Harcourt, Brace, Jovanovich, 1950.

Park, Albert L. *Building a Heaven on Earth: Religion, Activism, and Protest in Japanese-Occupied Korea*. Honolulu: University of Hawai'i Press, 2015.

Park, Hyunhee. *Soju: A Global History*. Cambridge: Cambridge University Press, 2021.

Park, Seong-Rae. "Modern Scientific Terms in East Asia: Their Births and Modifications." *International Area Studies Review* 7(1) (2004): 131–44.

Perry, G. "Nomenclatural Stability and the Botanical Code: A Historical Review." In *Improving the Stability of Names: Needs and Options: Proceedings of an International Symposium*, Kew, 20–23 February 1991, Königstein/Taunus Germany: The International Association for Plant Taxonomy, 1991, 79–93.

Pickering, Andrew. *Constructing Quarks: A Sociological History of Particle Physics*. Chicago, IL: University of Chicago Press, 1984.

Plutschow, Herbert E. *Philipp Franz Von Siebold and the Opening of Japan*. Folkestone: Global Oriental, 2007.

Prakash, Gyan. *Another Reason*. Princeton, NJ: Princeton University Press, 1999.

Pratt, Mary Louise. *Imperial Eyes: Travel Writing and Transculturation*. London: Routledge, 2008.

Pyenson, Lewis. "The Ideology of Western Rationality: History of Science and the European Civilizing Mission." *Science & Education* 2(4) (1993): 329–43.

Raj, Kapil. *Relocating Modern Science: Circulation and the Construction of Knowledge in South Asia and Europe, 1650–1900*. New York: Palgrave Macmillan, 2010.

Raj, Kapil. "Beyond Postcolonialism and Postpositivism: Circulation and the Global History of Science." *Isis* 104(2) (2013): 337–47.

Richards, Robert J. *The Romantic Conception of Life Science and Philosophy in the Age of Goethe*. Chicago, IL: University of Chicago Press, 2002.

Roberts, Lissa, Simon Schaffer, and Peter Dear eds. *The Mindful Hand*. Amsterdam: Koninkliijke Nederlandse Akademie van Wetenschappen, 2007.

Robinson, Michael Edson. *Cultural Nationalism in Colonial Korea, 1920–1925*. Seattle: University of Washington Press, 1988.

Rowe, William T. *Saving the World: Chen Hongmou and Elite Consciousness in Eighteenth-Century China*, Stanford, CA: Stanford University Press, 2001.

Sachs, Aaron. *The Humboldt Current: Nineteenth-Century Exploration and the Roots of American Environmentalism*. New York: Viking, 2006.

Schaffer, Simon, ed. *The Brokered World: Go-Betweens and Global Intelligence, 1770–1820*. Sagamore Beach, MA.: Science History Publications, 2009.

Schiebinger, Londa L. *Plants and Empire*. Cambridge, MA: Harvard University Press, 2004.

Schiebinger, Londa, and Claudia Swan. *Colonial Botany*. Philadelphia: University of Pennsylvania Press, 2005.

Schmid, Andre. *Korea between Empires, 1895–1919*. New York: Columbia University Press, 2002.

Scott, James C. *Seeing Like a State: How Certain Schemes to Improve the Human Condition Have Failed*. Yale Agrarian Studies. New Haven, CT: Yale University Press, 1998.

Secord, Anne. "Pressed into Service: Specimens, Space, and Seeing in Botanical Practice." In David N. Livingstone and Charles W. J. Withers eds., *Geographies of Nineteenth-Century Science*. Chicago, IL: Chicago University Press, 2011, 283–310.

Secord, James A. "Knowledge in Transit." *Isis* 95(4) (2004): 654–72.

Seow, Victor. "A Tradition of Invention: The Paradox of Glorifying Past Technological Breakthroughs." *East Asian Science, Technology and Society: An International Journal* 16(3) (2022): 349–66.

Setoguchi, Akihisa. "Control of Insect Vectors in the Japanese Empire: Transformation of the Colonial/Metropolitan Environment, 1920–1945." *East Asian Science, Technology and Society: an International Journal* 1(2) (2007): 167–81.

Shapin, Steven. *The Scientific Revolution*. Chicago, IL: University of Chicago Press, 1996.

Shapin, Steven and Simon Schaffer. *Leviathan and the Air-Pump: Hobbes, Boyle, and the Experimental Life*. Princeton, NJ: Princeton University Press, 1985.

Shen, Grace Yen. "Taking to the Field: Geological Fieldwork and National Identity in Republican China." *Osiris* 24(1) (2009): 231–52.

Shin, Dongwon. "How Commoners Became Consumers of Naturalistic Medicine in Korea, 1600–1800." *East Asian Science, Technology and Society* 4(2) (2010): 275–301.

Sivasundaram, Sujit. "Sciences and the Global." *Isis* 101(1) (2010): 146–58.

Sivin, Nathan. "Why the Scientific Revolution Did Not Take Place in China-or Didn't It?" *Chinese Science* 5 (1982): 45–66.

Smith, Pamela. *From Lived Experience to the Written Word: Reconstructing Practical Knowledge in the Early Modern World*. Chicago, IL: University of Chicago Press, 2022.

Spary, Emma C. *Utopia's Garden*. Chicago, IL: University of Chicago Press, 2000.Stafleu, Frans A. *Linnaeus and the Linnaeans: The Spreading of their Ideas in Systematic Botany, 1735–1789*. Utrecht: International Association for Plant Taxonomy, 1971.

Strasser, Bruno J. "Who Cares about the Double Helix?" *Nature* 422 (2003): 803–4.

Stevens, Peter F. *The Development of Biological Systematics: Antoine-Laurent De Jussieu, Nature, and the Natural System*. New York: Columbia University Press, 1994.

Suh, So Young. "Korean Medicine between the Local and the Universal: 1600–1945." Ph. D. Thesis, UCLA, 2006.

Tanaka, Stefan. *Japan's Orient: Rendering Pasts into History*. Berkeley: University of California Press, 1993.

Thiede, Arnulf, Yoshiki Hiki and Gundolf Keil. *Philipp Franz Von Siebold and His Era: Prerequisites, Developments, Consequences and Perspectives*. New York: Springer, 1999.
Thijsse, Gerard. "A Contribution to the History of the Herbaria of George Clifford III (1685–1760)." *Archives of Natural History* 45(1) (2018): 134–48.
Thomas, Julia Adeney, *Reconfiguring Modernity Concepts of Nature in Japanese Political Ideology*. Berkeley: University of California Press, 2002.
Toby, Ronald P. *State and Diplomacy in Early Modern Japan Asia in the Development of the Tokugawa Bakufu*. Stanford, CA: Stanford University Press, 1991.
Totman, Conrad D. *The Origins of Japan's Modern Forests: The Case of Akita*. Honolulu, HI: University of Hawaii Press, 1985.
Tsukahara, Togo. "Introduction to Feature Issue: Colonial Science in Former Japanese Imperial Universities." *East Asian Science, Technology and Society: An International Journal* 1(2) (2007): 147–52.
Tsuru, Shuntaro. "Irrigation Pumps in Late Colonial Taiwan: Farmers' Utilization of Technology and the Transition to Rice Cultivation." *Modern Asian Studies* 57(6) (2023): 1–37.
Turrill, William B. "John Christopher Willis. 1868–1958." *Biographical Memoirs of Fellows of the Royal Society* 4 (1958): 353–59.
Uchida, Jun. *Brokers of Empire: Japanese Settler Colonialism in Korea, 1876–1945*. Cambridge, MA: Harvard University Asia Center, 2011.
Walle, V. F. Vande, ed. *Dodonaeus in Japan: Translation and the Scientific Mind in the Tokugawa Period*. Leuven; Kyoto: Leuven University Press, 2001.
Walraven, Boudewijin. "The Natives Next-door: Ethnology in Colonial Korea." In Jan van Bremen and Akitoshi Shimizu eds., *Anthropology and Colonialism in Asia and Oceania*. Richmond, Surrey: Curzon Press, 1999, 219–44.
Weber, Andreas. "Encountering the Netherlands Indies: Caspar G.C. Reinwardt's Field Trip to the East (1816–1822)." *Itinerario* 33(1) (2009): 45–60.
Wigen, Kären. "Discovering the Japanese Alps: Meiji Mountaineering and the Quest for Geographical Enlightenment." *The Journal of Japanese Studies* 31(1) (2005): 1–26.
Wiley, Edward O. and Bruce S. Lieberman, eds. *Phylogenetics: Theory and Practice of Phylogenetics Systematics*. Hoboken, NJ: Wiley-Blackwell, 2011.
Wittner, David G. and Philip C. Brown, eds. *Science, Technology, and Medicine in the Modern Japanese Empire*. New York: Routledge, 2016.
Yang, Daqing. *Technology of Empire: Telecommunications and Japanese Expansion in Asia, 1883–1945*. Cambridge, MA: Harvard University Asia Center, 2010.
Yoon, Carol Kaesuk. *Naming Nature: The Clash between Instinct and Science*. New York: W. W. Norton & Company, 2009.
Young, Louise. *Japan's Total Empire: Manchuria and the Culture of Wartime Imperialism*. Berkeley: University of California Press, 1999.

Index

Note: Page numbers followed by "n" refer to end notes.

For Product Safety Concerns and Information please contact our EU representative GPSR@taylorandfrancis.com Taylor & Francis Verlag GmbH, Kaufingerstraße 24, 80331 München, Germany

Batch number: 10406157

Printed by Printforce, the Netherlands